The Digital Age in Agriculture

The Digital Age in Agriculture presents information related to the digital age in the agriculture sector. Agriculture is an essential activity for the continuity of life, yet is very labor-intensive and faces a wide variety of challenges. In the struggle against these difficulties, the superior features offered by technology provide important benefits. These technologies require expertise in various technical disciplines, and *The Digital Age in Agriculture* provides information to readers allowing them to make more informed decisions and giving them the opportunity to improve agricultural productivity.

Written by Mehmet Metin Ozguven, an expert who has conducted field studies and with a working technical knowledge of various topics pertaining to the agriculture age, this book covers many subjects important to the age of digital agriculture, including precision agriculture and livestock farming, using agricultural robots and unmanned aerial vehicles in agriculture practices, and image processing and machine vision. It is an essential read for researchers, agriculture sector workers, and agricultural engineers.

T0229522

The Digital Age in Agriculture

Mehmet Metin Ozguven

CRC Press
Taylor & Francis Group
Boca Raton London New York

CRC Press is an imprint of the
Taylor & Francis Group, an **informa** business

First edition published 2023
by CRC Press
6000 Broken Sound Parkway NW, Suite 300, Boca Raton, FL 33487-2742

and by CRC Press
4 Park Square, Milton Park, Abingdon, Oxon, OX14 4RN

CRC Press is an imprint of Taylor & Francis Group, LLC

© 2023 Taylor & Francis Group, LLC

Library of Congress Cataloging-in-Publication Data
Names: Özgüven, Mehmet Metin, author.
Title: The digital age in agriculture / Mehmet Metin Ozguven.
Description: First edition. | Boca Raton : CRC Press, 2023. | Includes bibliographical references and index. | Summary: "The Digital Age in Agriculture presents information related to the digital age in the agriculture sector. Agriculture is an essential activity for the continuity of life, yet is very labor-intensive and faces a wide variety of challenges. In the struggle against these difficulties, the superior features offered by technology provide important benefits. These technologies require expertise in various technical disciplines, and The Digital Age in Agriculture provides information to readers allowing them to make more informed decisions and giving the opportunity to improve agricultural productivity. Written by Mehmet Metin Özgüven, an expert who has conducted field studies and with a working technical knowledge of various topics pertaining to the agriculture age, this book covers many subjects important to the age of digital agriculture including precision agriculture and livestock farming, using agricultural robots and unmanned arial vehicles in agriculture practices, and image processing and machine vision. It is an essential read for researchers, agriculture sector workers, and agricultural engineers"-- Provided by publisher.
Identifiers: LCCN 2022035911 (print) | LCCN 2022035912 (ebook) | ISBN 9781032385778 (hardback) | ISBN 9781032385808 (paperback) | ISBN 9781003345718 (ebook)
Subjects: LCSH: Agricultural innovations. | Agriculture--Data processing. | Precision farming.
Classification: LCC S494.5.I5 O94 2023 (print) | LCC S494.5.I5 (ebook) | DDC 338.10285--dc23/eng/20230103
LC record available at https://lccn.loc.gov/2022035911
LC ebook record available at https://lccn.loc.gov/2022035912

ISBN: 978-1-032-38577-8 (hbk)
ISBN: 978-1-032-38580-8 (pbk)
ISBN: 978-1-003-34571-8 (ebk)

DOI: 10.1201/b23229

Typeset in Times
by Deanta Global Publishing Services, Chennai, India

Contents

Preface

Agriculture consists of two important agricultural production areas: Plant and animal production. Agricultural science includes activities with a wide range of specializations. Areas of specialization are not limited to one department or branch of science. There are many study subjects that require expertise in the same departments or branches of science. Although there are some changes in the names of faculties and departments in various universities, basically similar education and training activities are carried out. The departments in the faculties of agriculture where agricultural education is given, branches of science that are sub-units of these departments, and programs with subunits of these branches of science are as follows:

- Department of Biosystems Engineering
  - Division of Machinery Systems in Agriculture
  - Division of Energy Systems in Agriculture
  - Division of Land and Water Resources
  - Division of Agricultural Structures
- Department of Horticulture
  - Division of Vineyard Growing and Breeding
  - Division of Fruit Growing and Breeding
  - Division of Vegetable Growing and Breeding
  - Division of Ornamental Plant Growing and Breeding
  - Division of Greenhouse Growing
  - Division of Postharvest Physiology
- Department of Field Crops
  - Division of Grains and Legumes
  - Division of Industrial Plants
  - Division of Grassland, Meadow, and Forage Plants
  - Division of Biotechnology
- Department of Plant Protection
  - Division of Entomology
    - Program of Anatomy and Morphology
    - Program of Physiology
    - Program of Taxonomy and Systematics
    - Program of Applied Entomology
  - Program of Ecology and Epidemiology
  - Division of Phytopathology
    - Program of Virology
    - Program of Bacteriology
    - Program of Disease Ecology and Epidemiology
    - Program of Herbology
    - Program of Mycology
    - Program of Applied Phytopathology
- Department of Agricultural Machinery and Technologies Engineering
  - Program of Agricultural Machinery
  - Program of Energy in Agriculture
  - Program of Agricultural Tractors

- Department of Agricultural Structures and Irrigation
  - Program of Irrigation
  - Program of Drainage and Land Reclamation
  - Program of Agricultural Construction
- Department of Agricultural Economics
  - Division of Agricultural Management
  - Division of Agricultural Politics and Publications
- Department of Soil Science and Plant Nutrition
  - Division of Soil
    - Program of Soil Physics
    - Program of Soil Microbiology and Biochemistry
    - Program of Soil Mineralogy
    - Program of Soil Survey and Mapping
    - Program of Soil Genetics and Classification
    - Program of Soil and Water Conservation
  - Program of Soil Chemistry
  - Division of Plant Nutrition
    - Program of Soil Fertility
- Department of Zootechnics
  - Division of Biometrics and Genetics
  - Division of Animal Breeding
  - Division of Feeds and Animal Nutrition
- Department of Fisheries
  - Division of Fisheries Fishing and Processing Technology
    - Program of Fishing Technology
  - Program of Processing Technology
  - Division of Marine-Inland Water Sciences and Technology
  - Division of Aquaculture
- Department of Dairy Technology

In addition to the fact that there are many fields of study that require expertise in agricultural science, agriculture consists of complex events involving biological, physical, and chemical processes. Furthermore, it can be adversely affected by many factors such as weather, climate, soil characteristics, diseases and pests, and environmental pollution that cannot be controlled during agricultural production. For this reason, for efficient agricultural production, expertise alone is not sufficient; cultivation conditions must also be suitable for demand. The most efficient production will be achieved if the growing conditions meet the optimum demands, provided that the species and breed of the plant and animal grown are also suitable. Providing optimum conditions can be very difficult and costly. Therefore, for a successful agricultural business, the changes in each component should be determined accurately and in a timely manner, and the applications to be made should be decided by considering the possible interactions.

Technology continues to develop rapidly, and new applications continue to emerge with new methods and tools every day. To be able to apply these technologies, it is necessary to start from somewhere. On the one hand is learning, and on the other hand is the application of what has been learned. Although the topics in my *Precision Agriculture* book, which was published in Turkish in 2018, still maintain their validity and innovation, I have created chapters for this book with these new topics that I have been trying to learn and had the opportunity to apply in the last three years. I care very much about the application of these new techniques in agricultural production. I hope that the widespread use of these techniques will increase the quality and efficiency of agricultural production. Many entrepreneurs, whom I have met on various occasions and who want to do these businesses, had various questions. Entrepreneurs' skills in this field were suitable for doing the jobs they asked about. Since the subjects were very technical, it was necessary to explain where to start first. When we tried to explain the topics and were asked to suggest sources, there was no tidy literature on this subject. One of the reasons for this is that these subjects consist of literature belonging to different branches of science. I decided to prepare this book because I consider it almost impossible for my colleagues who are not experts in the field to understand the literature from different disciplines. For example, when I wanted to prepare articles, papers, or lecture notes on agricultural robots as an academician, I saw that the existing literature generally centered around industrial robots or sample agricultural applications. The fact that I had difficulty in finding literature that I could refer to in order to explain the subjects and that I thought that the needed information could be found more easily guided me in preparing the content of this book. For this reason, I created the content of the book with the subjects that I saw as needed. I have prepared this book in accordance with the format I have created by examining hundreds of books and hundreds of articles. I think that the content of the book is so important that it will shape our future as humanity. I believe the book will open new horizons and be an important handbook not only in agriculture but also in many fields. I hope it is useful.

Mehmet Metin Ozguven

Author

Mehmet Metin Ozguven is an Associate Professor in the Department of Biosystem Engineering at Tokat Gaziosmanpaşa University, Turkey. His research interests are in the areas of precision agriculture, precision livestock farming, artificial intelligence, deep learning, image processing, agricultural robots, unmanned aerial vehicles, autonomous vehicles, electronic applications in agriculture, and agricultural machinery. He has written four textbooks, two separate book chapters, published more than 40 peer-reviewed journal articles, presented more than 30 papers at national and international professional conferences, and has been awarded one best reviewer award by the *Journal of Environmental Informatics*. He has also been the translation editor of the *Handbook of Agricultural Engineering, Volume VI: Information Technology*. He worked as a project coordinator for 12 projects and as a researcher for four projects. He has two patent applications.

Precision Agriculture Technology, Smart Agricultural Technologies, Electric and Electronics, Sensors and Biosensors, Agricultural Automation, Use of Information Technologies in Agriculture, Precision Livestock Farming Applications, Instrumentation in Biosystems Engineering, Greenhouse Mechanization, and Plant Protection Machinery are some of the courses he teaches.

1 Precision Agriculture

1.1 INTRODUCTION

By evaluating the hardware, algorithms, and software that emerged with the development of technology together with the existing knowledge and experience in agriculture, it has been possible to facilitate agricultural operations and to bring alternative solutions to the problems that await solution or improvement. The use of technological methods, models, and tools that manage the processes of obtaining, processing, storing, transferring, and using information, as well as portable computers and hardware with high processing and computational power, can be found easily in the field, and their use in field applications has increased. Depending on the experience gained during the increasing use, the development has enabled it to be applied to different areas (Özgüven et al., 2020).

As a result of the rapid developments in information technology following the mechanization, automation, and control technologies during the development period of agricultural production, today, intelligent machines and production systems that control machines have begun to take over traditional production methods. Information technology consists of hardware, algorithms, and software developed for the management of the collection, processing, storage, transfer, and use of information processes. The implementation of present knowledge and experiences in agriculture together with machine learning, deep learning, artificial intelligence, modeling, and simulation applications enabled the development of real-time and automated expert systems, autonomous tractors or agricultural machines, and agricultural robotics applications (Ozguven, 2018).

Precision agriculture is the idea of doing the right thing, in the right place, at the right time. This idea is as old as agriculture. However, during the mechanization of agriculture in the 20th century, there was strong economic pressure to cultivate large fields with uniform agricultural practices. Precision agriculture provides a way to automate site-specific applications using information technology, thus making the site-specific application practical in commercial agriculture. Precision agriculture includes all agricultural production practices that use information technologies (e.g., variable rate application, yield monitors, remote sensing) to either tailor the use of inputs to achieve desired results or monitor those results (Bongiovanni and Lowenberg-Deboer, 2004).

Precision agriculture technologies, combined with control, electronics, computer, and databases with account data, present an advanced system approach. Using global positioning system, geographic information system, variable rate application, and remote sensing technologies, precision agriculture technologies, contrary to common fixed-level application methods which are applied all the same way to the whole land, use variable-level application methods (based on the application of fertilizer and chemicals to each section to its own needs, tillage at different levels, planting at different norms, irrigation and drainage at different levels) determining land and plant characteristics of small sections (soil moisture, nutrient level of soil, soil structure, product requirements, yield, etc.). As a result, precision agriculture technologies are agricultural production and management methods whose targets are more economic and more environmentally sensitive production (Özgüven and Türker, 2010).

In traditional agriculture, agricultural operations are carried out in a fixed norm all over the field. In precision agriculture, on the other hand, agricultural operations are carried out in variable norms specific to the field. While deciding at what level the variable norms will be for any application, it is necessary to determine the amount of need of the area for that application. This determination processing can be done in real time before or during the application. The decision-making phase can be made depending on the data coming from a sensor, or in some cases, it can be done by evaluating all the available information about the area, such as yield maps, soil maps, soil analysis results, soil pH,

DOI: 10.1201/b23229-1

soil compaction, electrical conductivity, plant growth status, disease and weed maps, soil moisture, and topography, together spatially. Application maps can be created for each agricultural process before or during the application. The benefits of precision agriculture can be listed as follows:

- Increased production efficiency;
- Improved product quality;
- The use of more effective chemicals and other inputs;
- Energy saving;
- Soil and groundwater protection.

Precision agriculture practices are not agricultural practices that are applied in a single year and left the next year. For accurate analysis in precision agriculture, it is recommended to use a yield map of at least three years. In precision agriculture, for example, with yield maps, regions with low yields and the factors limiting yield in these regions can be determined. In the production to be made in the following year, the management decisions to be taken by considering these factors also allow an increase in productivity. If there are land deficiencies such as low soil pH or soil compaction which limit the yield, one must try to eliminate them first. In cases where the deficiency cannot be eliminated, the amount of input that cannot be used by the plant can be reduced, thus contributing to economic production by preventing the use of unnecessary inputs and preventing these inputs from harming the environment. Precision agriculture practices begin with the acquisition of data by using various sensors and remote sensing technologies and continue with soil tests of the production area and determination of soil properties. Application maps are created by associating all information such as yield values, fertilizer and pesticide application norms, climate data, topographic data, weed density, and disease status of previous production seasons with their actual location in the production area. Then, by using appropriate hardware and software, variable-level applications are made in the field (Özgüven, 2018).

1.2   VARIABILITY

Variability is the most important feature that leads to the emergence of precision agriculture. Understanding the concept of variability will facilitate the understanding of precision agriculture. In Figure 1.1, photographs of the same cotton field are given. Looking at the two photos therein, it can be seen that the yield is high in some areas and the yield is low in some areas. Looking at the photos, it is seen that the cotton yields in the field are not the same and the yield is variable in different areas (Özgüven, 2018).

There are three types of variability in precision agriculture (Sındır, 2008):

- *Spatial variability*: It consists of differences on the land, including physical (soil properties, yield, etc.), chemical (nutrients, pH, etc.), and biological (plant growth and development, diseases, pests, weeds, etc.).
- *Temporal variability*: It is the variability seen in yield, soil, etc. properties during the plant development process from year to year.
- *Predictive variability*: It is the variability in total yield, pests and diseases, product prices, climatic forecasts, etc. management decisions that do not match the reality.

The variability that has significant effects on agricultural production can be divided into six groups (Zhang et al., 2002):

- *Yield variability*: Historical and current yield distributions.
- *Area variability*: Area topography (elevation, slope, direction, and terrace), proximity to area boundary and flows, etc.

FIGURE 1.1   Variability in cotton yield (Özgüven, 2018).

- *Soil variability*: Soil fertility (N, P, K, Ca, Mg, C, Fe, Mn, Zn, and Cu), soil fertility provided by fertilizer, soil physical properties (texture, density, mechanical strength, moisture content, and electrical conductivity), soil chemical properties (pH, organic matter, salinity, and cation exchange capacity), soil structure (available water-holding capacity and hydraulic conductivity), soil depth.
- *Plant variability*: Plant density, plant height, plant nutrient stresses for N, P, K, Ca, Mg, C, Fe, Mn, Zn, and Cu, plant water stress, plant biophysical properties (leaf area index, captured photosynthetically active radiation and biomass), plant leaf chlorophyll content, plant grain quality.
- *Variability in abnormal factors*: Weed infestation, insect infestation, nematode infestation, disease infestation, wind damage, and hay damage.
- *Management variability*: Tillage application, plant hybrid, seed rate sown, crop rotation, fertilizer application, pesticide application, irrigation pattern.

1.3   VARIABLE RATE APPLICATION

With precision agriculture practices, areas with different characteristics seen in agricultural areas and grown plants are determined. Applications made depending on these determined variations are called variable rate applications (VRAs). Since a VRA is the implementation phase, it is the most important phase of precision agriculture. By changing the application norms in VRAs, only as much input as needed can be applied at variable rates. While various benefits are obtained from all VRAs, the importance of variable rate fertilization and pesticide applications especially is in the first place in terms of providing financial gain and reducing environmental pollution (Özgüven, 2018). When the nozzles on the boom are examined individually in Figure 1.2, it is seen that liquid fertilizer or spraying was applied to a VRA.

FIGURE 1.2   Variable rate liquid fertilizer or pesticide application (Trimble, 2022a).

The main idea in the application of precision agriculture techniques is to first determine the variable crop potentials of the land and the factors that cause yield changes for that area with point studies made in the previous harvest period or the same planting period. Then, the necessary application (variable spraying, fertilizer application, etc.) is done in the next or the same growing period. Before applying variable rate input (Türker et al., 2003):

- Variability should be identified and quantified;
- Factors causing variability should be identified;
- To improve and enrich production, management and operating decisions should be determined for the correction of problems;
- The economic returns of these applications should be analyzed;
- Business and management effectiveness that will provide definite economic benefits and returns must be determined.

1.3.1   VARIABLE RATE APPLICATION METHODS

VRAs include a computerized control unit and associated equipment to deliver inputs such as fertilizer, seeds, pesticides, and water to the field at variable rates. There are two application methods: Map-based and sensor-based VRAs. These two systems can be applied together due to some advantages over each other, and thus more successful results can be obtained. Figure 1.3 shows the schematic of the spinner spreader making a VRA according to the map-based system.

In the map-based system, application maps are prepared in the GIS environment or with special software by the following analysis and the evaluation of the obtained data together by the experts (Özgüven, 2018):

- Maps prepared in the GIS environment by matching the results of soil sampling made to determine the levels of nutrients in the soil with spatial information with the help of GPS;
- Yield maps prepared in the GIS environment by matching the yield values obtained with yield sensors during crop harvesting with spatial information with the help of GPS;

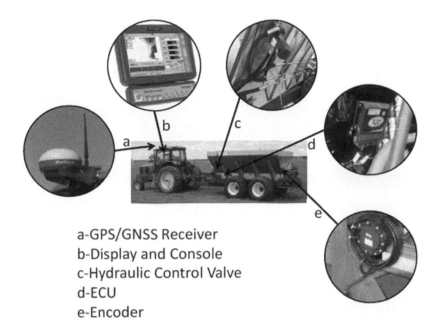

a-GPS/GNSS Receiver
b-Display and Console
c-Hydraulic Control Valve
d-ECU
e-Encoder

FIGURE 1.3 Components of the spinner spreader with automatic section control (Fulton et al., 2010).

- Maps such as plant stress maps obtained with aerial photographs obtained by remote sensing techniques;
- Maps prepared by matching the information obtained with sensors such as electrical conductivity sensors with spatial information with the help of GPS;
- All available data such as business records, soil type, and land characteristics.

In the sensor-based system, instant data obtained from plants and soil are sent to the computer with the help of sensors placed on vehicles. With the simultaneous processing of the data by the computer, the application rates are determined, and instant application is made by sending electrical signals to the rate adjuster by the control unit. Figure 1.4 shows the WeedSeeker® 2 system as an example of a sensor-based system.

Trimble's Field-IQ™ crop input control system provides productive and efficient functionality in planting, nutrient, and pest management operations (Figure 1.5). This system is a flow and application control system with advanced precision agriculture features such as avoiding seed and fertilizer overlapping, controlling material application rate, monitoring seed delivery and fertilizer blockage, and controlling the height of spray booms. It can provide automatic section control at the inch level in up to 48 individual sections. The application rate of six different materials such as granular seed, granular fertilizer, liquid, and anhydrous ammonia in different combinations can be controlled simultaneously.

A GreenSeeker®-brand NDVI (Normalized Difference Vegetation Index) sensor can be used to determine the amount of nitrogen needed by the plant with the help of the radiation values reflected from the leaves. GreenSeeker® is a commercially available sensor-based system used for the VRA of nitrogen. This sensor indirectly evaluates the chlorophyll (greenness) level and biomass amount by calculating the NDVI. Generally, NDVI values range from −1 to 1 and give negative values when the sensor measures bare soil. Also, NDVI values close to 0 indicate poor biomass while 1 indicates high biomass. By applying an equation, a given NDVI value can be converted to a site-specific nitrogen rate based on plant needs at the time of application. The equation used by the on-the-go VRA nitrogen system is developed through extensive research requiring NDVI measurements from an area with no nitrogen limitation, the number of days from planting to NDVI measurements,

FIGURE 1.4   Sensor-based system (Image provided by Trimble Inc.).

FIGURE 1.5   Trimble's Field-IQ™ crop input control system (Trimble, 2022b).

and the plant yield potential (Norwood et al., 2009). Figure 1.6 shows the study for the detection of *Cylindrocladium* black rot (CBR) disease in a peanut field using the GreenSeeker®-brand NDVI sensor. The GreenSeeker®-brand NDVI sensor produces infrared and near-infrared light. To determine the NDVI value of the plant, the reflected light is analyzed and the calculated NDVI value is recorded by transferring it to the handheld computer together with the location information obtained by GPS. Mapping is also done using software such as FarmWorks or ArcGIS.

The Yara N-Sensor and fertilizer spreader system is a sensor that measures the nitrogen needs in real time while passing through the field, and the VRA fertilizes when the tractor comes to the

FIGURE 1.6 Detection of *Cylindrocladium* black rot (CBR) disease in a peanut field with NDVI sensor.

FIGURE 1.7 Yara N-Sensor.

measured place. By using this system, VRA of the correct and optimum fertilizer ratio is ensured in each part of the field (Figure 1.7). The nitrogen sensor determines the nitrogen requirement by measuring the chlorophyll content of the plant covering a total area of approximately 50 m^2 and the light reflection rate in certain wavebands related to the biomass. Measurements are taken every second at normal operating speeds.

WeedSeeker® 2 is the next-generation spot spray system from Trimble Agriculture (Figure 1.8). The WeedSeeker® system has advanced optics and processing power for weed detection and elimination. When a weed passes underneath the sensor while on the move in real time, it signals its linked spray nozzle to precisely deliver herbicide and kill the weed, and the chemical is applied. In this way, by applying herbicide only to weeds, the amount of applied chemical is reduced up to 90%.

In the WeedSeeker® system, each sensor unit consists of a light source and an optical sensor. The sensors are mounted on a rod or spray arm in front of the spray head and directed to the ground

FIGURE 1.8 Weed control by selective spraying with the sensor-based WeedSeeker® 2 system (Image provided by Trimble Inc.).

(Figure 1.9). The microprocessor interprets the data read by the sensor. It senses the presence of weeds when a threshold signal is passed and sends a signal from the controller to the solenoid valve to deliver a certain amount of herbicide to spray only the weeds without spraying bare soils. Thus, spraying is done by opening the spray nozzle. It is an effective solution in areas where weeds occur intermittently.

1.3.2 COMPARISON OF VARIABLE RATE APPLICATION METHODS

Both methods have their own advantages and disadvantages. These strengths and weaknesses are described in the following (Morgan and Ess, 2003; Keskin and Görücü Keskin, 2012):

Advantages of map-based systems:
- The amount of input to be applied can be determined before the application is made. This information is important for planning how much input is needed in advance.
- The time elapsed between data collection and VRA benefits in terms of better data processing and analysis and higher accuracy.
- Since the necessary changes in the application amount are known in advance, it can be planned in advance to adjust the speed of the application machine when it is necessary to change the application amount.

Weaknesses of map-based systems:
- A GNSS is needed to determine the instant geographical position of the vehicle while it is moving.
- Sampling data should be collected, stored, and processed using computer and software (GIS) for a certain period of time before the application, and the application map should be obtained. This requires significant time and labor.

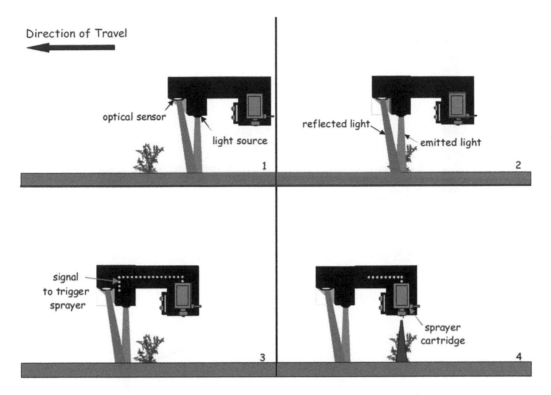

FIGURE 1.9 Operation of the Weedseeker® system (Anonymous, 2017).

- Special software is required to create a VRA map.
- Positioning errors may cause the application to be placed in the wrong place.
- Since samples collected from the field to determine variability usually represent a specific area and interpolation is applied to the data, the application for intermediate zones may not be sensitive enough.
- Soil and crop conditions may change during the time period between data collection and application.

1.3.3 Components in Variable Rate Application

Map-based and sensor-based VRAs are basically the same, although there are some differences in their components. For example, although the use of GNSS is not mandatory in sensor-based systems, it is preferred to use the applications made during the processes for recording purposes. A VRA system includes the use of a computer with the appropriate field application software for the measuring mechanism, the controller, a DGPS receiver, and a control mechanism (hydraulic valve, motor, etc.):

- *Computer*: A laptop, PDA, or computer systems produced by other equipment manufacturers can be used (e.g., John Deere, etc.). This component acts as the user interface and capability to run the application control software. Figure 1.10 shows (tractor, combine harvester and sprayer, etc.) a computer in the form of a monitor is seen on it, which provides monitoring and control for the entire VRA.
- *Controller*: Processes and controls the application rate. It can be a system separate from the software and control mechanism. It uses the setpoint ratio in the software and allows

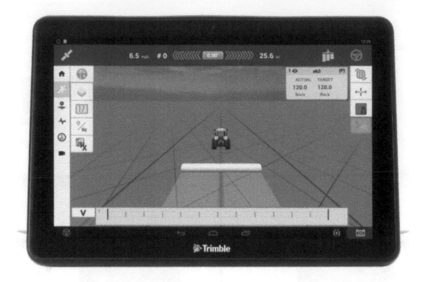

FIGURE 1.10 Trimble GFX-750 display (Trimble, 2022c).

FIGURE 1.11 Field-IQ ISOBUS liquid control system (Trimble, 2022d).

the control mechanism (motor or actuator) to output the appropriate rate. It uses feedback from ground speed radar or other speed sensors to compensate for speed changes. It also uses velocity or position feedback from the control mechanism to ensure it is rotated at the proper speed or positioned correctly (Figure 1.11).

- *Software*: Allows viewing application maps, determining desired application rate based on field position, and recording. It reads the loaded application map and reports the desired application rates to the controller (Figure 1.12).
- *GNSS receiver*: Provides location information used by the control software to set rates based on the application map. It also provides the ability to spatially log the actual rates applied to generate the applied maps. A GPS receiver with differential correction (WAAS, Starfire, OmniStar, RTK, etc.) is used in applications (Figure 1.13).
- *Hydraulic motor and valve*: It is the component that controls the measuring unit. It can be a motor, linear actuator, or another control device. In the disc-type fertilizer spreader, the fertilizer or pesticide particles fall on the rotating disc and are accelerated by the disc blades. VRA is achieved by changing the conveyor or chain speed, adjusting the opening

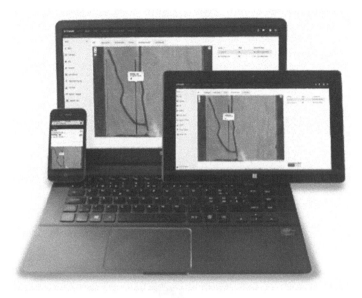

FIGURE 1.12   Trimble Farmer Core software (Image provided by Trimble Inc.)

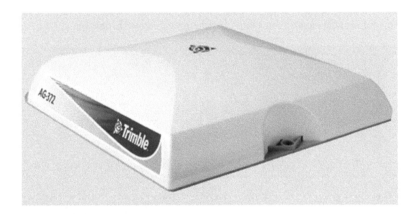

FIGURE 1.13   GNSS receiver (Trimble, 2022e).

of the cover, and changing the application speed. The drive mechanism used to control the conveyor or chain speed can be a pneumatic, electric, or hydraulic motor. These motors are also used to adjust the sowing norm. Since the row spacing is fixed in variable rate sowing machines, the sowing norm is made by changing the row spacing by adjusting the speed of the planter layouts. Figure 1.14 shows the current Trimble electric row clutch used for variable rate seeding or input application.

- *Control and/or measuring mechanism*: It is the component that directly controls the feed rate. The crawler chain (Figure 1.15) used to distribute solid manure to metering mechanisms, seed disc, or liquid injection system can be given as an example.

1.3.4   VARIABLE RATE APPLICATION SYSTEMS

In VRAs, the most commonly used input applications are fertilizer, pesticide, and sowing applications. In these applications, the inputs are in the form of powder, granule, and liquid in fertilization and spraying, and the seeds in sowing are granular. Powder and granular chemical fertilizers

FIGURE 1.14 Current Trimble electric row clutch (Image provided by Trimble Inc.).

FIGURE 1.15 Controlling the feed rate of the apron chain (Fulton et al., 2009).

and pesticides are distributed to the soil with disc-type and pneumatic-type fertilizer spreaders and spraying machines, and liquid fertilizers are distributed to the soil with variable rate sprayers. Pneumatic fertilizer and pesticide dispensers have transmission pipes like spraying booms. Powder and granular fertilizer and drug particles are conveyed to the soil by being carried in the pipe system with the effect of compressed air. The VRA is realized by adjusting the airflow rate of the fan by changing the operating speed of the hydraulic-driven fan and by controlling the section in the transmission pipes and opening and closing the cover settings. In addition, suspension systems and transmission pipes are kept in balance by making them parallel to the ground.

A VRA in liquid fertilizer distribution machines is performed by adjusting the pressure and flow rate in the hydraulic system, changing the speed of movement, section control in booms, or turning on and off the nozzles independently. Section control is provided by the control of boom valves and/or nozzle solenoids. Figure 1.16 shows a sensor-based VRA scheme for liquid fertilizer and pesticide application.

Advanced irrigation management and application technologies combined with advanced detection, modeling, and control technologies are used to ensure the best performance in VRA irrigation. With VRA irrigation, plant yield is optimized through systematic information collection and processing about the plant and area, and field-specific targets are set. In this way, reducing the use of inputs prevents waste of resources, minimizing adverse environmental impact. In precision irrigation, the goal is not only the yield increase, but also the practices that will allow savings in the use of inputs so as not to result in yield loss. With this system, data obtained continuously and effectively from the field is analyzed using advanced hardware and software, and more accurate decisions are made regarding farming. Furthermore, by using these techniques, it is possible to facilitate managers' decision-making in the effective management of irrigation networks (Figure 1.17) (Özgüven and Karaman, 2012).

Depending on soil compaction, variable rate soil tillage including variable depth, variable width, and variable speed applications can be made (Figure 1.18) (Keskin and Görücü Keskin, 2012).

1.3.5 CONTROL TECHNOLOGY IN VARIABLE RATE APPLICATION

The boom can be automatically turned on and off through boom control to prevent spraying to selected areas, such as waterways within the boundaries of the field, by mapping these areas.

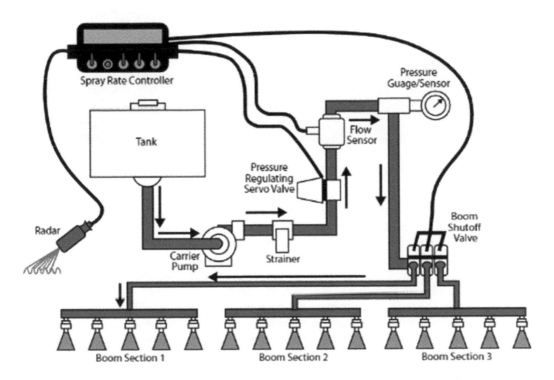

FIGURE 1.16 Sensor-based VRA scheme for liquid fertilizer applications (Grisso et al., 2011).

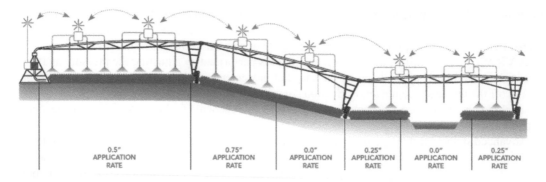

| 0.5″ APPLICATION RATE | 0.75″ APPLICATION RATE | 0.0″ APPLICATION RATE | 0.25″ APPLICATION RATE | 0.0″ APPLICATION RATE | 0.25″ APPLICATION RATE |

FIGURE 1.17   Individual sprinkler control allowing variable rate irrigation (Anonymous, 2011a).

FIGURE 1.18   Variable depth tillage (Görücü Keskin and Keskin, 2010).

In addition, if the boom section is located in a pre-applied area (Figure 1.19), the controller can turn off the boom section and eliminate overlaps. It also eliminates omissions by turning on boom sections after leaving the previously applied area behind.

An example of a spinner spreader without automatic section control or guidance system is given in Figure 1.20a. The operator has full control of deciding when to switch the spreader on-off and maintain parallel, adjacent transitions. However, this determination could be difficult as it may be visibly challenging to see where manure has been previously applied. Therefore (a) skips and (b) overlaps can occur frequently. A spinner spreader equipped with a guidance system and automatic section control is shown in Figure 1.20b. The operator does not have to guess where previous manures have been applied. Because of this, skips or overlaps are greatly reduced. Overlaps (Figure 1.20b–a) at headlands or field ends is minimized compared to operating without an automatic control system.

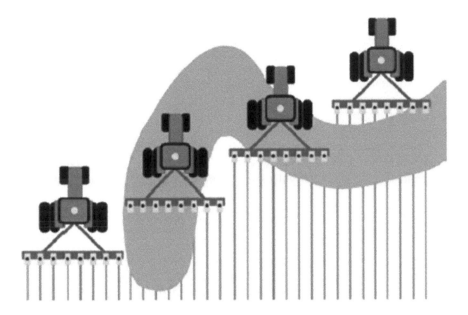

FIGURE 1.19 Electronic boom control to eliminate overlaps available for fertilization, spraying, and sowing (Grisso et al., 2011).

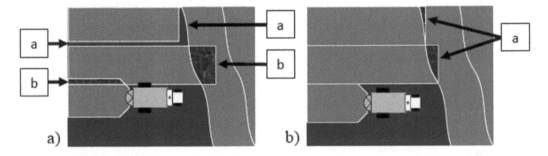

FIGURE 1.20 Use of automatic control system in fertilization: (a) skips and (b) overlaps (Fulton et al., 2010).

1.4 BENEFITS OF PRECISION AGRICULTURE

As a result of the rapid technological advances after the Second World War, up to 100% yield increase was achieved thanks to the agricultural techniques called "Green Revolution", started to be implemented in agriculture. The use of chemical fertilizers and pesticides in vegetable production and the use of hormones and antibiotic-like feed additives in animal production increased the yield and reduced the cost but caused a very rapid deterioration of the ecosystem. Some crops and animal products, produced with these techniques that contain residues, have adversely affected human health, and caused significant increases in health problems, particularly cancer (Ak, 2004). The increasing use of soil, irrigation, and agricultural chemicals played a great role in the growth of agricultural production during the Green Revolution. However, it is accepted that the gains often have adverse effects on the natural resources used in agriculture, including degradation of land, salting of irrigated areas, excessive extraction of groundwater, formation of pest resistance, and erosion of biodiversity. Agriculture has harmed the wider environment through deforestation, greenhouse gas emissions, and nitrate pollution in water supplies (FAO, 2011a).

Climate-smart agriculture aims to sustainably increase food safety and revenues, reduce potential risks, adapt to climate change while capturing benefits, and gain flexibility. Protected agriculture links other innovations such as agricultural ecology, agricultural forestry, and the development of crop varieties with greater tolerance for pests, diseases, drought, flooding, and salinity (FAO, 2013). In climate-smart agriculture, agroforestry systems are an important tool for food production in a sustainable way, while maintaining the ecosystem, especially in regions prone to environmental degradation. It also promotes sustainable livestock production through mixed plant-livestock systems that fully integrate environmental goals and production goals. Examples include the rotation of pasture and forage plants to improve soil quality and reduce erosion, and the use of livestock manure to maintain soil fertility (FAO, 2017).

Sustainable agriculture targets such as the long-term income of the producer in agriculture, elimination of environmental problems caused by pesticide pollution, chemical fertilizer pollution, and wrong irrigation can be achieved by the application of precision agriculture technologies. The benefits of precision agriculture are listed in the following sections.

1.4.1 Effective Fertilizer Usage

Fertilizer is considered one of the most important inputs of agricultural production. Increasing productivity in agriculture requires the use of more inputs per unit area, and therefore fertilizer and fertilization come to the fore since they play an important role in plant nutrition. Improper and excessive use of fertilizers causes soil and water pollution. Chemical substances that do not dissolve easily in water, such as nitrate and phosphate, can be mixed into surface and groundwater, resulting in harmful effects on human health (Zengin, 2008).

To decide on the most suitable fertilizer form, dosage, and application method, soil samples should be taken from the land to be fertilized, and necessary soil analyses should be performed in accordance with the technique. On the other hand, a homogeneous fertilization program applied to large agricultural areas will often result in waste of plant nutrients, breakdown of balance between nutrients, pollution of natural resources, and economic losses. As a result of homogeneous fertilization applied according to soil analysis results from a mixed sampling taken from different parts of the land, some parts of the land will be over-fertilized and some parts will get lesser fertilizer than needed, and this will cause fertilizer accumulation in the soil, or will cause decreased yield in areas with an inadequate amount of fertilizer. This will also lead to the disruption of the proportional balance between the nutrients (Karaman et al., 2012).

Thanks to the GPS, RS, GIS, and VRA technologies in precision agriculture, the variability in different parts of the land can be easily determined, variability is mapped, and field-specific applications are performed depending on the variability. Figure 1.21 shows soil maps of a field prepared according to soil analysis results.

Variability in potassium, phosphorus, magnesium, and pH values is seen in Figure 1.21, as a result of soil analysis. When the scales on the maps given as examples are examined, variability in different parts of the field can be seen. In precision agriculture, variability values are taken into consideration during fertilization, and fertilizer is applied at a variable rate needed by field-specific applications. Thanks to the VRA, it is possible to do fertilization, spraying, and planting in specific amounts needed in the area, as well as to perform desired operations during the application, thanks to the sensor and control systems developed.

1.4.2 Effective Pesticide Usage

Pesticide is a generic name given to all compounds used against various agents, such as plant diseases, harmful insects, and weeds that can cause a decrease in agricultural products. In Turkey, there are more than 240 pests, more than 80 diseases, and about 30 weeds that cause pre-harvest and post-harvest crop loss at the economic level in cultured plants. In the absence of agricultural control against these pests, it is estimated that around 35% loss of crops will occur on average each year. Adverse effects of pesticides on humans and the environment are listed below (Anonymous, 1999):

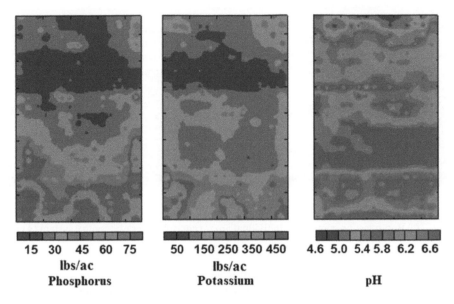

15 30 45 60 75   50 150 250 350 450   4.6 5.0 5.4 5.8 6.2 6.6
     lbs/ac   lbs/ac
   Phosphorus   Potassium   pH

FIGURE 1.21   Varying soil maps (Davis et al., 1998).

- *General effects*: Pesticides begin to disintegrate into different chemicals from the moment they are sprayed. Some of these disintegration products can sometimes be more toxic and persistent than the essential pesticide.
- *Effects on people*: Pesticides enter the human body by mouth, skin, or respiratory tract. Due to the chemical structure of pesticides or the breakdown products called metabolites, some of them accumulate in the body and some of them deteriorate nerve cells, resulting in very dangerous consequences.
- *Effects on waters*: Pesticides can pose dangers for aquatic organisms such as fish and human health by mixing with dam waters, groundwaters, or rivers that provide drinking water and irrigation water.
- *Effects on soil*: Pesticides applied to insects in the soil, nematodes, and seeds are mixed directly into the soil, and have a negative effect on soil flora, which has a useful activity in the soil.
- *Effects on atmosphere*: Some pesticides that can evaporate create environmental pollution by mixing into the air we breathe.
- *Effects on beneficial insects, nutrients, farm animals, and birds*: Pesticides that spread to nature in a variety of ways can damage a wide variety of living things.

Thanks to VRAs, it is possible to do spraying in the specific amount needed in the area, as well as to perform the desired spraying thanks to the sensor systems and control systems developed in precision agriculture (Figure 1.22).

Since a manually controlled sprayer cannot stop spraying immediately when it encounters a trench as in Figure 1.22, it causes contamination of the trenches. Thanks to the GPS/GNSS-based automatic control unit developed in precision agriculture, spraying is stopped with the boom section or individual nozzle control when encountering a trench, and the spraying continues afterward. In this way, the unwanted places, such as the trenches in the field, are not sprayed.

1.4.3   EFFECTIVE IRRIGATION

The pressures on natural resources are driven not only by changes in demand but also by climate changes. It is estimated that rainfall and temperatures will become more variable with climate

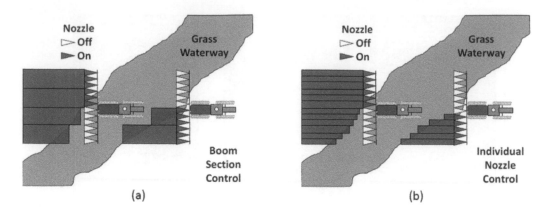

FIGURE 1.22   (a) Boom section control, (b) individual nozzle controlled sprayer control system (Anonymous, 2011b).

change, which will lead to more drought. It will have particularly severe effects on rain-fed small farming systems in mountainous areas and the tropics, which make up 80% of the world's farmland and account for about 60% of global agricultural production. Water consumption for agriculture represents 70% of all water consumption. The FAO estimates that more than 40% of the rural population lives in river basins with water scarcity (FAO, 2011b).

Water scarcity, defined in terms of access to water, is a critical restriction for agriculture in many parts of the world. More than 1.2 billion people, one-fifth of the world's population, live in areas of physical water scarcity and are unable to find enough water for their demands. About 1.6 billion people live in watersheds with water shortages, where human capacity or financial resources will be insufficient to develop adequate water supplies (Figure 1.23). Today's water scarcity will possibly cause complexity in the population in the coming years. A growing population is an important factor, but the main causes of water problems are due to other reasons. These include lack of interest in water and poverty, inadequate and incomplete investments, inadequate human resources, ineffective institutions, and mismanagement (IWMI, 2007).

Figure 1.23 shows that areas of physical and economic water scarcity are divided into four sections in the world (IWMI, 2007):

- *Little or no water scarcity*: Abundant water resources, of which less than 25% of the water from rivers is used for people's needs.
- *Physical water scarcity* (water resources development, approaching or exceeding sustainable limits): Use of more than 75% of the river flow for agricultural, industrial, and domestic purposes (including the recycling of return flows). Defining water availability by associating it with water demand means that dry areas do not necessarily have water scarcity.
- *Approaching physical water scarcity*: More than 60% of the river flows are retracted. These basins will experience physical water scarcity in the near future.
- *Economic water scarcity* (although water in nature is locally available to meet human demands, humanitarian, institutional, and financial capital limits access to water): Water supplies are plentiful compared to water use. Although less than 25% of the water in the rivers is used for human purposes, there is still malnutrition.

Undoubtedly, the most important factor in increasing crop production per unit area is irrigation and fertilization. There is an absolute need for soil water in the transformation of fertilizers into the form that can be taken by plants (Doğan, 2001). Irrigation for intensive agricultural activities can lead to unsustainable water use. Land drainage can lead to lower groundwater levels and also to the destruction of wetlands, which are the habitat of many species. In addition, pesticides and nitrates

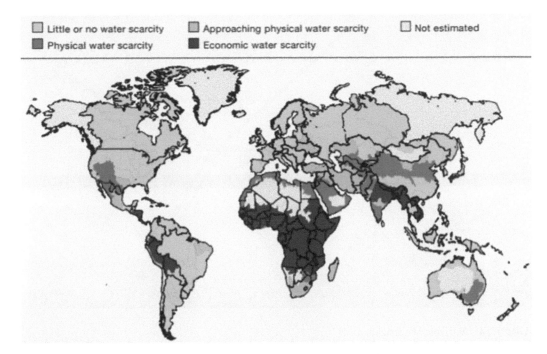

FIGURE 1.23   Areas of physical and economic water scarcity (IWMI, 2007).

from fertilizer use may leak into groundwater, and similarly, nitrogen and phosphorus resulting from fertilizer use may mix with surface water. Irrigation and drainage can affect groundwater levels and cause salinity in the soil (Dişbudak, 2008).

Advanced irrigation management and application technologies combined with advanced detection, modeling, and control technologies are used to ensure the best performance in precision irrigation systems. With precision irrigation technologies, plant yield is optimized through systematic information collection and processing about the plant and area, and field-specific targets are set. In this way, reducing the use of inputs prevents waste of resources, minimizing adverse environmental impact. In precision irrigation, the goal is not only the yield increase, but also the practices that will allow savings in the use of inputs so as not to result in yield loss. With this system, data obtained continuously and effectively from the field is analyzed using advanced hardware and software, and more accurate decisions are made regarding farming. Furthermore, by using these techniques, it is possible to facilitate managers' decision-making in the effective management of irrigation networks (Özgüven and Karaman, 2012). In precision irrigation, the amount of water is applied according to soil type, topography, product type, or barriers in the field. Figure 1.24 shows the individual sprinkler-controlled pivot system allowing variable rate irrigation.

1.4.4   EFFECTIVE MANAGEMENT

More complex finer adjustments in decision-making and input practices do not necessarily increase farm profits or wealth in overly vague, variable, and uncontrolled activities, such as agriculture. The impact of risk factors, such as the weather and price, may exceed the potential benefits of more precise production decisions and methods (Knight and Malcolm, 2009). Precision agriculture has many potential benefits, such as increased crop yield, cheaper input costs through improved process control, and reduced environmental impact (Weiss, 1996).

As a result of the advantages provided by precision agricultural technology, an agricultural enterprise will be able to establish a registration system by recording detailed information in GIS, such as

FIGURE 1.24   Variable rate irrigation.

yield, disease, climate information, the types of agricultural products produced, the processes such as spraying, fertilization, types of inputs such as seeds, pesticides, fertilizers, etc. In this way, efficiency will be increased by ensuring that more effective business decisions are taken. The income from agricultural production is increased through increased crop yield, higher-quality products, and reduced input expenses, such as fertilizer, pesticides, seeds, and water. Increased income makes the agricultural business more stable, leading to the opportunity to increase employment. In this way, it contributes to rural development.

1.4.5   Contribution to Food Safety

Food safety is the use of resources and strategies to ensure that foods are properly produced, processed, and distributed so they are safe for consumption. Food safety is related to the presence of food-borne hazards such as chemical, physical, and biological hazards in food at the point of consumption (Jevsnik et al., 2008). Food security is often shaped by four dimensions: Food availability, food access, food utilization and use, and food stability. These dimensions form the general framework of the definition established by the Food and Agriculture Organization of the United Nations (FAO) (United Nations, 2017). Food security exists when everyone, at all times, has physical, social, and economic access to adequate, safe, and nutritious food that meets their nutritional needs and dietary preferences for an active and healthy life (FAO, 2016).

The world's population is expected to increase to about ten billion by 2050, and agricultural demand (in a modest economic growth scenario) is expected to increase by about 50% compared to 2013. High-input, resource-intensive farming systems that cause large amounts of deforestation, water scarcity, soil depletion, and higher amounts of greenhouse gas emissions cannot provide sustainable food and agricultural production. Innovative systems that protect and enhance the natural resource base, while increasing productivity, are required (FAO, 2017). Today, different approaches, such as plant breeding, genetically modified foods, in vitro planting, and the spread of closed ecological systems, are implemented as a solution to increase food availability. In addition to these methods, the use of modern technologies under the name of precision agriculture is also recommended as a way of ensuring food safety. Using various methods such as remote sensing and geographic information systems, early intervention is provided to prevent unwanted events through monitoring vegetation at every stage of production from soil processing to the harvest. Different methods of precision agriculture can provide the following benefits in terms of food safety (Talebpour et al., 2015):

- Crop quality:
- Increasing crop quality using spatially variable fertilizer management;
- Reducing the impact of climate change;
- Timely control of pests;
- Increasing irrigation efficiency when water resources are scarce;
- Protection of crop quality with precision post-harvest application.
- Sustainability and rural development;
- Traceability;
- Risk management.

1.4.6 TRACEABILITY

According to EC 178/2002, "traceability" is the ability to trace and follow a food, feed, food-producing animal, or substance intended or expected to be incorporated into a food or feed, through all stages of production, processing, and distribution (Art. 3.15), and "stages of production, processing and distribution" is any stage, including import, from and including the primary production of a food up to and including its storage, transport, sale, or supply to the final consumer and, where relevant, the importation, production, manufacture, storage, transport, distribution, sale, and supply of feed (Anonymous, 2002). While these definitions cover product, input, and process traceability, traceability can also be realized for different categories and purposes (Cebeci, 2006):

- Product traceability;
- Process traceability;
- Input traceability;
- Genetic traceability;
- Disease and residue traceability;
- Measuring/measurement traceability.

Precision agriculture can provide opportunities to track crops using a system. Among these opportunities is the process that describes all the applications made to produce the final product. As a result, crop monitoring and traceability capability is becoming one of the main topics in precision agriculture research, especially in farm management monitoring (McBratney et al., 2005). By using the geographic information system as a tool in precision agriculture, traceability information can be obtained by associating soil conditions, local features, and basin areas with the agricultural environment. Therefore, the characteristics of the immediate and indirect environment of the field can be determined to document the history of the events in the production area from sowing to harvest (Oger et al., 2010). Another important application related to traceability in precision agriculture is the use of radio frequency identification (RFID) technology. The widespread use of RFID in precision agriculture makes it possible to increase the effectiveness, efficiency, and profitability of agricultural systems while avoiding undesirable impacts on the environment. Obtaining real-time information allows farmers to adjust strategies at any time, providing a solid foundation to identify differences and change operations accordingly. Examples of where RFID tags are used to ensure traceability include (Ruiz-Garcia et al., 2011):

- Animal identification and monitoring;
- Measurement of greenhouse temperature, relative humidity, and lighting conditions in greenhouses, where spectral imaging and environmental detection are used together with RFID technology;

- Measurement of soil temperature using wireless sensors;
- In orchards, in the boxes for matching trees corresponding to the fruits harvested during harvest;
- A cold chain that uses RFID tags with embedded temperature sensors.

In precision agricultural technology, machines and tools have intelligent processing units that can communicate with each other, with other units, and with the farm management unit. In this way, intelligent machines can collect multi-purpose information at high spatial and temporal resolution. As seen in Figure 1.25, the ISO11783 standard, called ISOBUS, provides a manufacturer-independent plug-and-play solution for each combination of tractors and equipment. Integrated into the farm business, this technology makes an important contribution to the documentation and traceability of products and processes, including changes in field operations (Auernhammer and Speckmann, 2006).

1.4.7   Risk Management

It has been observed that there has been a significant increase in the number and severity of meteorological natural disasters because of global warming and the greenhouse effect, which have been more felt in the world in recent years. In the agricultural sector, which is defined as the "open-top factory", the great financial losses caused by the effect of natural risks threaten the world economy to a great extent (Dinler, 2003). First of all, in the management of natural risks in agriculture, there are technical protection measures. In addition to early warning systems, the use of systems that provide minimum water consumption against the risk of drought, the tendency toward plant patterns that consume less water, and dwarf fruit growing are some of them. Cultivation of frost-, hail-, and storm-resistant varieties gains the ability to resist natural risks as part of risk management (Dinler et al., 2005).

Managing productivity risk is always a challenge for farmers and farm managers. They may have the best prices, but if they have little or nothing to sell, the result can be disastrous. Production or yield risk arises from a variety of factors, such as weather-related events (inadequate rainfall, hail, etc.), insects, diseases, crop variety selection, soil characteristics, or production techniques. Precision farming can enable a producer to manage the spatial component of return risk, but it does not always help him cope with weather risks (Gandonou, 2005).

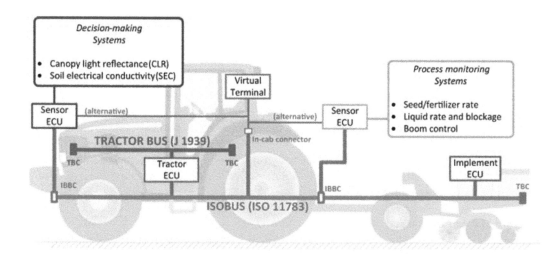

FIGURE 1.25   ISOBUS setup of a tractor and a towed implement (Paraforos et al., 2019).

Adopting precision farming to make decisions about risk management, improving control of crop growing conditions, and providing more and better information: The purpose of this approach to risk management is the site-specific treatment of trouble areas to reduce the probability of low yields and returns according to temporal variability in yields and net returns while also taking into account spatial variability. Although high risks of using precision farming, like not eliminating the probability of crop loss, financial risk due to investment in bad crop season, and human and technological risks, early prediction of yield using remote sensing (RS) data leads to an increase in farmers' confidence in early marketing. It also makes contracting easier by pushing agriculture closer to "producing to specification". This means that if farmers have more control over input application and yield quality, buyers will be more willing to contract early to ensure their returns (Lowenberg-DeBoer, 1999).

1.5 COMPONENTS OF PRECISION AGRICULTURE

The success to be obtained from precision agriculture technologies depends on the level of technology used. Therefore, there is a need to use the best available technology. The basic elements of technologies used in the field of precision agriculture technology can be grouped into three groups (Figure 1.26). With the use of data collection technologies in the first group, necessary spatial and temporal basic data are provided for farmers. Data processing and decision-making technologies in the second group include technologies that make raw data fit for purpose and interpret it. Application technologies in the third group enable the planned operations to be carried out in the production area by making use of the maps and comments created (Keskin and Görücü Keskin, 2012). Technologies and processes related to precision agriculture are summarized in Figure 1.27.

The use of remote sensing satellites, which are one of the basic components of precision agriculture, in monitoring crops, land use, vegetation indices, plants and their characteristics, diseases and pests, physical and chemical properties of soils, fertilizer application, water status, water resources, etc. and implementing management decisions has increased recently. Significant increases are seen in both the number and imaging resolutions of these satellites. A case study of agricultural monitoring application with remote sensing is shown in Figure 1.28.

With the use of technologies such as artificial intelligence, robotics, Internet of Things, autonomous vehicles, drones, advanced computers, cloud computing, and big data in today's precision agriculture applications, the transition to the Agriculture 4.0 period has been made. Thus, with the use of these technologies in agriculture, more competitive, efficient, and sustainable agricultural practices are possible. In this way, the opportunity to instantly obtain and evaluate data in agricultural applications has emerged. Thus, more conscious decisions can be made for better-quality production in agricultural production. Figure 1.29 shows the transition from precision agriculture to Agriculture 4.0 with the use of IoT.

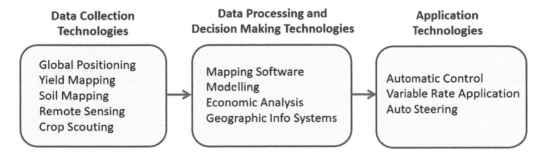

FIGURE 1.26 Basic elements of precision agriculture technologies (Keskin and Keskin, 2012).

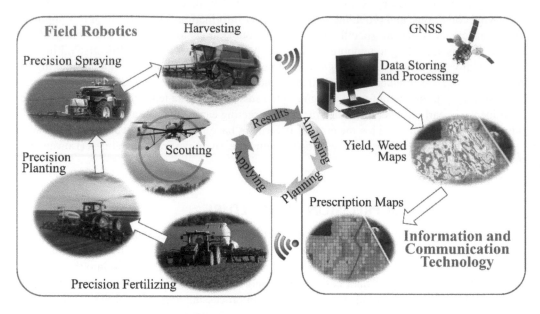

FIGURE 1.27 Technologies and processes involved in precision agriculture (Gonzalez-de-Santos et al., 2020).

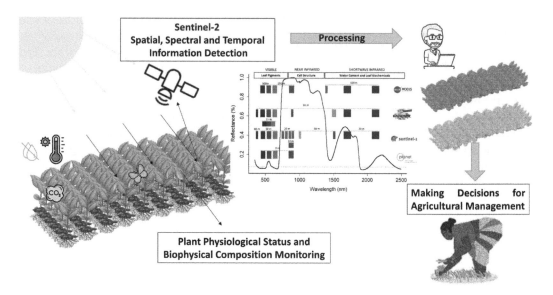

FIGURE 1.28 A case study of agricultural monitoring application with remote sensing (Segarra et al., 2020).

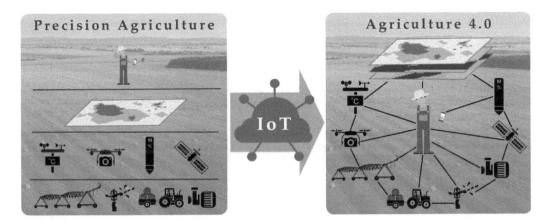

FIGURE 1.29 Transition from precision agriculture to Agriculture 4.0 with the use of IoT (Monteleone et al., 2020).

REFERENCES

Ak, İ., 2004. Apolyont Doğal Tarım ve Hayvancılık Projesi. I. Uluslararası Organik Hayvansal Üretim ve Gıda Güvenliği Kongresi. 28 Nisan–1 Mayıs, 2004, s.144. (Turkish).

Anonymous, 1999. Türkiye'nin Çevre Sorunları. Türkiye Çevre Vakfı Yayınları no: 131. Ankara. (Turkish).

Anonymous, 2002. Regulation (EC) No 178/2002 of the European Parliament and of the Council of 28 January 2002 Laying Down the General Principles and Requirements of Food Law, Establishing the European Food Safety Authority and Laying Down Procedures in Matters of Food Safety. *Official Journal of the European Communities*, L 31/1, [Accessed 1.2.2002].

Anonymous, 2011a. Irrigation Advances Conserving Water, Energy and Labor. 20(1) Spring 2011.

Anonymous, 2011b. Automatic Section Control (ASC) Technology for Agricultural Sprayers. Alabama Cooperative Extension System. Precision Agrkiculture Series Timely Information. Agriculture, Natural Resources & Forestry. August 2011.

Anonymous, 2017. www.ntechindustries.com/rowcrop.html. [Accessed 07.11.2017].

Auernhammer, H. and Speckmann, H., 2006. Section 7.1 Dedicated Communication Systems and Standards for Agricultural Applications, pp. 435–452 of Chapter 7 Communication Issues and Internet Use, in CIGR Handbook of Agricultural Engineering Volume VI Information Technology. Edited by CIGR-The International Commission of Agricultural Engineering; Volume Editor, Axel Munack. St. Joseph, Michigan, USA: ASABE. (Çevirmenler: Yaşar, E., Tarhan, S. ve Ozguven, M.M.; Çeviri Editörleri: Tarhan, S. ve Ozguven, M.M.).

Bongiovanni, R. and Lowenberg-Deboer, J., 2004. Precision Agriculture and Sustainability. *Precision Agriculture*, 5, 359–387.

Cebeci, Z., 2006. Gıda İzlenebilirliğinde Bilgi Teknolojileri. Ulusal Tarım Kurultayı, 15–17 Kasım 2006, Çukurova Üniversitesi, Adana. Bildiriler s. 189–195. (Turkish).

Davis, G., Casady, W. and Massey, R., 1998. Precision Agriculture: An Introduction. University of Missouri Extension, WQ450.

Dinler, T., 2003. Tarımda Meteorolojik Karakterli Doğal Afetler ve Risk Yönetim Teknikleri, III. Atmosfer Bilimleri Sempozyumu, İstanbul. (Turkish).

Dinler, T., Yaltırık, A., Çetin, B., Özkan, B., Gülçubuk, B., Sürmeli, E., Ekmen, E., Saner, G., Akçaöz, H., Karahan Uysal, Ö., Karaaslan, S. ve Kıymaz, T., 2005. Tarımda Risk Yönetimi ve Tarım Sigortaları. ZMO Türkiye Ziraat Mühendisliği VI. Teknik Kongresi, Cilt.2, s.1209-1232, Ankara. (Turkish).

Dişbudak, K., 2008. Avrupa Birliği'nde Tarım-Çevre İlişkisi ve Türkiye'nin Uyumu. AB Uzmanlık Tezi. T.C. Tarım ve Köyişleri Bakanlığı Dış İlişkiler ve Avrupa Birliği Koordinasyon Dairesi Başkanlığı. Ankara. (Turkish).

Doğan, O., 2001. Toprak-Su-Gübre-Bitki İlişkilerinde Araştırmanın Önemi. T.C. Tarım ve Köyişleri Bakanlığı, Tarımın Yeniden Yapılanmasında Toprak-Su Politika Toprak Muhafaza ve Sulama Politikaları Sempozyumu, 30–31 Ocak 2001, Ankara. (Turkish).

FAO, 2011a. Land Tenure, Climate Change Mitigation and Agriculture. Mitigation of Climate Change in Agriculture (MICCA) Programme. June 2011. Rome.

FAO, 2011b. The State of the World's Land and Water Resources for Food and Agriculture (SOLAW). www .fao.org/nr/solaw/solaw-home.

FAO, 2013. Climate-Smart Agriculture Sourcebook. Rome.

FAO, 2016. Food Security Indicators. http://www.fao.org/economic/ess/ess-fs/ess-fadata/en/.

FAO, 2017. The Future of Food and Agriculture – Trends and Challenges. Rome. ISBN: 978-92-5-109551-5.

Fulton, J., Brodbeck, C., Winstead, A. and Norwood, S., 2009. Overview of Variable-Rate Technology. Alabama Cooperative Extension System. Precision Agriculture Series Timely Information. Agriculture, Natural Resources & Forestry. January 2009.

Fulton, J., Brodbeck, C., Winstead, A. and Norwood, S., 2010. Automatic Section Control (ASC) for Spinner Disc Spreaders. Alabama Cooperative Extension System. Precision Agriculture Series Timely Information. Agriculture, Natural Resources & Forestry. December 2010.

Gandonou, J.M.A., 2005. Essays on Precision Agriculture Technology Adoption and Risk Management. University of Kentucky Doctoral Dissertations. Paper 227. http://uknowledge.uky.edu/gradschool_diss /227.

Gonzalez-de-Santos, P., Fernández, R., Sepúlveda, D., Navas, E., Emmi, L. and Armada, M., 2020. Field Robots for Intelligent Farms—Inhering Features from Industry. *Agronomy*, 10, 1638. https://doi.org/10 .3390/agronomy10111638.

Görücü Keskin, S. and Keskin, M., 2010. Advanced Technologies in Soil Tillage (Lecture). 9th International Symposium on the Results of Turkish-German Research Studies in Agriculture. 22–27 March 2010, Mustafa Kemal University, Antakya, Hatay, Turkey, pp. 38–48.

Grisso R., Alley M., Thomason, W., Holshouser, D. and Roberson, G.T., 2011. Precision Farming Tools: Variable-Rate Application. Virginia Cooperative Extension. Virginia Tech. Publication 442–505.

IWMI (International Water Management Institute), 2007. *A Comprehensive Assessment of Water Management in Agriculture*. Edited by D. Molden. Water for Food, Water for Life: A Comprehensive Assessment of Water Management in Agriculture Earthscan. www.earthscan.co.uk

Jevsnik, M., Hlebec, V. and Raspor, P., 2008. Consumer Interpretation of the Term Food Safety. *Acta Alimentaria*, 37(4), 437–448.

Karaman, M.R., Brohi, A.R., Müftüoğlu, N. M, Öztaş, T. ve Zengin, M., 2012. Sürdürülebilir Toprak Verimliliği. Koyulhisar Ziraat Odası Kültür Yayınları No:1. (Turkish).

Keskin, M. and Görücü Keskin, S., 2012. Precision Agriculture Technologies (Hassas Tarım Teknolojileri). Book (In Turkish), Mustafa Kemal University, 212 pages. ISBN: 978-975-7989-33-2.

Knight, B. and Malcolm, B., 2009. A Whole-Farm Investment Analysis of Some Precision Agriculture Technologies. *AFBM Journal*, 6(1), 41–54.

Lowenberg-DeBoer, J. 1999. Risk Management Potential of Precision Farming Technologies. *Journal of Agricultural and Applied Economics*, 31(2), 275–286.

McBratney, A., Whelan, B., Ancev, T. and Bouma, J., 2005. Future Directions of Precision Agriculture. *Precision Agriculture*, 6(1), 7–23.

Monteleone, S., Moraes, E.A.D., Tondato de Faria, B., Aquino Junior, P.T., Maia, R.F., Neto, A.T. and Toscano, A., 2020. Exploring the Adoption of Precision Agriculture for Irrigation in the Context of Agriculture 4.0: The Key Role of Internet of Things. *Sensors*, 20, 7091. https://doi.org/10.3390/s20247091.

Morgan, M. and Ess, D., 2003. *The Precision-Farming Guide for Agriculturists*. 2nd Edition. John Deere Publishing, Moline, IL.

Norwood, H., Ortiz, B., Winstead, A. and Fulton, J., 2009. On-the-go Crop Sensing. Alabama Cooperative Extension System. Alabama A&M and Auburn Universities. March 2009.

Oger, R., Krafft, A., Buffet, D. and Debord, M., 2010. Geotraceability: An Innovative Concept to Enhance Conventional Traceability in the Agrifood Chain. *Biotechnologie, Agronomie, Société et Environnement*, 14(4), 633–642.

Ozguven, M.M., 2018. The Newest Agricultural Technologies. *Current Investigations in Agriculture and Current Research*, 5(1), 573–580. https://doi.org/10.32474/CIACR.2018.05.000201.

Özgüven, M.M., 2018. Hassas Tarım. Akfon Yayınları, Ankara. ISBN: 978-605-68762-4-0. (Turkish).

Ozguven, M.M. and Türker, U., 2010. Application of Precision Farming in Turkey, Comparative Analysis of Wheat, Cotton and Corn Production. *Journal of Agricultural Machinery Science*, 6(2), 127–135.

Özgüven, M.M. ve Karaman, S., 2012. Hassas Sulama Teknolojileri. 27. Tarımsal Mekanizasyon Ulusal Kongresi, 5–7 Eylül Samsun, S:288-297. (Turkish).

Özgüven, M.M., Türker, U., Akdemir, B., Çolak, A., Acar, A.İ., Öztürk, R. ve Eminoğlu, M.B., 2020. Tarımda Dijital Çağ. Türkiye Ziraat Mühendisliği IX. Teknik Kongresi. Ocak 2020, Ankara. Bildiriler Kitabı-1, s.55–78. (Turkish).

Paraforos, D. S, Sharipov, G.M. and Griepentrog, H.W., 2019. ISO 11783-Compatible Industrial Sensor and control Systems and Related Research: A Review. *Computers and Electronics in Agriculture*, 163, 104863. https://doi.org/10.1016/j.compag.2019.104863.

Ruiz-Garcia, L. and Lunadei, L., 2011. The Role of RFID in Agriculture: Applications, Limitations and Challenges. *Computers and Electronics in Agriculture*, 79(1), 42–50.

Segarra, J., Buchaillot, M.L., Araus, J.L. and Kefauver, S.C., 2020. Remote Sensing for Precision Agriculture: Sentinel-2 Improved Features and Applications. *Agronomy*, 10, 641. https://doi.org/10.3390/agronomy10050641.

Sındır, K.O., 2008. Tarımda Yeni Teknolojiler. Gıda Tarım Kongre ve Konferansları. Adana. (Turkish).

Talebpour, B., Türker, U. and Yegül, U., 2015. The Role of Precision Agriculture in the Promotion of Food Security. *International Journal of Agricultural and Food Research*, 4(1), 1–23.

Trimble, 2022a. Field-IQ™ Crop İnput Control System Datasheet. [Accessed 06.11.2022].

Trimble, 2022b. https://agriculture.trimble.com/product/field-iq-system/. [Accessed 06.11.2022].

Trimble, 2022c. https://agriculture.trimble.com/blog/choosing-the-best-display-for-your-farm/. [Accessed 06.11.2022].

Trimble, 2022d. https://agriculture.trimble.com/product/field-iq-isobus-liquid-control-system/. [Accessed 06.11.2022].

Trimble, 2022e. https://positioningservices.trimble.com/wp-content/uploads/2019/02/AG-372_receiver_HR _500x500.png. [Accessed 06.11.2022].

Türker, U., Güçdemir, İ. ve Karabulut, A., 2003. Alansal Değişkenliğin Hassas Tarım Teknolojilerinden Yararlanarak Belirlenmesi Üzerine Bir Araştırma. Tarımsal Mekanizasyon 21. Ulusal Kongresi, 3–5 Eylül 2003, Konya. (Turkish).

United Nation, 2017. The Role of Science, Technology and Innovation in Ensuring Food Security by 2030. United Nations. Economic and Social Council. 27 February 2017.

Weiss, M., 1996. Precision Farming and Spatial Economic Analysis: Research Challenges and Opportunities. *American Journal of Agricultural and Applied Economics*, 78, 1275–1280.

Zengin, E., 2008. Küreselleşme Sürecinde Tarım'da Sürdürülebilirlik ve Çevre Sorunları. *Alatoo Academic Studies*, 3(2), S:44–54. (Turkish).

Zhang, N., Wang, M. and Wang, N., 2002. Precision Agriculture-A Worldwide Overview. *Computers and Electronics in Agriculture*, 36(1), 113–132.

2 Precision Livestock Farming

2.1 IMPORTANCE OF ANIMAL PRODUCTION

Animal production is an important sector due to its great contribution to the economy such as the production of meat, milk, and other animal products necessary for a balanced diet for people, the creation of new employment in the fields of leather and textile, medicine, feed, and equipment industries based on animal husbandry, the realization of rural development, and the production of plants and plant residues that are not consumed as human food. Therefore, it has an important place and value in human life and the national economy (Özcen, 2019). Livestock farming activities, which are important for the sustainability of agricultural production, are carried out together with plant production, so that pasture plants, other roughage, agricultural products, and food industry residues that cannot be consumed as human food are evaluated. Thus, it is ensured that agricultural enterprises can continue their activities economically in every period of the year (Yığmatepe and Özgüven, 2020). Livestock farming is less affected by extreme changes in climatic conditions than crop production. This is achieved through both physiological and behavioral adaptation mechanisms, as well as the greater adaptability or tolerance of animals to temperature changes. For this reason, it is the guarantee of the farmer's family when it comes to extreme climatic conditions (Ertuğrul, 2012).

Adequate and balanced nutrition plays an important role in the protection and development of the health of individuals; meat and meat products, milk and dairy products, vegetables and fruits, and bread and cereals are divided into four groups. In the milk and dairy products group, there are foods made from milk such as yogurt, cheese, and milk powder. These foods are important sources of many nutrients, including protein, calcium, phosphorus, vitamin B2, and vitamin B12 (Ünal and Besler, 2012). Animal products (meat, milk, eggs, honey, and by-products) are indispensable in human nutrition and cannot be substituted for other nutrients due to their biological properties. Eight amino acids, which are also important for human growth, development, staying healthy, and brain development, are only found in proteins of animal origin. Protein amounts in animal foods are 15–20% in meat, 19–24% in fish, 12% in eggs, 3–4% in milk, and 15–25% in cheese. For these reasons, animal products such as red meat, white meat, milk, and eggs should be consumed regularly. However, 72% of the daily protein consumed in our country is from plant-based foodstuffs (Anonim, 2015).

2.2 ANIMAL NUTRITION

Animals need nutrients such as water, carbohydrates, protein, fat, vitamins, and minerals to survive and produce various products. Animals obtain these nutrients from the many feeds and feed additives used in animal feeding and the water they drink. In addition to providing nutrients, feeds also have the functions of forming fillers and giving desired color, odor, and taste to animal products (Filya and Canbolat, 2015). Balanced and economical nutrition of animals is possible for them with balanced and healthy ration practices. The ration is the total amount of feed that meets the daily energy, protein, mineral, and vitamin requirements of an animal. Considering that 50–70% of the total production costs in livestock enterprises are composed of feed costs, it is very important for an economic business to decide which feeds to choose and to what extent they will be used in the rations while meeting the nutrient requirements of the animals (Alçiçek and Kırkpınar, 2015).

It is possible to classify the feeds used in animal nutrition in many ways. According to the international official feed classification, it is possible to organize feeds into seven main groups. The first

DOI: 10.1201/b23229-2

three groups of this classification are forages, the next two groups are dense (concentrated) feeds, and the last two groups are feed additives (Toker et al., 1994):

- Dry roughage;
- Meadow-grassland green fodder;
- Silo feeds;
- Energy source feeds;
- Protein source feeds;
- Mineral supplements;
- Vitamin supplements.

Forages are plant-based feeds with high fiber content (raw cellulose) used as animal feed in fresh, dried, and silage forms. Grass-eating makes up the main part of animal rations. The raw cellulose content of the feeds in this group is high, but the protein and energy levels are low. Forages are important for grass-eating animals, not only because they meet their nutritional requirements, but also because they regulate digestive system movements. The cellulose, protein, mineral substances, and vitamin contents of roughage vary considerably. Although legume roughage can contain up to 20% crude protein, this rate drops to 3–4% in straw. Animals fed with high-density feed must necessarily include roughage in their rations. In addition, roughage has the effect of developing rumen muscle tissue and epithelial cells and stimulating intra-rumen fermentation. Dense feeds are rich in energy and protein content, and the degree of digestibility of nutrients is high. Some of the dense feeds have higher energy content and some have higher protein content. Energy feeds with a cellulose content of less than 18% are used to increase the energy level of the ration. Feeds with a protein level of more than 20% are called protein supplements. Protein-rich feeds are very important in animal nutrition. Feed additives, on the other hand, are substances that provide the highest level of benefit from the nutrients consumed by animals with feed, do not impair their health, and enable them to develop and produce better. Commonly used feed ingredients in animal nutrition are preservatives, antioxidants, flavors, enzymes, probiotics, prebiotics, toxin binders, methane formation inhibitors, flatulence inhibitors, buffers, pellet binders, and vitamin-minerals (Filya and Canbolat, 2015).

2.3   MECHANIZATION IN LIVESTOCK

Mechanization in livestock farming, production of feeds necessary for balanced and adequate nutrition of animals and their presentation to animals, cleaning and storage of manure accumulated in the shelter environment to provide a clean living space for animals, providing suitable climatic environmental conditions (temperature, humidity, clean air) for animals, and collection of animal products are all the realization of many processes such as evaluation (Yıldız et al., 2008). They are aimed to realize both economic production and optimum health conditions with the triple relationship between animal–human–machine with mechanization in livestock farming. In addition, it is also important that the animal, the working person, and the product produced are healthy. Animal husbandry and related mechanization can be examined under three main groups (Ayık et al., 2015):

- *Feed mechanization*: All of the methods to be applied for the storage, preparation, and feeding of the feed obtained according to the harvest method (shredded, long, bale) are only valid for the livestock enterprises that produce their own feed. The type and amount of feed to be produced differ according to the type and size of the enterprise. The necessity of harvesting in a limited time depending on the climatic conditions necessitates the establishment of a good and complete mechanization (harvest + transport + storage) chain, especially in large enterprises.
- *Manure mechanization*: These are wide-ranging applications that include the collection of solid and liquid residues in the barns, their removal from the barn, their storage by accumulating and transporting them to the field after maturation, and their distribution as

farm manure. In addition, if the production of biogas from barn manure is also intended, the facilities for biogas mechanization should be compatible with manure mechanization facilities in terms of technical and economic management.

- *Obtaining the product*: In milk production enterprises, facilities and arrangements suitable for the type of barn should be provided for milking and temporary storage in the enterprise. There is no need for such a structure in fattening enterprises. The necessity of increasing work efficiency necessitated the development of milking places and milking techniques. In milking parlors, milker road time is minimized. On the other hand, milking systems that are commanded according to milk flow have been developed. In addition, it is ensured that the milking heads are automatically retrieved after the end of the milking work.

The optimum ambient temperature for cattle varies between 10–25°C for dairy cows. The temperature dropping to zero degrees in open barns is not a big problem. The effect of relative humidity on cattle is considered together with temperature. Cows are more uncomfortable with cold and humid air than with cold and dry air. Within the optimum temperature limits, the relative humidity should be 60–80%. It is necessary to ventilate the barns to remove excess heat, humidity, and odors in the barn and to provide fresh air. The hourly amount of air required for one cattle per unit animal is 240 m^3 in summer and 114 m^3 in spring and autumn. One of the most important environmental conditions required in barns is light. For adequate natural lighting, the window surface area should be $\dfrac{1}{15} - \dfrac{1}{20}$ of the barn floor area (Yavuzcan, 1995).

Milk production mechanization is required for milk production: 25% for manure removal, 15% for feeding, and 10% for other works in dairy cow farming, which is one of the most important animal husbandry branches. Milk yield, udder health, and milk quality depend on careful milking and the conscious use of machinery and equipment. Milking and in-house storage of milk should be carried out in a clean and healthy way for quality milk production. In addition, the psychology and anatomy of the animal should be considered during milking (Ayık et al., 2015). To preserve the natural properties of the milk and to increase the storage period, it must be cooled. Microorganisms in milk, especially bacteria, reproduce rapidly at the temperature at which milk is milked, causing various fermentations. As a result, the milk turns sour and spoils. To obtain healthy milk, the milk must be cooled immediately within two hours after milking. The temperature of a newly expressed milk is around 36–37°C, which is the body temperature of the animal. The milk to be delivered to the dairy must not be hotter than 12°C. Therefore, milk should be cooled down to 8–12°C, depending on the shortness or length of transportation time (Yavuzcan, 1994).

2.4 PRECISION LIVESTOCK PRODUCTION

The first desired condition in animal production is breeding of breeds with high meat and milk yield. The second one is to make sure that the highest level of individual potential of animals is achieved through adequate and balanced nutrition. The third is to take preventive health measures against diseases that cause major losses in animal production and to minimize the use of drugs with the early detection of diseases and the necessary intervention. Precision livestock farming (PLF) practices have contributed significantly to the solution of the problems experienced in animal breeding and in increasing the desired yield and quality in meeting the increasing animal food needs (Özgüven, 2017a). In PLF, the animals are constantly monitored individually and as a herd, along with their performance, behavior, physiological parameters, welfare status, housing environment, product, reproduction, and health status, and are recorded for the purpose of making necessary business decisions and creating a database. Warning systems are established to detect sudden changes in general situations. For this purpose, sensors are used to automatically obtain computerized imaging technologies, RFID tags, movement activity, rumination number, ruminal pH, pulse signal, body temperature, respiratory rate, body weight, body condition score, location information, and acoustic information.

As animal production systems continue to increase in scale and complexity, it has become difficult to comprehensively monitor the condition of the animals and the housing and operating environment that are critical for the animals' welfare. For example, in poultry farming, each farm has more than one building and each building houses tens of thousands of broilers or hundreds of thousands of laying hens. Routine daily care for such large flocks is limited to inspecting the house once or twice a day, removing dead or diseased chickens, and checking to ensure that equipment such as feeders, drinkers, and ventilation systems are working during the inspection. However, there is now an increasing trend to use monitoring systems on commercial farms to assess and manage animals and their environments in real time (Mench, 2018).

To make the automatic welfare assessment system functional, it should meet five main conditions: (1) Cows should be automatically identified; (2) monitoring and recording of housing environment parameters should be carried out; (3) monitoring of cow behaviors, physiological parameters, production levels, and other animal-based parameters reflecting welfare status should be done; (4) software for processing and interpretation of the data obtained and models for welfare prediction should exist; and (5) data on housing environments, animal-based parameters, diseases, and welfare scores should be kept in a database and complemented in real time. In particular, data collected on cows for breeding recommendations and health checks can be used for automatic assessment of animal welfare. An important factor of automatic welfare monitoring is the arrangement of data exchange between different parts of the integrated system (Figure 2.1). This is influenced by (1) the structure of the network; (2) the unification of the interfaces and protocols; and (3) the management

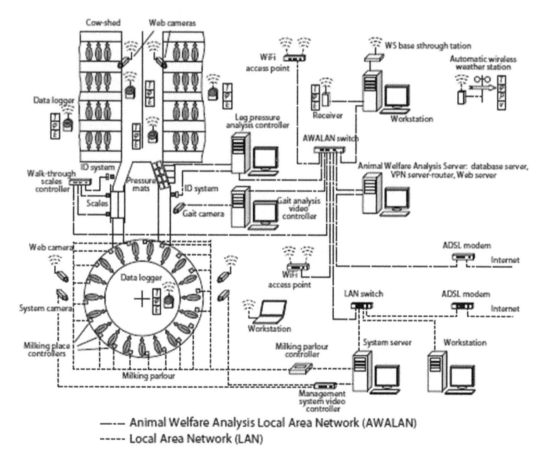

—·— Animal Welfare Analysis Local Area Network (AWALAN)

----- Local Area Network (LAN)

FIGURE 2.1 Configuration example of local area and animal welfare analysis networks in PLF (Poikalainen et al., 2013).

information system's openness for data exchange. In addition, manual data entry and special tools for experiment control (for example, some of the health status information is entered manually) are required (Poikalainen et al., 2013).

2.5   PRECISION LIVESTOCK FARMING APPLICATIONS

Although there are examples of precision livestock farming (PLF) practices in ovine and poultry breeding, they have more applications in cattle breeding. The main reason for this is that cattle breeding is generally carried out in barns by animals from breeds with high meat and milk yields and businesses with large herds, and there are suitable environments for the use of developed systems and devices. To obtain optimum efficiency from animals in large herds with high material values and where individual animal observation is more difficult and less frequent, enterprises prefer sensitive animal production practices. PLF practices enable both the herd and individual animals to be summarized and to report situations that require intervention. In this way, the benefit to be obtained reaches the maximum level. With electronic animal identification systems, animals are identified, information is collected with developed devices and systems, and data is recorded by storing data in other environments such as computer or internet environments with wired–wireless data transfer methods (Özgüven, 2018).

Ever since man began domesticating animals a few thousand years ago, we have always relied on our intuition, common knowledge, and sensory signals to make effective animal breeding decisions. So far, this has helped us make important gains in animal livestock farming and farming. Together, increasing food demand and the progress in sensing technology have the potential to make animal farming more centralized, large-scale, and efficient. Sensors, big data, AI, and machine learning play a role in helping livestock farmers to reduce production costs, increase productivity, improve animal welfare, and grow more animals per hectare. At the same time, despite the difficulties and limitations of the technology, new opportunities are being explored with newly developed techniques (Neethirajan, 2020). Figure 2.2 shows the data monitored to establish optimum growth conditions in dairy cattle in the PLF.

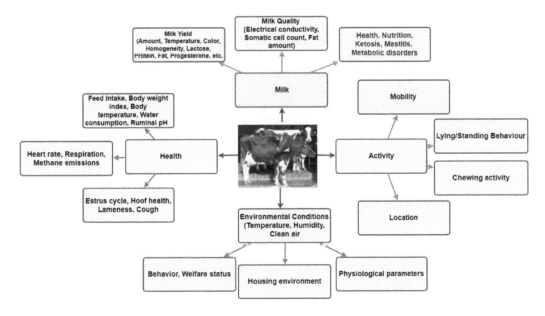

FIGURE 2.2   Data monitored in PLF.

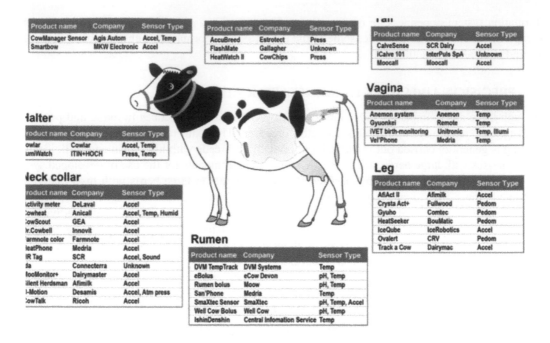

Product name	Company	Sensor Type
CowManager Sensor	Agis Autom	Accel, Temp
Smartbow	MKW Electronic	Accel

Product name	Company	Sensor Type
AccuBreed	Estrotect	Press
FlashMate	Gallagher	Unknown
HeatWatch II	CowChips	Press

Tail

Product name	Company	Sensor Type
CalveSense	SCR Dairy	Accel
iCalve 101	InterPuls SpA	Unknown
Moocall	Moocall	Accel

Vagina

Product name	Company	Sensor Type
Anemon system	Anemon	Temp
Gyuonkei	Remote	Temp
iVET birth-monitoring	Unitronic	Temp, Illumi
Vel'Phone	Medria	Temp

Halter

Product name	Company	Sensor Type
Cowlar	Cowlar	Accel, Temp
RumiWatch	ITIN+HOCH	Press, Temp

Neck collar

Product name	Company	Sensor Type
Activity meter	DeLaval	Accel
Cowheat	Anicall	Accel, Temp, Humid
CowScout	GEA	Accel
Dr.Cowbell	Innovit	Accel
Farmnote color	Farmnote	Accel
HeatPhone	Medria	Accel
IR Tag	SCR	Accel, Sound
Ida	Connecterra	Unknown
MooMonitor+	Dairymaster	Accel
Silent Herdsman	Afimilk	Accel
U-Motion	Desamis	Accel, Atm press
CowTalk	Ricoh	Accel

Leg

Product name	Company	Sensor Type
AfiAct II	Afimilk	Accel
Crysta Act+	Fullwood	Pedom
Gyuho	Comtec	Pedom
HeatSeeker	BouMatic	Pedom
IceQube	IceRobotics	Accel
Ovalert	CRV	Pedom
Track a Cow	Dairymac	Accel

Rumen

Product name	Company	Sensor Type
DVM TempTrack	DVM Systems	Temp
eBolus	eCow Devon	pH, Temp
Rumen bolus	Moow	pH, Temp
San'Phone	Medria	Temp
SmaXtec Sensor	SmaXtec	pH, Temp, Accel
Well Cow Bolus	Well Cow	pH, Temp
IshinDenshin	Central Infomation Service	Temp

FIGURE 2.3 Commercially available wearable wireless sensors in cattle (Yoshioka et al., 2019).

Data provides support for decision systems: Individual or herd sheep or cows, farmers, veterinarians, experts with knowledge of farming, professional associations, research institutes, certification institutions, etc. The data collection elements that feed the decision process can be divided into three groups: (1) Sensors attached to animals or in their environment capture the behavior of the system; (2) user interfaces allow the farmer to observe the system; (3) revealing mechanisms extract information from the analysis of the system by experts. Data and information from these elements are stored in several databases and can be classified and then structured through an experience feedback process and summarized in a historical database (heart rate, body temperature, etc.). Expert opinion is either used directly in the decision process or combined with historical technological data from the experience feedback process to create business rules (IF the sheep looks weak AND the temperature is above 40°C THEN call the veterinarian). These rules are used in the decision process. Structured data is the rules generated by the experience feedback process from historical technological data and the revealing expert knowledge (Villeneuve et al., 2019). Figure 2.3 shows commercially available wearable wireless sensors for early disease detection in cattle.

2.5.1 ELECTRONIC ANIMAL RECOGNITION SYSTEMS

In identification, an identification number is assigned to each animal, and the material (e.g., tag) bearing this identification number is placed on the animal. In this way, it is possible to mark all animals in the enterprise in a way that can be separated from the others and thus be monitored continuously. Today, traditional methods (such as ear tags, tattooing, branding, or stamping), electronic methods containing RFID (such as rumen boluses, ear tags, and injectable transponders), and biometric methods (such as retinal scanning, nose prints, and DNA) are used for identification and tracking of animals (Bowling et al., 2008). RFID is a technology that enables identification information to be transmitted via radio waves (Figure 2.4). RFID electronic identification systems are widely used in animal husbandry due to their advantages such as being able to read all the characteristics of many animals at once within the herd software system, allowing remote reading and writing of even the

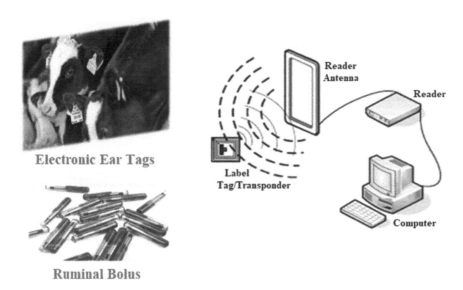

FIGURE 2.4 Electronic animal identification systems (Özgüven, 2017b).

characteristics of animals in motion, not needing an open field of view, and making changes easily when desired (Özgüven, 2018). Rumen boluses are used to determine rumen temperature, pressure, and pH in animals to prevent the occurrence of health problems such as subacute rumen acidosis and swelling, which cause animal deaths and loss of productivity. Measurements can be made by placing a bolus with a temperature sensor, processor, and radio transmitter in the rumen to monitor rumen temperature, rumen pressure, and rumen pH (Singh et al., 2014).

2.5.2 Automatic Feed Measuring Systems

Animals are fed in different rations and amounts according to their developmental status and yield to obtain optimum efficiency individually during the feeding of the animal. To adjust and give feed amounts in different rations, automatic coarse-concentrate feed mixers and dispensers with electronic scales, automatic feeder systems, and drinker systems measuring water consumption have been developed (Özgüven, 2018). To obtain maximum milk yield, optimum feed is fed according to the yield status of each animal. When the animals enter the feeding cabin, the animal identification information is detected by the system and the feed to be given during the day is poured in front of them in a controlled manner (Kuşçu and Arın 2003). In the preparation of feed rations, corn silage, barley and oat straw, cotton and sunflower meal, dried alfalfa grass, concentrate feed, and mineral feed materials are mixed. In this way, it is tried to ensure that the daily nutritional needs of the animals are fully met (Yıldız and Özgüven, 2018).

The usage of PLF can help farmers to improve management tasks such as monitoring animal performance and health and optimizing feeding strategies. A key component of PLF is precision livestock nutrition, which consists of providing individuals or a group of animals with the amount of nutrients in real time that maximizes nutrient utilization without loss of performance. Precise feeding of animals can reduce protein intake by 25% and nitrogen release into the environment by 40%, and increase profitability by approximately 10%. Feeding success depends on the automatic and continuous collection of data, data processing and interpretation, and control of farm processes. With the advancement of PLF feeding, new nutritional concepts, and the development of mathematical models, we can predict individual animal nutritional requirements in real time. Further progress for these technologies will require the coordination of different experts and stakeholders such as nutritionists, researchers, engineers, technology suppliers, economists, farmers, and consumers

(Pomar et al., 2019). Figure 2.5 shows the system developed for the automatic determination of individual feed consumption. In addition to determining the amount of feed, these systems can also monitor eating habits and eating frequency with software developed based on artificial intelligence.

There are many factors that affect how much feed an individual animal consumes and feed efficiency. These factors can be divided into two important categories as individual animal variations and environmental factors including management decisions. Figure 2.6 shows the system developed for the automatic determination of individual feed consumption. In addition to determining the

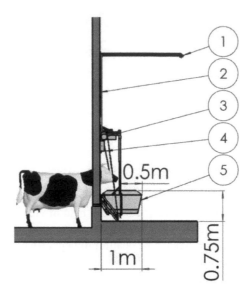

FIGURE 2.5 Developed system for automatic determination of feed consumption: (1) camera, (2) supporting beam, (3) load cell, (4) electronics box, (5) feeding container (Halachmi et al., 2019).

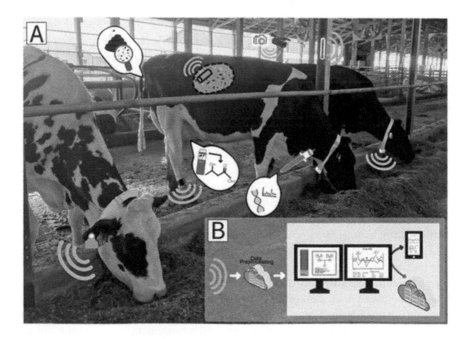

FIGURE 2.6 An improved system for automatic determination of feed consumption (Siberski-Cooper and Koltes, 2022).

amount of feed, these systems can also monitor eating habits and eating frequency with software developed based on artificial intelligence. In the study by Siberski-Cooper and Koltes (2022), various technologies such as milk spectral data, activity, rumen measurements, and image-based phenotypes were associated with feed intake.

Campos et al. (2018) presented a method for classifying different grass-eating behaviors by the surface electromyography (sEMG) signal of the masseter muscle. The main hypothesis tested in the study is whether rumination and food eaten can be distinguished from sEMG signal features using machine learning techniques. The three scenarios examined were differentiation (IR) between ruminant and grazing, food identification (FC) for four different foods, and both conditions combined (FCR). In the study, a new segmentation technique was developed and applied to automatically subdivide the chewing motion signal, evaluated by eight features extracted from seven classifier signals (LDA, QDA, SVM, MLP-NN, RBF-NN, KNN, and MLP-NN), and combined into five sets. Despite the similar characteristics of grasses, it has been reported that the results of food recognition were found to be reasonable and that feeding and rumination were distinguished with relatively high accuracy.

2.5.3 DRINKER SYSTEMS MEASURING WATER CONSUMPTION

The amount of water needed by dairy cattle cannot be determined precisely; it is between 80 and 120 liters depending on the type of feed, ration of dry matter, ambient air temperature, and humidity. They need 4–5 kg of water to make 1 kg of milk (Toker et al., 1994). A lactating cow, of whose body milk is 87% and water is approximately 55–65%, consumes more water than its body weight. As a matter of fact, the daily water consumption of a highly productive cow can exceed 150 liters. In addition, the decrease in cow's water consumption causes a decrease in milk production. The water consumption demands of the cows are directly related to their feed consumption and the amount of milk milked. Cows consume 30% to 50% of the water they need within one hour of leaving the milking parlor. In order for the cows to consume enough water, the drinker should be wide, the water should be stagnant, and the barn environment should be calm. It is possible for the cow, which exhibits natural water drinking behavior in a calm environment, to produce more milk while feed consumption is encouraged due to drinking more water (Delaval, 2015). Significant variation in water consumption can be a warning about the health status of dairy cattle. For this reason, automatic drinkers have been developed to calculate the amount of water consumption (Tarhan et al., 2015).

Tarhan et al. (2020) developed an automatic water consumption measurement system that allows only one cow to drink water at a time. It consists of three main units: Mechanical unit, electronic unit, and data acquisition/processing unit. The mechanical unit holds different parts together and provides durability with its mounting to the ground. It consists of side barriers that allow only one cow to drink water at a time, a bowl drinker, plumbing fittings, a storage box in which the electronic unit and a battery are kept under lock, and an RF antenna pole with a power supply cabinet (Figure 2.7). In the field trial where dairy cows were used to determine the performance of the developed system, it was reported that each cow's daily water intake was between 80.55 and 164.41 liters, the daily water intake durations were between 581 and 2,870 seconds, and the daily drinking events were between 13 and 40.

2.5.4 MILK QUALITY MEASUREMENT SYSTEMS

To compare production records of dairy cows, it is necessary to standardize the records on the same basis. While standardizing production records, lactation period, the number of milkings per day, calving age, and calving date are considered. The standard lactation period used to compare production records is 305 days. PLF practices must be used to produce high-quality milk. Characteristics of high-quality milk include being free from dirt and other deposits, low bacterial count, no chemical contamination, low somatic cell count, no added water, and good taste. In addition, management

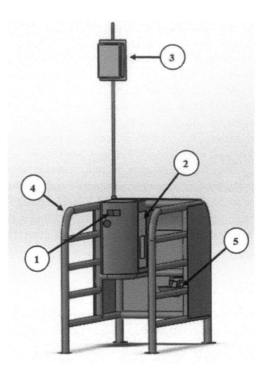

FIGURE 2.7 Mechanical unit: (1) storage box, (2) RFID reader, (3) RF transmitter pole and power supply cabinet, (4) side barriers, (5) bowl drinker with muzzle paddle (Tarhan et al., 2020).

practices to help produce high-quality milk include cleaning the cows, herds, barns, milking parlors, dairy houses, and milking equipment, keeping the cows healthy, cooling the milk properly, using the right cleaning and sanitation methods on the equipment, keeping chemicals away from the milk, and preventing flavors in the milk (Flanders and Gillespie, 2015). Milk compositions such as milk yield, milk flow amount, milking time and duration, milk temperature, electrical conductivity of milk, milk color, milk homogeneity, lactose, protein, fat, progesterone, lactate dehydrogenase, urea, and beta hydroxyl butyrate can be determined with milk quality measurement systems mounted on the automatic milking system. In addition, the number of somatic cells is automatically determined, all data obtained are recorded, and the system gives warnings when necessary. By analyzing the milk, negativities such as udder health, nutrition, ketosis, and metabolic disorders can also be determined.

AfiLab brand milk analyzer is a device that performs real-time analysis of the milk components fat, protein, and lactose and gives an indication of blood and somatic cell counts (Figure 2.8). The working technique is based on spectroscopy and therefore does not interfere with the milk flow in the line or alter the milk in any way. A device is installed at each stall and each cow's milk is analyzed during each milking. The acquisition of milk component data takes place in the same configuration and time frame as the milk weight data measured by the electronic milk meter (Arazi et al., 2011). Besides measuring milk yield, conductivity and lactose content are scanned, and changes in these components accurately indicate mastitis. Ketotic cows have increased fat levels and lower protein levels in their milk. The AfiLab milk analyzer measures milk components and AfiFarm software detects when fat-to-protein ratios exceed 1.4, thus giving an early diagnosis of ketosis. By using the combined data of milk yield, conductivity, and milk components, important information such as subclinical ketosis, subclinical mastitis, digestive problems, and genetic development is obtained (Afimilk, 2021).

Arazi et al. (2011) compared the data obtained from the AfiLab milk analyzer, which was recorded three times a day during the milking of 800 milking cows in a commercial dairy herd,

FIGURE 2.8 Assembled AfiLab milk analyzer used in the milking parlor.

with the reference laboratory data. At the end of the study, small differences in daily calculated bulk tank oil and protein were found between the analyzer and laboratory data (–0.05% to +0.28% and +0.01% to +0.05%, respectively), high-quality (protein>% 3.2), which determined that there was a 5.2% difference in milk volume. They also reported that the analyzer can be used at individual, group, and herd levels.

2.5.5 MONITORING OF MOVEMENT ACTIVITIES IN DAIRY CATTLE

Automatic systems can easily identify when a cow is eating, standing, or lying down and can inform the operator of significant changes in the cow's behavior. The system can also determine the cow's heat and health information. Having the ability to monitor this type of information is profitable for businesses and overall better comfort for their cows. Therefore, these systems are effective in improving the deficiencies of human detection by observation, reducing birth losses due to human detection errors, and detecting estrus in cows (Özgüven and Tan, 2017).

Lying behavior is considered important for the welfare of dairy cattle. It is variable in individual cows, however, and there is interest in determining the factors that influence this variation in commercial dairy farms. However, measuring lying behavior in a commercial setting is difficult and requires continuous observation over a long period of time, particularly to obtain information about aspects of lying behavior such as the number, frequency, and pattern of lying movements. Ankle-mounted and neck-mounted accelerometers are used for automatic and continuous recording of lying behavior. The data obtained from accelerometers are used to evaluate a range of factors potentially associated with variations in lying behavior, such as lameness, leg lesions, body condition, transition diseases, bedding, flooring, feed distribution schedules, fecal contamination of barns, and animal density (Mench, 2018).

Solano et al. (2016) used electronic data loggers to measure lying behavior in 40 Holstein cows on each of 141 dairy farms to determine associations with lameness. Figure 2.9 shows the variation in the prevalence of lameness on farms and the variation in the daily lying times of the herds on those farms. It has been reported on farms that lame cows had longer lying times than healthy cows and with fewer and longer and more variable lying times than non-lame cows.

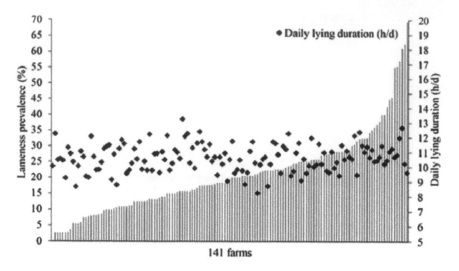

FIGURE 2.9   Average prevalence of lameness (bars) and average daily lying time (hours/day) on each of the 141 dairy farms (Solano et al., 2016).

Sepúlveda-Varas et al. (2016) reported that cows with mastitis had changes in the amount of feed they ate and their motility, and they slept less and stood up more than healthy cows. In the study, the eating behaviors of Holstein cows were determined by using electronic feeders that allow the cows to eat individual feed, measuring the amount of feed eaten, and tracking the RFID identification numbers of the cows. The daily dry matter intake of cows with mastitis started to decrease five days before the day of diagnosis, and a decrease of 1.2 kg/day occurred. One day after the diagnosis and the start of the treatment, the time to eat feed increased from 117 minutes/day to 189.2 minutes/day, the number of visits to the manager increased from 25.4 to 39.1, and the rate of eating increased from 85.4 g/day to 92.1 g/day. It has been reported that partial decreases occur in the daily feeding time and the number of visits to the feeder per day on the second and third days after the start of the treatment, with an increase in feed intake rates up to 102.1 g/day, and the cow's eating behavior improves rapidly with the start of the treatment.

Benaissa et al. (2019) compared leg-mounted and neck-mounted accelerometers to automatically classify cows' behavior and examined the effect of sampling rate and the number of accelerometer axes logged on classification performances. Lying, standing, and feeding behaviors of 16 different lactating dairy cows were recorded with 3D accelerometers for six hours, while behaviors were recorded simultaneously using visual observation and video recordings as references, and different features were extracted from the raw data and machine learning algorithms were used for the classification.

As a result of the study, classification models using combined data of leg-mounted and neck-mounted accelerometers classified three behaviors with high precision (80–99%) and sensitivity (87–99%). For leg-mounted accelerometers, lying behavior was classified as high precision (99%) and sensitivity (98%). The feed was classified more accurately by the neck-mounted accelerometer than the leg-mounted accelerometer (precision 92% vs. 80%; sensitivity 97% vs. 88%) and standing was the most difficult behavior to classify when only one accelerometer was used. The classification performances were not highly influenced when only X, X, and Z, or Z and Y axes were used instead of three axes in the classification, especially for the neck-mounted accelerometer. In addition, the accuracy of the models was reduced by about 20% when the sampling rate was reduced from 1 Hz to 0.05 Hz.

In open areas such as pastures, the movements of animals can be monitored continuously (or at certain time intervals, for example, every five minutes) by attaching a GPS-mounted collar to each

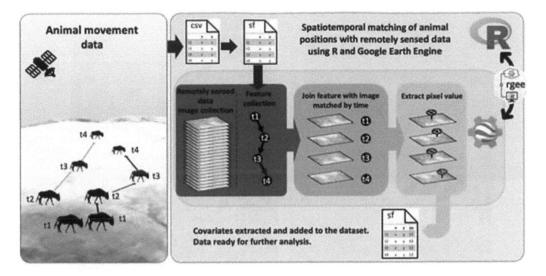

FIGURE 2.10   Monitoring animal movements in a bordered grassland (Crego et al., 2021).

animal individually or as a herd. The data to be obtained in this way is transferred to the GIS database in real time and necessary analyses are made. These analyses are an indicator of animal/herd movements (recorded speeds at different times of the day and in grazing areas, lying, standing, and walking times, etc.). These data will enable the determination and mapping of the most productive grazing times of the day, so that an effective grazing plan can be realized by determining the grazing areas that the animals most prefer and where they stay the longest (Kahveci, 2014). Trotter et al. (2010) determined the most preferred grazing areas of cattle using GPS data. There is a stream bed in the north of this area. It is also possible to plot the time-speed graphs of the herd (night speed and duration, morning speed and duration, evening speed and duration, etc.) using GNSS data (Trotter et al., 2010). Figure 2.10 shows an example of a study in which animals were tracked using GPS.

2.5.6 DETECTION OF LAMENESS

When cattle are raised in meadows and pastures, which are their natural habitats, foot problems are almost non-existent. However, lameness and foot problems are frequently experienced in heavy-bodied, highly productive culture breed cattle, which are forced to spend most of their lives indoors on concrete floors. Foot diseases lead to serious treatment costs, food, and environmental pollution as well as meat, milk, and fertility losses that determine the quality of breeding. Severe pain that occurs in foot diseases blocks the hypothalamus and pituitary systems of the brain and overturns all life and productivity functions of the animal. Therefore, early diagnosis and timely foot care are of great importance (Mulaoğlu, 2019). Lameness is classified according to its severity value, and for this purpose, there are studies as in Figure 2.11, which are based on lameness degrees and posture and walking positions (Yaylak, 2008; Çeçen, 2014).

Experts decide the degree of lameness by looking at these positions of the cows. Animals with a score of three or more in the movement score should be observed and the problem should be investigated, and solutions should be sought. Lameness can be determined automatically with the image processing technique as in Figure 2.12 and foot pressure–sensitive mats as in Figure 2.13. These systems are usually placed at the entrance or exit of the milking facility. The way lameness is determined by the image processing method is like what experts do. Here, the images are evaluated with the visualization of the posture and walking positions and the developed models, and the presence of lameness and the degree of lameness are decided. In the detection of lameness with

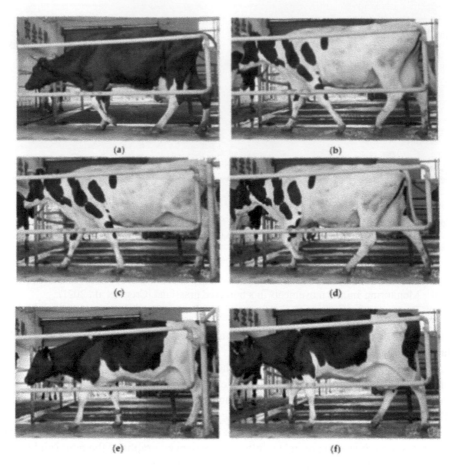

FIGURE 2.11   Examples of posture walking positions in degrees of lameness (Kang et al., 2021).

foot pressure–sensitive mats, the lame cow cannot step on its lame foot as much as it normally should due to pain. Lameness can be detected automatically by measuring this difference, which is normally required, with foot pressure–sensitive mats and evaluating the measured values with the developed models.

2.5.7   DETECTION OF ESTRUS

The aim of effective reproductive management is to achieve pregnancy in the earliest biologically possible time period for each cow. A missed estrus causes an average loss of 21 days per animal. For this purpose, it is necessary to determine the estrus accurately. Accurate determination of estrus increases the success of artificial insemination and calving rate and increases the profitability of the enterprise by allowing longer milk production from cows and at higher rates. Insufficient or incorrect determination of estrus causes artificial insemination not to be done on time and successfully, resulting in delayed insemination, decreased pregnancy rate, and prolongation of the birth interval, and reduces the milk and calf production potential of enterprises. Estrus is a condition that occurs just before ovulation, causes physiological and behavioral changes, and occurs in a period of 4 to 24 hours. Due to the short and variable estrus period, errors are experienced for various reasons in the accurate determination of estrus, especially in large-scale enterprises with many animals. Animals in estrus move more than other animals. For this reason, trying to determine the heat by measuring

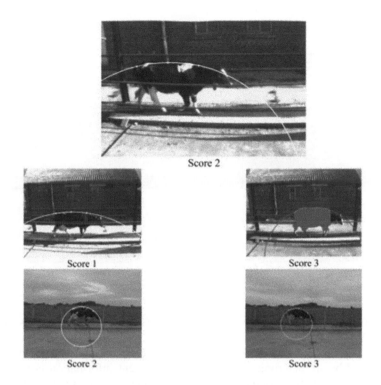

FIGURE 2.12   Detection of lameness with image analysis systems (Poursaberi et al., 2010).

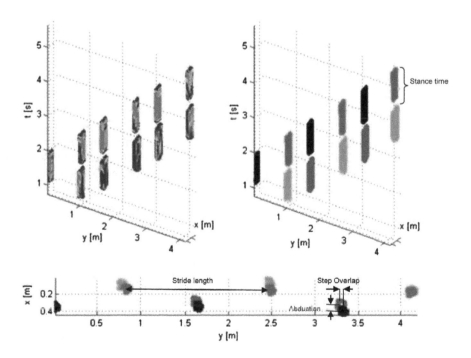

FIGURE 2.13   Automatic detection of lameness with GAITWISE footbed pressure-sensitive mats (Maertens et al., 2011).

FIGURE 2.14   Leg-mounted pedometer.

the step taken using a pedometer (Figure 2.14) is the most preferred method because of its high success rate and easy use (Özgüven, 2018).

Motion sensors in the pedometer detect movement and generate an impulse, and the generated impulses are counted by an electronic counter. This movement information is transferred to the computer at certain intervals by various data transfer methods. The computer software compares the activity information of the animal with the old activity information and decides whether it is estrus. It allows timely insemination by giving a warning for the animals for which estrus is predicted (Özgüven, 2018). Another sign of estrus is the act of mounting or attempting to mount other cattle. Even if cows are observed two or three times for 30 minutes every day to detect estrus behaviors, there is only a 12–19% chance of catching them. This is because more than 60% of the mounting occurs between night and morning. For this reason, automatic systems are used for estrus detection in cows (Chen and Lin, 2015).

If there is a decrease in milk yield while there is an increase in the motility values, or when there are sudden changes in the electrical conductivity measurement values of the vaginal mucosa, it shows that there is estrus. Herd management software also uses information such as the estrus warning date of the animal, the date of estrus, the number of days after estrus, the next estrus, and the day until the next estrus in the evaluation made when detecting estrus. In addition, the software allows the recording of important features and animals in terms of herd management such as veterinary affairs (diseases, drugs, treatment, interpretation, etc.), warning lists, milking, reproduction, activity, health, automatic animal separation, automatic weight measurement, and feeding information (Özgüven and Tarhan, 2019).

2.5.8   Measurement of Body Characteristics and Gait Analysis

Negretti et al. (2011) conducted a study on 36 Comisana sheep and 50 Saanen goats to make predictions about morphological characteristics and weight in sheep and goat production with image processing applications (Figure 2.15). For this purpose, the distance has been fully computerized using a laser measurement application to automatically calculate the morphological parameters, thus making it possible to take new angular and surface measurements. In addition, animal weight assessments were made to establish regression equations for the indirect determination of live weight. As a result of the study, a surface parameter (surface area of side profile, SLP) was measured, which was shown to be highly correlated with the live weight (LW) of buffaloes, sports horses, and milk and beef cattle, which could be determined by image processing techniques, and

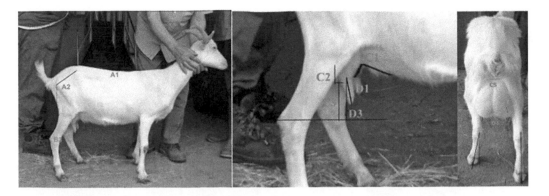

FIGURE 2.15 Examples of measurement points in Saanen goats (Negretti et al., 2011).

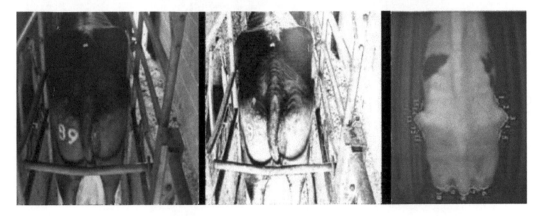

FIGURE 2.16 Determination of body condition score with image analysis systems (Coffey and Bewley, 2014).

in sheep and goats correlations between SLP and LW were 0.95 and 0.96, respectively ($P < 0.01$). The regression equation (LW = −59.25 + 0.03 × SLP) has been used in sheep, while (LW = −3.5 + 0.014 × SLP) has been used to estimate live weight in goats.

A body condition score is an evaluation made with a score between one and five, which is used to determine the level of fatness in cows. The body condition score has effects on health problems that may occur during or immediately after birth and on reproductive and milk production ability during lactation (Yıldız and Özgüven, 2018). Figure 2.16, Figure 2.17, and Figure 2.18 show the study of body condition score determination with image analysis systems.

Livestock animal identification is of great importance to achieve precision animal production, as it is a prerequisite for modern livestock management and automated behavioral analysis. Regarding cow identification, computer vision-based methods have been widely considered due to their non-contact and practical advantages. Hu et al. (2020) propose a new non-contact cow identification method based on the fusion of deep part features. In the study, first, a set of side view images of the cows were captured, and the YOLO object detection model was applied to locate the cow object in each original image, which was then divided into three parts, head, trunk, and legs, with a part segmentation algorithm using frame differentiation and segmentation span analysis. Next, three independent CNNs were trained to extract deep features from these three parts, and a feature fusion strategy was designed to fuse the features. Finally, an SVM classifier trained with the fused features was used to identify each cow. The proposed method achieved 98.36% cow identification accuracy in a dataset containing side view images of 93 cows, outperforming existing studies. Experimental results showed the effectiveness of the proposed cow identification method and good potential for

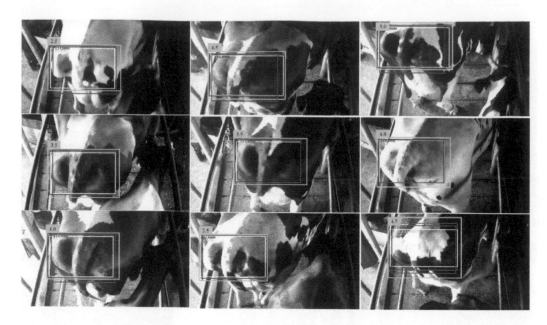

FIGURE 2.17 Examples of SSD, original SSD, and cow tail detection by YOLO-v3 method (Huang et al., 2019).

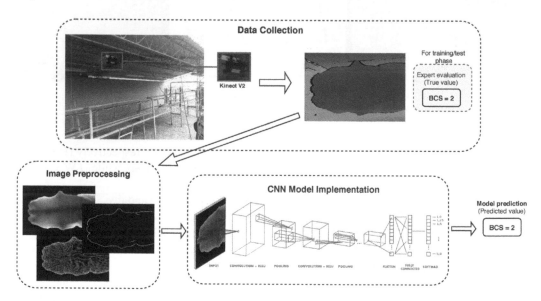

FIGURE 2.18 Determination of body condition score with image analysis systems (Rodríguez Alvarez et al., 2019).

this method in the individual identification of other farm animals. Figure 2.19 shows some parts of interest for cow identification and Figure 2.20 shows the flow chart of the proposed cow identification method.

2.5.9 Monitoring of Broiler Houses and Early Warning System

It has become common to have up to 100,000 broilers on broiler farms. However, it has become difficult to observe such a large number of chickens. In addition to feeding, maintenance, and

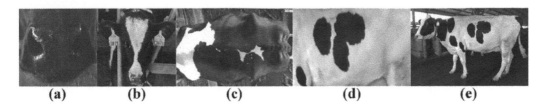

FIGURE 2.19 Some parts of interest for cow identification: (a) nose, (b) face, (c) back, (d) trunk, and (e) general cow object (Hu et al., 2020).

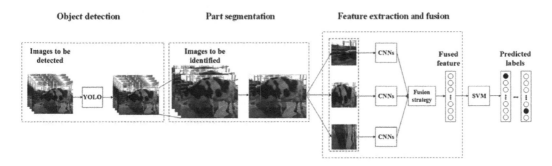

FIGURE 2.20 Flow chart of the proposed cow identification method (Hu et al., 2020).

preparation of suitable environmental conditions, diseases and various malfunctions can be experienced frequently. Some of these activities can be done with automation. However, maintenance and control operations are labor-intensive. Since these activities are difficult, time-consuming, important, and require early intervention, the coops are automatically monitored with computer vision-based methods and early warning systems have begun to be used.

A good example of how precision animal production technology is valuable to the farmer is the eYeNamic system used to monitor common problems in broiler farming. The eYeNamic system consists of three or four cameras mounted on the ceiling that provide pictures of the distribution of broilers (Figure 2.21 and Figure 2.22). The system uses an algorithm that compares the actual measured distribution of animals with a predicted value at that time of day, giving the farmer an alarm when the actual measured value differs by more than 25% from the predicted value. The only parameter used in the operation of the system is the change in the time of distribution of broilers to the available area. The success of the system demonstrates once again that it is necessary to measure animal responses continuously and that it is not necessary to measure many variables to obtain systems that add value. Using the system means that detecting most problems can save the farmer the long hours of work normally spent on checks. Farmers are advised to enter the coop and disturb the broilers only to solve problems and do not need to disturb them if there are no problems (Berckmans, 2017).

Proper spatial distribution of chickens is indicative of a healthy flock. For this reason, Guo et al. (2020) developed and tested a machine vision-based system that can automatically monitor the floor distribution of chickens in their study. For the new method developed for recognizing the bird distribution in the images, the coop floor is virtually defined/divided into drinking, feeding, and resting/exercise zones (Figure 2.23). Each pen was monitored with an HD camera mounted on the ceiling (2.5 m above the floor) to capture video. Videos were saved as AVI files in a video recorder. As broilers grew, images collected each day were analyzed individually to avoid biases due to body weight/size changes over time. About 7,000 images were used to construct a BP neural network model for ground distribution analysis and 200 images were used to validate the model. Broiler distribution defined by the BP model is shown in Figure 2.24. It is concluded that missed detections

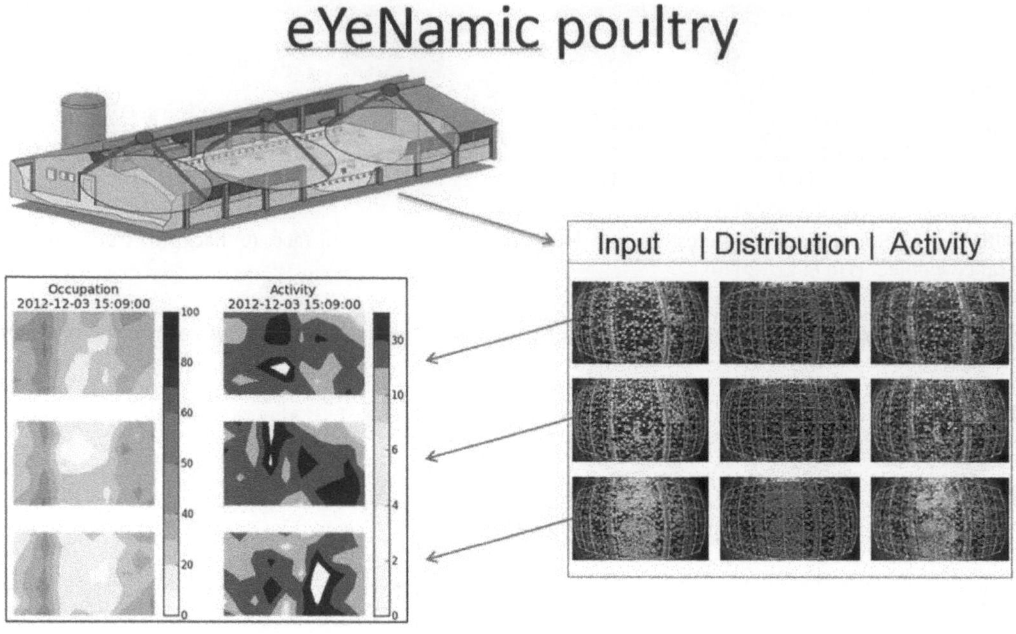

eYeNamic poultry

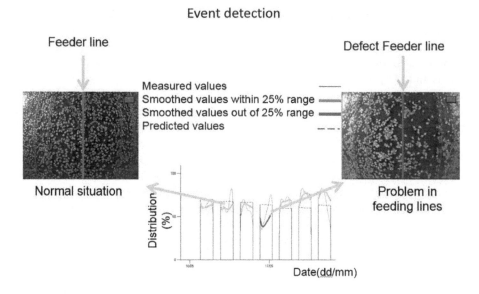

FIGURE 2.21   Real-time image analysis of broiler behavior with three top view cameras (Berckmans, 2017).

FIGURE 2.22   Image taken from broilers analyzed in real time by the eYeNamic system (Berckmans, 2017).

are due to interference with equipment such as the feeder hanging chain and waterline. Researchers have reported that their work continues to eliminate these problems and that this system can be used in commercial facilities.

2.5.10   MONITORING HEALTH STATUS WITH VOICE ANALYSIS

Acoustic monitoring of farm animals can serve as an effective management tool to improve animal health, welfare, and farm efficiency. Jahns (2008) developed a call recognizer that determines the

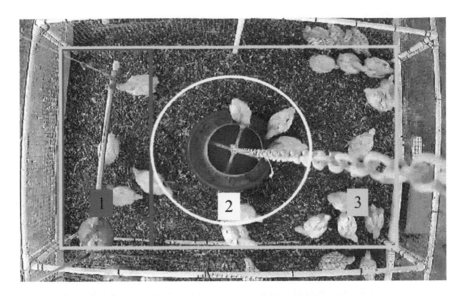

FIGURE 2.23 Division of the coop floor. The red box (1) represents the drinking area, the yellow circle (2) represents the feeding area, and the cyan box (3) represents the rest/exercise zone (Guo et al., 2020).

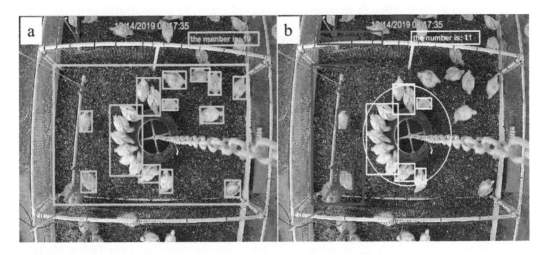

FIGURE 2.24 Broiler distribution defined by the BP model: (a) total broilers monitored and identified, (b) distribution of broilers in drinking and feeding areas. (Guo et al., 2020).

meaning of cows' voices and presents this meaning to the farmer. The researcher stated that the voice recognition of animals can be regarded as a statistical paradigm; during the learning or training phase, feature vectors from known calls are calculated from the feature vectors of sounds with the same meaning, reference patterns are created and stored, the feature vectors from an unknown sound are calculated in the same way for recognition, and the system then determines the reference pattern that is most similar to the feature vector to be recognized and outputs its meaning. The study revealed that in speech recognition latent Markov models (HMMs) have proven to be very efficient with double stochastic processes (Figure 2.25), and as a result, HMMs are very suitable for animal voice recognition.

Because animal sounds in breeding contain information about animal welfare and behavior, automatic sound detection has the potential to facilitate a continuous acoustic monitoring system for use in a range of PLF applications. Bishop et al. (2019) presented a multi-purpose livestock voice classification

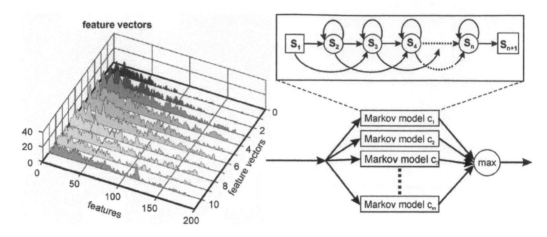

FIGURE 2.25 Voice recognition using hidden Markov models (Jahns, 2008).

algorithm using voice-specific feature extraction techniques and machine learning models. In the study, to test the multi-purpose nature of the algorithm, three separate datasets of farm animals consisting of sheep, cattle, and shepherd dogs were created, voice data were extracted from continuous recordings made in situ at three different farming operations reflecting the actual distribution conditions, Mel-frequency cepstral coefficients and discrete wavelet transform–based (DWT) features were made, and the classification was determined using an SVM model. At the end of the study, high accuracy was obtained for all datasets (sheep: 99.29%, cattle: 95.78%, dog: 99.67%), but the classification performance alone was insufficient to determine the most appropriate feature extraction method for each dataset. Computational timing results of DWT-based features were significantly faster (14.81–15.38% reduction in execution time) and a highly sensitive livestock vocalization classification algorithm has been developed, which forms the basis of an automated livestock vocalization detection system.

Mahdavian et al. (2021) stated that research in PLF can be used as a biomarker of chicken sound health status, and one of the most important steps for this purpose is to examine ways to use acoustic characteristics as criteria for disease diagnosis. The researchers have worked to establish a database to determine the health status of chicken sounds and to determine important acoustic properties. They collected five acoustic characteristics of 150 healthy and diseased commercial broilers in three groups: Control, bronchitis, and Newcastle disease. In the study, the growth environment conditions in terms of relative humidity, temperature, daylight hours, and feeding were controlled according to the standard protocol, in which chickens were raised with the same protocol as at Tarbiat Modares University and the University of Minnesota. An eye drop method was applied to inoculate the ten-day-old birds with the virus. Serological tests were performed by taking blood samples three times in total to confirm the health status of the chickens, and chicken body dissection was also performed to confirm health status. Results of data analysis showed that among the five acoustic features studied, wavelet entropy (WET) had the best performance and was able to detect bronchitis on the third day after inoculation with 83% accuracy while incorrectly detecting sick birds as healthy was less than 14% and 6% on the third day and fourth day, respectively. In the case of Newcastle disease, although WET and Mel-frequency cepstral coefficients (MFCC) exhibited similar accuracy (80% and 78% respectively on the fourth day), the difference was that WET was more reliable in detecting healthy birds while MFCC had better performance detecting challenged birds.

Du et al. (2018) developed a surveillance system that uses Kinect microphone arrays to automatically monitor the abnormal vocalizations of birds during the night (Figure 2.26). In the study, Kinect sensor direction predictions were found to have high accuracy, based on the principle of time difference of arrival (TDOA) of the sound source localization (SSL) method. The system had an accuracy rate of 74.7% in laboratory tests and 73.6% in small bird group tests for differential area

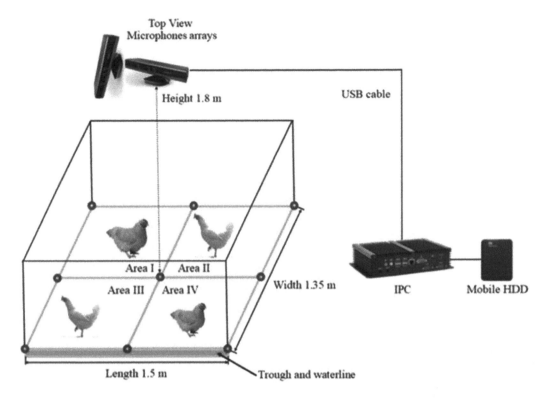

FIGURE 2.26 Schematic of the poultry health monitoring platform (Du et al., 2018).

voice recognition. Also, in small-group tests, the flocks produced an average of 40 sounds per bird during the feeding period. On average, each normal bird was found to make more than 53 sounds during the day (noon to 6:00 p.m.) and less than one sound at night (11:00 p.m.–3:00 a.m.). The researchers reported that this system could be used to detect abnormal poultry status for the study of animal behavior and welfare at night.

Ferrari et al. (2013) reported that respiratory diseases cause death and loss of productivity in intensive pig farming, and cough is an important symptom for screening and diagnosis. To compare the acoustic properties of different types of cough sounds, analyze their acoustic properties, and automatically identify coughs, they worked with an algorithm-based alarm system to provide farmers with early warning about the health status of their herds. The properties examined in the study are peak frequency, sound duration, energy envelope, and time constant. The study, an automated online recognition and localization procedure for sick pig cough sounds where the instantaneous energy of the signal was used to detect and extract individual sounds and their duration was used as a pre-classifier. They stated that automatic regression analysis was used to calculate an estimate of the sound signal, the parameters of the estimated signal were evaluated to identify the sounds, and a localization technique based on the time difference of arrival was of acceptable accuracy for this application. It is suggested that the presented application can be used for online monitoring of welfare status in pig farming and for early diagnosis of a cough hazard and faster treatment of sick animals. Figure 2.27 shows the comparison graph between coughing sounds and grunting in terms of power spectral density and frequency.

2.5.11 Automatic Temperature Measurement with Thermal Camera

Hoffmann and Schmidt (2015) measured the body temperatures of ten cows (Holstein-Friesians aged three to nine years) and nine calves (Holstein-Friesians, eight to 35 weeks old) using a thermal

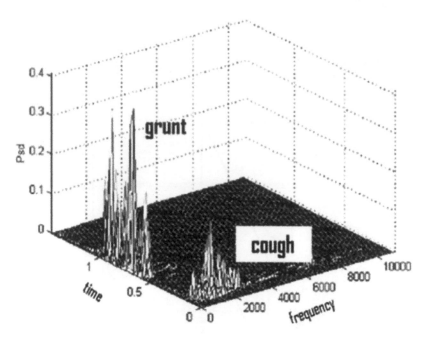

FIGURE 2.27 Comparison between pig cough sound and grunt in terms of power spectral density and frequency (Ferrari et al., 2013).

camera. The measurement was taken at a certain distance (approximately 100 cm) between the camera lens and the body surface and with a fixed measurement angle. For measurements of cows, the thermal imager was placed laterally in front of an automated milking system to film the head and forepart of the cows during milking. In the calf barn, the camera is placed laterally in the automatic calf feeder to film the head and back of the calves as they are fed. A plate with a reference temperature (preset to 40.0°C) was set up and used as a comparison value for the thermal imager. At the end of the study, thermal image temperatures of cows and calves were generally in a very wide range, and the thermal imager temperatures of cows were between 36.0 and 38.7°C in the body area (range: 2.7 K), between 35.5 and 37.5°C (range: 2.0 K) in the head area, between 36.4 and 38.2°C (range: 1.8 K) in the body area, and between 36.8 and 38.4°C (range: 1.6 K) in the head area in calves. Therefore, they reported that it is important to focus on the individual relationship between reference temperature and thermal temperature for each animal.

Wang et al. (2021) conducted a study to obtain a sensitive and accurate method for measuring the body temperature of cattle. In this study, the effect of environmental factors such as wind speed, ambient temperature, and humidity on infrared thermography (IRT) was taken into account and a new calibration method was proposed (Figure 2.28). The proposed calibration method reduced the influence of ambient temperature and humidity on the IRT results, thus increasing the accuracy of the IRT temperature. The differences in mean value and standard deviation value between recorded rectal reference temperature and IRT temperature were 0.04°C and 0.10°C, respectively, and the proposed system significantly improved the measurement consistency of IRT temperature and cattle body temperature reference.

The heat stress of broilers in commercial broiler houses reduces productivity and thus farm profitability. Climate control systems use sensors that measure the temperature around the broilers, which may differ from the actual body temperature of the broilers. Bloch et al. (2019) designed and validated a method for estimating broiler body temperature for commercial broiler houses. As a method, a real-time calibrated low-cost thermal camera, thermal camera image processing algorithm, and lasso regression estimation model were used. The prototype house (Figure 2.29)

FIGURE 2.28 Measurement of body temperatures of cattle with a thermal camera and auxiliary sensors (Wang et al., 2021).

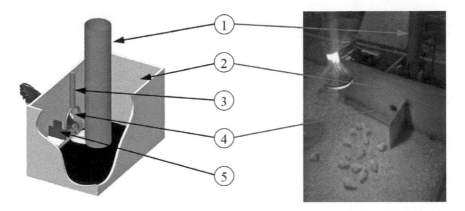

FIGURE 2.29 A prototype of the core temperature measurement system (left side) operating in an experimental broiler house (right side) before implementation in commercial broiler houses. (1) Feeder, (2) box, (3) RFID antenna, (4) hole for a chicken, (5) thermal camera (Bloch et al., 2019).

developed for the application of this method was tested for 15 broilers for 21 days (ages 14–35 days). The researchers have reported that the estimated body temperature was compared with the actual body temperature measured by temperature recorders implanted in the abdominal cavity, and the accuracy was found to be ± 0.27°C, measured in a broiler every 16 minutes on average. An experiment in a commercial broiler house showed a discrepancy between climate control activation and broiler estimated body temperature.

2.5.12 Automatic Animal Weighing, Sorting, and Marking Systems

The measurement takes place while the cow passes over the automatic scale placed at the exit of the milking parlor. The measured values are combined with the cow identification information and sent to the herd software system using radio frequency. In cases such as treatment, insemination, group change, and decrease in milk yield, cows are separated with automatic separation gates placed at the exit of the milking parlor. It is also possible to mark with paint in different colors when desired (Figure 2.30) (Tarhan et al., 2015).

FIGURE 2.30 Automatic animal weighing and sorting gates.

REFERENCES

Alçiçek, A. ve Kırkpınar, F., 2015. Rasyon Hazırlama ve Yemleme Yöntemleri. Hayvan Besleme (Ak, İ.). Anadolu Üniversitesi Yayınları No:2244. Eskişehir. (Turkish).

Anonim, 2015. Hayvancılık Sektör Raporu. TİGEM. (Turkish).

Afimilk, 2021. https://www.afimilk.com/parlor-automation. [Accessed 28.01.2021].

Arazi, A., Pinski, N., Schcolnik, T., Aizinbud, E., Katz, G. and Maltz, E., 2011. Innovations Arising from Applied Research on a New On-line Milk Analyzer and a Behavior Meter. *New Trends for Innovation in the Mediterranean Animal Production.* (Editors: Bouche, R. Derkimba, A. and Casabianca, F.) EAAP publication No. 129. Wageningen Academic Publishers, Wageningen.34–43 ISBN: 978-90-8686-170-5.

Ayık, M., Çilingir, İ. ve Onurbaş Avcıoğlu, A., 2015. Hayvancılıkta Mekanizasyon. Ankara Üniversitesi Ziraat Fakültesi Yayınları No: 1624, Ders Kitabı No: 576. Ankara. (Turkish).

Benaissa, S., Tuyttens, F.A.M., Plets, D., de Pessemier, T., Trogh, J., Tanghe, E., Martens, L., Vandaele, L., Van Nuffel, A., Joseph, W. and Sonck, B., 2019. On the use of on-Cow Accelerometers for the Classification of Behavioursi Dairy Barns. *Research in Veterinary Science*, 125, 425–433.

Berckmans, D., 2017. General Introduction to Precision Livestock Farming. *Animal Frontiers*, 7(1), 6–11. https://doi.org/10.2527/af.2017.0102.

Bishop, J.C., Falzon, G., Trotter, M., Kwan, P. and Meek, P.D., 2019. Livestock Vocalisation Classification in Farm Soundscapes. *Computers and Electronics in Agriculture*, 162(2019), 531–542. https://doi.org/10.1016/j.compag.2019.04.020.

Bloch, V., Barchilon, N., Halachmi, I. and Druyan, S., 2019. Automatic Broiler Temperature Measuring by Thermal Camera. *Biosystems Engineering.* https://doi.org/10.1016/j.biosystemseng.2019.08.011.

Bowling, M.B., Pendell, D.L., Morris, D.L., Yoon, Y., Katoh, K. and Belk, K.E., 2008. Review: Identification and Traceability of Cattle in Selected Countries Outside of North America. *The Professional Animal Scientist*, 24, 287–294.

Campos, D.P., Abatti, P.J., Bertotti, F.L., Hill, J.A.G. and da Silveira, A.L.F., 2018. Surface Electromyography Segmentation and Feature Extraction for Ingestive Behavior Recognition in Ruminants. *Computers and Electronics in Agriculture*, 153, 325–333. https://doi.org/10.1016/j.compag.2018.08.033.

Chen, C. and Lin, H., 2015. Estrus Detection for Dairy Cow Using Zigbee-Based Sensor Networks. *International Journal of Information and Electronics Engineering*, 5, 250–253.

Coffey, M. and Bewley, J., 2014. Precision Dairy Farming (PDF). https://ruralfuturesconf.agresearch.co.nz/mediawiki/images/e/e9/Precision_Dairying_NZ_MC_V2.ppt.

Crego, R.D., Masolele, M.M., Connette, G. and Stabach, J.A., 2021. Enhancing Animal Movement Analyses: Spatiotemporal Matching of Animal Positions with Remotely Sensed Data Using Google Earth Engine and R. *Remote Sensing*, 13, 4154. https://doi.org/10.3390/rs13204154.

Çeçen, G., 2014. Sığırlarda Topallık ve Ayak Hastalıkları. Sentez Yayınları,168s. Bursa. (Turkish).

Delaval, 2015. Water intake and cow comfort. DeLaval. http://www.delaval.com/en/-/Dairy-knowledge-and -advice/Cow-comfort/Drinking-areas/. [Accessed 28.03.2015].

Du, X., Lao, F. and Teng, G., 2018. A Sound Source Localisation Analytical Method for Monitoring the Abnormal Night Vocalisations of Poultry. *Sensors*, 18, 2906. https://doi.org/10.3390/s18092906.

Ertuğrul, M., 2012. Giriş. Hayvan Yetiştirme (Editör: Ertuğrul, M.). Anadolu Üniversitesi Yayınları No.:2255. Eskişehir. (Turkish).

Ferrari, S., Silva, M., Exadaktylos, V., Berckmans, D. and Guarino, M., 2013. The Sound Makes the Difference: The Utility of Real Time Sound Analysis for Health Monitoring in Pigs. *Livestock Housing Modern Management to Ensure Optimal Health and Welfare of Farm Animals*. (Editors: Aland, A. and Banhazi, T.) Wageningen Academic Publishers. 407–418 ISBN: 978-90-8686-217-7.

Filya, İ. ve Canbolat, Ö., 2015. Yemler ve Yem Katkı Maddeleri. Hayvan Besleme (Editör: Ak, İ.). Anadolu Üniversitesi Yayınları No:2244. Eskişehir.

Flanders, F.B. and Gillespie, J.R., 2015. *Modern Livestock and Poultry Production*. 9th Edition. Cengage Learning, Boston. ISBN: 978-1-133-28350-8.

Guo, Y., Chai, L., Aggrey, S.E., Oladeinde, A., Johnson, J. and Zock, G., 2020. A Machine Vision-Based Method for Monitoring Broiler Chicken Floor Distribution. *Sensors*, 20, 3179. https://doi.org/10.3390 /s20113179.

Halachmi, I., Levit, H. and Bloch, V., 2019. Current Trends and Perspective of Precision Livestock Farming (PLF) with Relation to IoT and Data Science Tools. FFTC Agricultural Policy Platform.

Hoffmann, G. and Schmidt, M., 2015. Monitoring the Body Temperature of Cows and Calves with a Video-Based Infrared Thermography Camera. *Precision Livestock Farming Applications*. (Editor: Halachmi, I.) Wageningen Academic Publishers, Wageningen, 231–238.

Hu, H., Dai, B., Shen, W., Wei, X., Sun, J., Li, R. and Zhang, Y., 2020. Cow Identification Based on Fusion of Deep Parts Features. *Biosystems Engineering*, 192, 245–256. https://doi.org/10.1016/j.biosystemseng .2020.02.001.

Huang, X., Hu, Z., Wang, X., Yang, X., Zhang, J. and Shi, D., 2019. An Improved Single Shot Multibox Detector Method Applied in Body Condition Score for Dairy Cows. *Animals*, 9, 470. https://doi.org/10 .3390/ani9070470.

Jahns, G., 2008. Call Recognition to Identify Cow Conditions-A Call-Recogniser Translating Calls to Text. *Computers and Electronics in Agriculture*, 62(2008), 54–58. https://doi.org/10.1016/j.compag.2007.09 .005.

Kang, X., Zhang, X.D. and Liu, G., 2021. A Review: Development of Computer Vision-Based Lameness Detection for Dairy Cows and Discussion of the Practical Applications. *Sensors*, 21, 753. https://doi.org /10.3390/s21030753.

Kahveci, M., 2014. Uydularla Konum Belirleme Sistemleri (GNSS)'nin Hassas Tarımda Kullanımı ve Sağladığı Katkılar. *Harita Teknolojileri Elektronik Dergisi Cilt*, 6(2), 35–48. (Turkish).

Kuşçu, H. ve Arın, S., 2003. Büyükbaş Hayvan Beslenmesinde Bilgisayar Kontrollü Otomatik Yemleme Sistemi Tasarımı. IJCI Proceedings of International Conference on Signal Processing, ISSN 1304-2386, Volume 1, No. 2, September. (Turkish).

Maertens, W., Vangeyte, J., Baert, J., Jantuan, A., Mertens, K., De Campeneere, S., Pluk, A., Opsomer, G., Van Weyenberg, S. and Van Nuffel, A., 2011. Development of a Real Time Cow Gait Tracking and Analysing Tool to Assess Lameness Using Pressure Sensitive Walkway: the GAITWISE System. *Biosystems Engineering*, 110(1), 29–39. https://doi.org/10.1016/j.biosystemseng.2011.06.003.

Mahdavian, A. Minaei, S., Marchetto, P.M., Almasganj, F., Rahimi, S. and Yang, C., 2021. Acoustic Features of Vocalization Signal in Poultry Health Monitoring. *Applied Acoustics*, 175, 107756. https://doi.org/10 .1016/j.apacoust.2020.107756.

Mench, J.A., 2018. Science in the Real World, Benefits for Researchers and Farmers (Section 6). *Advances in Agricultural Animal Welfare Science and Practice*. (Editor: Mench, J.A.) Woodhead Publishing is an imprint of Elsevier, Duxford.111–128. ISBN: 978-0-08-101215-4.

Mulaoğlu, Ş., 2019. Büyükbaş Hayvancılık (Sığırcılık). https://www.tarimorman.gov.tr/HAYGEM/Belgeler/ Hayvanc%C4%B1l%C4%B1k/B%C3%BCy%C3%BCkba%C5%9F%20Hayvanc%C4%B1l%C4%B1k /2019%20Y%C4%B1l%C4%B1/Buyukbas_Hayvan_Yetistiriciligi.pdf. [Accessed 04.06.2020]. (Turkish).

Neethirajan, S., 2020. The Role of Sensors, Big Data and Machine Learning in Modern Animal Farming. *Sensing and Bio-Sensing Research*, 29, 100367. https://doi.org/10.1016/j.sbsr.2020.100367.

Negretti, P., Bianconi, G., Bartocci, S., Terramoccia, S. and Noè, L., 2011. New Morphological and Weight Measurements by Visual Image Analysis in Sheep and Goats. *New Trends for Innovation in the Mediterranean Animal Production*. (Editors: Bouche, R. Derkimba, A. and Casabianca, F.) EAAP publication No. 129. Wageningen Academic Publishers, Wageningen.227–232. ISBN: 978-90-8686-726-4.

Özcen, D., 2019. Karma Yem Üretiminde Yem Üretiminde Soğutucudan Geçen Yemin Toz Dönüş Sürelerinde İyileştirme Yapılarak Yem Üretim Kapasitesinin Artırılması Ve Enerji Tasarrufu Sağlanması. Gaziosmanpaşa Üniversitesi Fen Bilimleri Enstitüsü Biyosistem Mühendisliği Anabilim Dalı Yüksek Lisans Tezi. Tokat. (Turkish).

Özgüven, M.M., 2017a. Hassas Hayvansal Üretim. Tarım Türk Dergisi Hayvancılık. Kasım-Aralık 2017, Sayı: 68, S: 28-31. (Turkish).

Özgüven, M.M., 2017b. Akıllı (Hassas) Tarım Uygulamaları. Muş Ovası Tarım ve Hayvancılık Çalıştayı 16–17 Mayıs 2017, Muş. (Turkish).

Özgüven, M.M., 2018. Hassas Tarım. Akfon Yayınları, Ankara. ISBN: 978-605-68762-4-0. (Turkish).

Özgüven, M.M. ve Tan, M., 2017. Kızgınlık Tespitinde Kullanılan Pedometrelerde Kablosuz Veri İletim Yöntemleri. Gaziosmanpaşa Bilimsel Araştırma Dergisi. ISSN: 2146-8168. Cilt: 6, Sayı: Özel Sayı (BSM-2017), Sayfa: 61–69. (Turkish).

Özgüven, M.M. ve Tarhan, S., 2019. Süt Sığırlarının Hareketlerinin İzlenmesi İçin Yeni Bir Pedometre Tasarımı. International Erciyes Agriculture, Animal Food Sciences Conference 24–27 April 2019. Erciyes University, Kayseri/Turkiye. (Turkish).

Poikalainen, V., Kokin, E., Veermäe, I. and Praks, J., 2013. Towards an Automatic Dairy Cattle Welfare Monitoring. *Livestock Housing Modern Management to Ensure Optimal Health and Welfare of Farm Animals*. (Editors: Aland, A. and Banhazi, T.) Wageningen Academic Publishers, Wageningen. 393–406. ISBN: 978-90-8686-217-7.

Pomar, C., van Milgen, J. and Remus, A., 2019. *Precision Livestock Feeding, Principle and Practice. Poultry and Pig Nutrition: Challenges of the 21st Century*. Wageningen Academic Publishers,Wageningen. 397–418.

Poursaberi, A., Bahr, C., Pluk, A., Van Nuffel, A. and Berckmans, D., 2010. Real-time Automatic Lameness Detection Based on Back Posture Extraction in Dairy Cattle: Shape Analysis of Cow with Image Processing Techniques. *Computers and Electronics in Agriculture*, 74, 110–119. https://doi.org/10.1016/j.compag.2010.07.004.

Rodríguez Alvarez, J., Arroqui, M., Mangudo, P., Toloza, J., Jatip, D., Rodriguez, J.M., Teyseyre, A., Sanz, C., Zunino, A., Machado, C. and Mateos, C., 2019. Estimating Body Condition Score in Dairy Cows from Depth Images Using Convolutional Neural Networks, Transfer Learning and Model Ensembling Techniques. *Agronomy*, 9, 90. https://doi.org/10.3390/agronomy9020090.

Sepúlveda-Varas, P., Proudfoot, K.L., Weary, D.M. and von Keyserlingk, M.A., 2016. Changes in Behaviour of Dairy Cows with Clinical Mastitis. *Applied Animal Behaviour Science*, 175, 8–13.

Siberski-Cooper, C.J. and Koltes, J.E., 2022. Opportunities to Harness High-Throughput and Novel Sensing Phenotypes to Improve Feed Efficiency in Dairy Cattle. *Animals*, 12, 15. https://doi.org/10.3390/ani12010015.

Singh, S.P., Ghosh, S., Lakhani, G.P., Aklank, J. and Biswajit, R., 2014. Precision Dairy Farming: The Next Dairy Marvel. *Veterinar Sci Technolo*, 5, 2. http://doi.org/10.4172/2157-7579.1000164.

Solano, L., Barkema, H.W., Pajor, E.A., Mason, S., LeBlanc, S.J., Nash, C.G.R., Haley, D.B., Pellerin, D., Rushen, J., de Passille´, A.M. and Vasseur, E., 2016. Associations Between Lying Behavior and Lameness in Canadian Holstein-Friesian Cows Housed in Freestall Barns. *Journal of Dairy Science*, 99, 2086–2101.

Tarhan, S., Ozguven, M.M. ve Ertuğrul, M., 2015. Süt Sığırı İşletmelerindeki Bilgi Teknolojileri Uygulamaları, GAP VII. Tarım Kongresi, 28–30 Nisan Şanlıurfa. (Turkish).

Tarhan, S., Yavuz, M., Zengin, K., Ozguven, M.M., Ertuğrul, M. ve Demirtaş, F., 2020. Modern Hayvancılık İşletmelerinde Hayvanların Sağlık Durumlarındaki Değişimlerin İzlenmesinde Kullanılabilecek Yeni Bir Otomasyon Sisteminin Geliştirilmesi, 1160332 Nolu TÜBİTAK 1001 Projesi. (Turkish).

Toker, E., Zincirlioğlu, M. ve Alarslan, Ö.F., 1994. Hayvan Yetiştirme (Yemler ve Hayvan Besleme). Ankara. (Turkish).

Trotter, M.G., Lamb, D.W., Hinch, G.N. and Guppy, C.N., 2010. GNSS Tracking of Livestock: Towards Variable Fertilizer for the Grazing Industry. 10th International Conference on Precision Agriculture, 18–21 July 2010, Denver, USA.

Ünal, R.D. ve Besler, H.T., 2012. Beslenmede Sütün Önemi. Sağlık Bakanlığı Yayınları. Yayın No: 727. Ankara. (Turkish).

Villeneuve, E., Akle, A.A., Merlo, C., Masson, D., Terrasson, G. and Llaria, A., 2019. Decision Support in Precision Sheep Farming. IFAC Papers OnLine 51–34 (2019) 236–241.

Wang, F.-K., Shih, J.-Y., Juan, P.-H., Su, Y.-C. and Wang, Y.-C., 2021. Non-Invasive Cattle Body Temperature Measurement Using Infrared Thermography and Auxiliary Sensors. *Sensors*, 21, 2425. https://doi.org /10.3390/s21072425.

Yavuzcan, G., 1994. Tarımsal Elektirifikasyon. Ankara Üniversitesi Ziraat Fakültesi Yayınları No: 1342, Ders Kitabı No: 390. Ankara. (Turkish).

Yavuzcan, G., 1995. İçsel Tarım Mekanizasyonu. Ankara Üniversitesi Ziraat Fakültesi Yayınları No: 1416, Ders Kitabı No: 409. Ankara. (Turkish).

Yaylak, E., 2008. Süt Sığırlarında Topallık ve Topallığın Bazı Özelliklere Etkisi. *Hayvansal Üretim*, 49(1), 47–56. (Turkish).

Yığmatepe, V.K. ve Ozguven, M.M., 2020. Sultansuyu Tarım İşletmesi Süt Sığırcılığı Faaliyetlerinde Girdi ve Maliyetlerin Belirlenmesi. *Turkish Journal of Agricultural Engineering Research*, 1(2), 339–353. https://doi.org/10.46592/turkager.2020.v01i02.010. (Turkish).

Yıldız, A.K. ve Özgüven, M.M., 2018. Hassas Hayvansal Üretim Uygulamaları ve Yozgat Hayvancılığında Uygulanabilirliği. Uluslararası Bozok Sempozyumu Bildiri Kitabı. s. 59–73. Yozgat-Türkiye. (Turkish).

Yıldız, Y., Karaca, C. ve Dağtekin, M., 2008. *Hayvancılıkta Mekanizasyon*. Hasad Yayıncılık, İstanbul. (Turkish).

Yoshioka, K., Mikami, O, Higaki, S. and Ozawa, T., 2019. Early Detection of Livestock Diseases by Using Wearable Wireless Sensors. FFTC Agricultural Policy Platform (FFTC-AP). https://ap.fftc.org.tw/article/1626.

3 Agricultural Robots

3.1 WHAT IS A ROBOT?

Robots are the most advanced automation systems. There are fundamental differences between automation and robots. In automation, often repetitive tasks are performed to evaluate the data coming from the sensors and to provide the desired output. The robot, on the other hand, must be able to move, perceive and define its environment during its movement, and make choices in order to perform a specified task. Therefore, a robot is a programmable mechanical device used to perform certain physical activities or tasks involving decision-making. In addition to the controlled robots, there are also robots that work autonomously (by themselves, without human intervention) (Özgüven, 2019a). Robots that can operate with remote control are used in places where it is difficult to work such as submarines or in dangerous work such as bomb disposal when it is necessary to intervene for security purposes in the face of unexpected or negative situations. It is useful to explain the concept of "autonomous" here. Although there are autonomous robots, not all robots work autonomously. The autonomous expression in the definition of a robot means to do things by itself at a certain level without human intervention. To be a fully autonomous robot, as in driverless vehicles, it is necessary to have the following capabilities: (1) providing motion control while the robot performs its task, (2) sensing the environment precisely so that it can avoid obstacles, and (3) accurately estimating its position to implement path planning and navigation.

Autonomous robots use various equipment and systems to determine their positions, create the desired route correctly, and accurately map the surrounding obstacles and objects while performing their duties. Radar sensors, laser scanners, LIDAR, GPS (Global Position System), INS (inertial navigation system), ultrasonic sensors, and cameras are used as hardware. Software is used to process this information obtained for the control of the movement and turn it into usable and useful information. The software includes the following features: Advanced decision mechanisms that perform operations such as image, audio, and video processing algorithms, artificial neural networks, machine learning, and statistical data analysis (Özgüven, 2018).

The robot must have a task. To accomplish his task, he must be able to act on his own and must have the intelligence to make a choice. A device that does not have a task, cannot move, and cannot make a choice cannot be defined as a robot. For the robot to perform its task, it must first have a body made up of joints suitable for the movements it needs to perform. There are mechanical and electronic parts placed in the body. There are manipulators for the robot to fulfill its task, actuators for the robot and manipulator to move, sensors for the detection of the robot and the environment, and control units where all the movement and operation of the robot are managed.

The robotics field covers expertise in a range of technical disciplines, including mechanical engineering, electrical and electronic engineering, computer science, applied mathematics, industrial engineering, cognitive science, psychology, biology, bio-inspired design, and software engineering. Therefore, a robotic system is made possible through the synthesis of theories and techniques from many fields. In particular, mechatronics engineering facilitates and enables the development of complex robotic systems from standard subsystems and has accelerated the progress of the robotics field in recent years (Kurdila and Ben-Tzvi, 2020). The technical disciplines that contribute to robotics are shown in Figure 3.1.

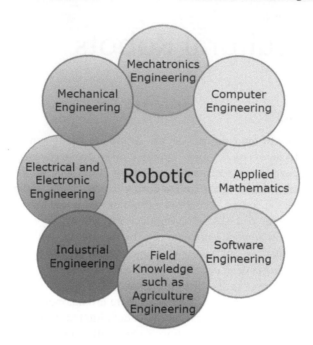

FIGURE 3.1   Areas that contribute to robotics.

3.2   ADVANTAGES AND DISADVANTAGES OF ROBOTS

Robots offer certain benefits that humans cannot. The use of robotic systems can improve the quality of life of workers by freeing them from dirty, boring, dangerous, and heavy labor. For this reason, it can be said that robots give people the opportunity to engage in work and perform better. In addition, it is thought that the work capacity and skill of humans cannot be compared with the speed, quality, reliability, and durability of a robotic system. Robots can be the solution in situations such as the need to increase production performance for the industry to remain competitive, pressure from the market to improve quality, increasing costs, and shortages of qualified workforce. Some people feel that the use of robotic systems has increased unemployment and prevented many from earning an income. It is true that robots can cause unemployment by displacing human workers, but robots also create new qualified jobs such as robot technicians, salesmen, engineers, programmers, and supervisors. The advantages and disadvantages of robots can be classified as follows (Gupta et al., 2017).

3.2.1   ADVANTAGES OF ROBOTS

The use of robots increases productivity, safety, efficiency, quality, and consistency of products.

- Robots can operate in hazardous environments without life support, comfort, or safety concerns.
- Robots do not need environmental comforts such as lighting, air conditioning, ventilation, and noise protection.
- Robots can work continuously without fatigue or boredom, there are no undesirable situations such as hangovers and late arrival to work, and they do not need health insurance or vacation.
- Robots always have repeatable precision unless something happens to them or wears out.
- Robots can be much more accurate and precise than humans.

3.2.2 Disadvantages of Robots

- Robots replace human workers, which creates economic problems such as loss of salary and social problems such as dissatisfaction and arguments among employees.
- Unless the situation is predicted and the system is conditioned, robots are incapable of responding in an emergency. Safety measures are needed to ensure that they do not harm the operators and the machinery working with them.
- Robots are costly due to initial equipment costs, setup costs, need for training, and need for programming.

3.3 ROBOT ACCIDENTS AND SAFETY

Robot accidents occur during programming, program updates, maintenance, repair, testing, installation, or adjustment rather than under normal operating conditions. Situations that cause injury to the operator, programmer, or corrective maintenance workers are the result of faulty operation but often occur while inside the robot's work area. In addition, accidents can occur in cases of mechanical failure. Therefore, all robotic systems must have a way to control the release of stored energy and turn off the power from outside the cover. A detailed risk assessment must be carried out to ensure the safety of workers operating, servicing, and maintaining the robotic system. Operators must be protected by mechanical or non-mechanical presence-sensing fixed barriers and locked barrier limiters (Gupta et al., 2017).

3.4 CLASSIFICATION OF ROBOTS

Many universities and research organizations around the world conduct active research in various fields of robotics. Some of the leading research organizations are MIT (Massachusetts Institute of Technology), JPL (Jet Propulsion Lab, NASA), CMU (Carnegie Mellon University), and Stanford University. These and many other organizations are involved in various fields of robotics. These robotics fields can be broadly classified as (Anonymous, 2007):

- *Robotic manipulator*: Robotic arms have become useful and economical tools in manufacturing, medicine, and other industries.
- *Wheeled mobile robots*: Wheeled mobile robots perform many tasks in industry and military fields.
- *Legged robots*: Movement on the ground can be accomplished by walking, rolling, sliding, and jumping.
- *Underwater robots*: Underwater robots equipped with cameras serve many purposes, including tracking fish and searching for sunken ships.
- *Robot vision*: Sight is our strongest sense. It provides us with a large amount of information about the environment and enables rich, intelligent interaction in dynamic environments. It is not surprising, therefore, that a great deal of effort has gone into providing machines and robots with sensors that mimic the capabilities of the human visual system. The first step in this process is the creation of sensing devices that capture the raw light information used by the human visual system. Two available technologies for building image sensors are CCD and CMOS. These sensors have certain limitations in terms of performance compared to the human eye.
- *Artificial intelligence*: Artificial intelligence is a branch of computer science and engineering that deals with intelligent behavior, learning, and adaptation in machines and robots. Artificial intelligence research is focused on building machines to automate tasks that require intelligent behavior such as control, planning and programming, diagnostics and the ability to answer consumer questions, handwriting, speech, and facial recognition.

- *Industrial automation*: Automation is the use of control systems such as computers to control industrial machines and processes, replacing human operators. It is a step beyond mechanization within the scope of industrialization. While mechanization provides machines that assist human operators with the physical requirements of the job, automation greatly reduces the need for human sensory and mental needs.

3.5 NEW TREND ROBOTS

3.5.1 Smart Autonomous (Robot) Vehicles

Automobiles form the transportation backbone of many economies, and significant financial investment has been made around the world to provide the road infrastructure necessary to carry traffic. As a direct result of the importance of the automobile, roads and highways are often congested due to the individual driving skills of the drivers, the fact that driving can be tiring and leads to fatigue and inattention, the inefficiency of communication modes between drivers, and the extreme vulnerability to error. It is thought that smart robot vehicles can provide much more safety than a human operator with mobile robot technology applications, and the following suggestions are presented to increase productivity (Dudek and Jenkin, 2010):

- Driving assistants that provide additional information and sensors to increase the performance of the human driver in terms of both safety and efficiency;
- Convoy systems that propose to develop automatic delivery convoys in which only the leading vehicle is driven by a human operator, but all other vehicles in the convoy move autonomously;
- Autonomous driving systems that suggest removing the operator completely from the loop and that each vehicle automatically moves on its own;
- Autonomous highway systems that propose to consider the entire highway as a system and autonomously control vehicle groups;
- Autonomous urban systems proposing to address problems related to urban transport through the application of autonomous vehicle technologies.

3.5.2 Autonomous Flying Robots

In recent years, there has been a rapid development in autonomous unmanned aerial vehicles (UAV) and micro aerial vehicles (MAV) equipped with autonomous control devices. These became known as "robotic aircraft" and their use became widespread. They can be classified according to their military or civilian use. There has been a remarkable improvement in UAVs and MAVs for military use. However, the extraordinary capabilities of many unmanned helicopters used for civilian agrochemical spraying are being exploited. UAVs offer great advantages when used for aerial surveillance, reconnaissance, and control in complex and dangerous environments. Low risk and high confidence in mission success are two strong motivations for the continued widespread use of UAVs. In addition, many other technological, economic, and political factors have stimulated the development and study of UAVs. First, technological advances make an important contribution. The latest sensors, microprocessors, and propulsion systems are smaller, lighter, and more capable than ever, allowing levels of durability, efficiency, and autonomy that exceed human capabilities. Secondly, UAVs are successfully used on the battlefield and successfully deployed in many missions. These factors have resulted in more financing and many production orders. Third, UAVs can operate in hazardous and polluted environments, and can also operate at both lower and higher altitudes where manned aircraft cannot fly. UAVs can successfully fulfill their duties without human intervention thanks to navigation sensors and microprocessors. Technologies in communication systems such as bandwidth, frequency, flexibility, adaptability, security, and cognitive controllability of information and data flows ensure smooth data connection. Ground station command, control, and communication are provided with non-vehicle infrastructures such as human–machine interfaces, multiple aircraft, target designation, downsizing of ground equipment, and voice control (Nonami, 2010).

3.5.3 HUMANOID ROBOTS

Humanoid robots are so called because they have a human-like physical appearance in which only articulated connections are used in terms of design, that is, they have a body, a head, two legs, and two arms with multi-fingered hands. Humanoid robots are being developed to work autonomously in various environments such as dwellings, offices, factories, and disaster areas to perform a wide range of physical tasks, physically contact and communicate with people without endangering them, and operate tools and manipulate objects designed for humans. In terms of design, they have a human-like physical appearance, i.e., a torso, a head, two legs, and two arms with multi-fingered hands. Only articulated joints are used. From a control point of view, a humanoid robot should have a hierarchical controller structure and a sensor subsystem that are needed to perform the operations. Operations include perception and cognition, learning, task sequence planning, locomotion trajectory planning and generation, walking control, whole-body manipulation planning with motion/force components, end-link motion/force trajectory generation, transformation, and tracking control, balance and posture control with optimal force distribution in the presence of external disturbances, and low-level actuator and joint space control (Nenchev et al., 2019).

3.5.4 WEARABLE ROBOTS

Robots today interact more closely with humans than ever before. Initially, robots were designed for use only in industrial settings to replace humans in tedious and repetitive tasks and tasks requiring precision, but today their use has expanded from merely information exchange (in teleoperation tasks) and service robots to close interaction involving physical and cognitive methods by increasing interaction with the human operator. A wearable robot can be defined as a technology that extends, complements, substitutes, or enhances human function and capability or empowers or replaces (a part of) the human limb where it is worn. Wearable robots can be robots that can work with human limbs to replenish lost ability or restore weak functions after a disease or neurological condition, as in orthotic robots or exoskeletons, or an electromechanical robot that replaces limbs lost after an amputation, as in prosthetic robots (Pons et al., 2008). Figure 3.2 shows a sensor integrated into a

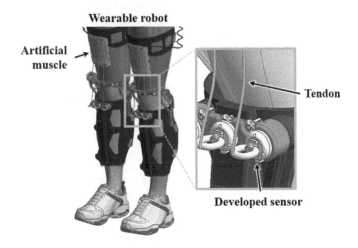

FIGURE 3.2   Tri-axis force/torque sensor developed for wearable robots (Jeong et al., 2021).

wearable robot. The artificial muscle contracts to assist the user's lower body strength and the force of contraction can be measured by the sensor through the tendon (Jeong et al., 2021).

The interface between the human and the robot can exchange signals to execute an action, provide feedback for human motor control, and monitor the state of the human–robot interface and its environment. In one application, wearability imposes several specific requirements on sensors, actuators, and energy storage technologies. To equip a wearable robot with a measuring system, the designer must inevitably accept tradeoffs between functional versatility and simplicity of implementation. Therefore, the most suitable one should be chosen while making the selection. For example, when defining reliable sensors for a wearable application, it can be useful to analyze a wide variety of candidate meters. Measurement requirements for a system include accurate tracking of motion or force, measuring the state of the human–robot interface, receiving a physiological signal for feedback, etc., which can be considered or combined. The design includes damping, resistance modulation, power-up, etc. at the level of human joints or limbs to respond to signals from the human body and the environment. Actuators may also be required. Surface electrodes are used on the skin to measure biological muscle and brain activity signals and internal electrodes are placed next to active cells to provide feedback and control to wearable robots. There are also studies dealing with the sensing of microclimate conditions at a human–robot interface, the fusion of inertial sensor data in a leg exoskeleton, and the biologically based design of a knee actuator system (Moreno et al., 2008).

3.5.5 UNDERWATER ROBOTS

Underwater robots have seen increasing interest from research and industry in recent years. Today, it is common to use manned underwater robotic systems to perform tasks such as seabed and pipeline exploration, cable maintenance, monitoring and maintenance of offshore structures, collection, and release of biological research. The difficult challenges of manned vehicle operation, the huge costs, and the risks of operating in such a harsh environment pave the way for underwater robot studies, gradually making it possible to perform such tasks in a fully autonomous manner. These studies are technologically and theoretically challenging as they cover a wide range of technical and research topics. Shipment of an autonomous vehicle in an unknown and unstructured environment with limited online communication requires some in-vehicle information and the vehicle's ability to respond reliably to unexpected situations. Techniques such as artificial intelligence, neural network, discrete events, and fuzzy logic can be useful in this high-level task control. The vehicle's sensor system must cope with a noisy and unstructured environment. Also, technologies such as GPS are not applicable due to the impossibility of underwater electromagnetic transmission, and vision-based systems are not entirely reliable due to poor visibility in general. The propulsion system usually consists of thrusters and control surfaces, which have a non-linear dynamic and are greatly affected by hydrodynamic effects. In this framework, the use of a manipulator mounted in an autonomous vehicle plays an important role. In terms of control, underwater robots are much more difficult to control than ground robots due to the unstructured environments found in underwater robots, mobile base, significant external disturbances, low bandwidth of sensors and propulsion systems, difficulty in estimating dynamic parameters, and high non-linear dynamics (Antonelli, 2006).

3.5.6 SOFT ROBOTS

Soft robots can be used in food processing plants based on agriculture, especially packaging, unpacking, palletizing, etc. in the production line. Almost all components of industrial robots, which are widely used in operations, are made of hard metal. Many problems such as weight restrictions, complex control mechanisms, and high cost of development are encountered in the production phase of these robots, which are now being called rigid robots. Therefore, the trend in robot construction today is to design by being inspired by the functional structures and shapes of living things in nature. In this context, research on the development of robots with pneumatic or hydraulic actuators, which are produced from polymers or plastic-derived materials for use in various fields of industry, and thus are much more agile, more biocompatible, lighter, and less energy-requiring, has increased in recent years, different types of soft-tissue robots have been developed, and soft robots have begun to be produced for various purposes (Figure 3.3). These robots, in which soft materials are used to obtain flexible structures that mimic nature, can be designed to deform in many different dimensions and to benefit from non-linear mechanics. With the potential to go beyond the bone and joint-like structures used in rigid robots, soft robots offer functionality that corresponds to muscles, musculoskeletal ligaments, and skin. These features will make soft robots suitable for working in areas close to people and rapidly changing environments where safety is kept at the highest level. These robots have been manufactured in recent years using 3D printers that can simultaneously process soft, rubber-like materials and hard materials (Özgüven et al., 2020).

3.5.7 INDUSTRIAL AND DELTA ROBOTS

There is a tendency toward automation systems where human labor is removed from the production line during the processing of agricultural products as food, and all processes in the factory, from product acceptance to storage and sales, are monitored and recorded automatically at every stage (Özgüven et al., 2020). When the shape of the agricultural product is regular and well placed on the production line, simple electromechanical solutions will suffice instead of robots. Poor line localization often involves visual servo robot procedures, especially if the product is also irregular. In addition, the most important advantage of robotic applications over integrated electromechanical systems is that operations can be reprogrammed for different tasks using simple, easy-to-apply procedures (Gray and Davis, 2013). Robots are machines that are intended to replace humans in the execution of tasks involving physical activities or decision-making. The sensors are part of the robot's electronic subsystem, which includes various electronics and other low-power components. Besides the sensors used to determine the configuration of the robot, the perception of the surrounding environment is usually done by means of visual and tactile sensors. Robot vision systems consist of several cameras and a processing unit (Kyriakopoulos and Loizou, 2006).

The use of robotics has increased productivity in almost every industrial sector. Initially used in the food industry, the robots are standard industrial robots used for end-of-line tasks including simple product placement, packaging, and palletizing (Figure 3.4a). However, Delta robots were later produced because of the demand for machines for faster and more efficient quick pick and place operations on individual food products (Figure 3.4b). These robots are optimized for fast work (typically 100–120 operations per minute) with light payloads (typically 1–2 kg) and have been particularly successful in processing food products. Future production trends are a human-operator-free workplace and a continued reduction in robot costs despite rising labor costs (Gray and Davis, 2013).

FIGURE 3.3 Examples of soft robot hands developed for food processing (SoftGripping, 2022).

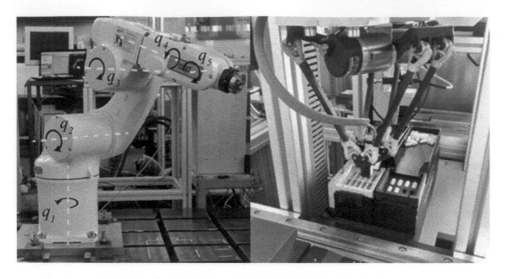

FIGURE 3.4   (a) Industrial robot (Bottin et al., 2020), (b) Delta robot (Fathi et al., 2021).

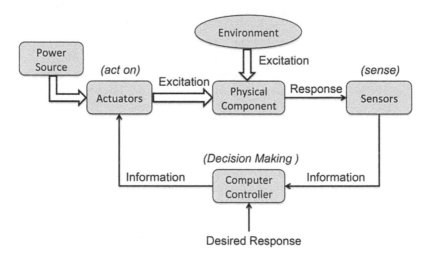

FIGURE 3.5   Structure of a robotic system (Perez et al., 2018).

3.6   COMPONENTS OF ROBOTS

The robot consists of the manipulator, which is specially designed for the robot to fulfill its task, the actuators that enable the robot and the manipulator to move, the sensors that detect the robot and its environment, and the control units that manage all the movement and operation of the robot (Özgüven, 2019b). In the structure of the robotic system, all the components of the system (mechanics, actuators, sensors, computers, and control) are initially considered as part of the design. Robotic systems have a mechanical component whose desired motion behavior is controlled using force actuators commanded by a computer control system that processes information from data produced by sensors (Figure 3.5). Decisions made in each component of the system often influence the decisions of other components, and the design requires multidisciplinary knowledge as they contribute together as factors affecting overall performance, economy, and safety. For example, decisions about mechanical design affect the dynamic properties of the mechanical component and thus the potential complexity of the motion control structure. On the other hand, the desired closed-loop behavior and the dynamics of the mechanical components determine the energy (Perez et al., 2018).

There is a diverse population of robots developed over the years that have roamed the air, land, or sea, from robot manipulators to mobile robots. These robots can imitate humans or animals or have new topologies to perform desired tasks. However, despite these differences, there are some common features that many robots share. Figure 3.6 shows several components of a typical robotic system (Kurdila and Ben-Tzvi, 2020).

3.6.1 Robot Chassis

The robot chassis should be in the shape and dimensions of the design suitable for performing the tasks expected from the robot that is being developed in the simplest way and should be made of materials that are durable enough to perform its duty without deteriorating in the ambient conditions it is in during operation. Electronic embedded system components of the robot such as motors, motor drivers, sensors, and batteries should be placed in a case to protect them from external factors. Recently, it has become widespread to use composite materials, which offer many advantages in terms of lightness and mechanical strength, as robot and drone covers (Özgüven et al., 2016). The dimensions of the chassis, the type of material used, the amount of power required, the mode of movement suitable for the working environment, the wheel selected depending on this, the sensor and engine used, the load-carrying capacity depending on the selection, and the vibration that will occur during operation are all conditions that must be considered in the chassis design. In addition, the design of the chassis should indicate how the robot should behave toward those who encounter it when it interacts with humans while performing its tasks. This is a situation that will help the robot to operate safely.

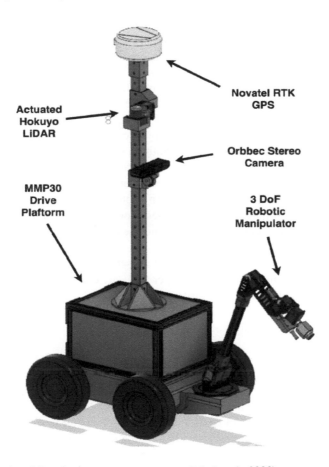

FIGURE 3.6 Typical mobile robotic system components (Iqbal et al., 2020).

The robot must have a base chassis or other structure before sensors, processor power, and actuators can be selected or installed. Considering the difficulties that may be experienced in the preparation of this, it should not be given up or considered unimportant. In unmanned aerial vehicles, for example, it is necessary for the robot to be able to fly long distances and to provide a solid chassis foundation for high-resolution cameras, radar, other sensors, and weapons. In this and other scenarios, materials science is a critical part of the equation. Considering the fundamental physics of many man-made materials, making a robot twice as tall often means quadrupling its mass. One humanoid robot, the Willow Garage PR2, is roughly 1.5 meters long and weighs about 180 kilograms. This weight brings with it many problems. The robot's portability is limited, its heavy attachments must be carefully managed for safety, and battery life suffers because it must carry that much mass. Such robots need to be lighter to gain wider appeal (Jordan, 2016).

3.6.2 Sensors

For the robot to perform its task, since it will interact with its environment while moving, it needs to have information about the environment, and sensors containing various sensing modes are used for this. Although sensors are very diverse, they are devices that provide information to the robot about the robot's environment. During the operation of the sensors, the values of the physical quantities measured from their inputs are given by converting them into electrical signals from their outputs. According to this incoming information, the robot software decides what the robot should do and runs the manipulator for motion or the actuators for the action to be performed to fulfill the relevant decision. A very important issue to be considered here is that there may be sensor malfunctions or an error in the data coming from the sensor. In such cases, the robot may not be able to perform its task successfully and may cause accidents. For this reason, trying to detect or correct the errors will enable the robot to perform its task successfully. For this purpose, the advantages of different sensors can be used together. Using sensors together is called sensor fusion.

An example of sensor fusion is the operation of an autonomous tractor. Due to the fact that an autonomous tractor works in an open area and in a wide variety of environments, some sensors may not receive information or errors may occur in the incoming information. For this reason, it is preferred to process data from different sensors together in the form of sensor fusion, and they are used together by transforming the unique advantages of different types of sensors used into a more comprehensive detection (Figure 3.7). In this way, information from the sensors provides more detailed information about the position of the tractor during movement, its surroundings, and surrounding objects. To process this information and turn it into useful information, advanced decision mechanisms are used, such as image, audio, and video processing algorithms, artificial neural networks, machine learning, statistical data analysis, and the autonomous tractor works successfully (Özgüven, 2018).

As can be seen in Figure 3.7, an autonomous robot is used by integrating a radar sensor, camera, and ultrasonic sensor. Thus, safe warning and danger areas are created at varying distances, covering all directions surrounding the autonomous tractor.

3.6.3 Controller

Controllers are microprocessors that can be programmed, usually in low-level programming languages, and used for describing robot tasks. Programs can be changed and reprogrammed at will. A microcontroller, computer, or PLC can be used instead of a microprocessor. More than one controller can be used in a robot, with each module being its own controller. In such cases, communication between controllers is necessary or can be used in a central computer that integrates the activities of several controllers. Using the data from the sensors, the controller decides to start and stop the actuators, motors, or end effectors according to the commands in its program and move or operate them by sending a signal. Without the controller, the robot cannot be moved, and the robot will not

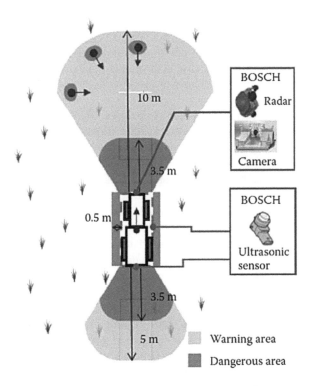

FIGURE 3.7 Sensor fusion used in an autonomous tractor (Noguchi, 2013).

work. In addition, the controller monitors the robot's surroundings and the robot's movements with the help of sensors and controls the features such as the position and speed of the movements. So, the controller, for example, knows when to move the robot arm and in which direction.

3.6.4 ACTUATORS

Actuators act as the muscles of the robotic system and move the robot joints. Actuators perform operations such as the movement of the robot from one place to another and the movement of an object from one place to another with the signals it sends because of the controller's decision. While selecting the actuator, first, it should be considered whether it is compatible with the controller, and it should be selected so that it provides physical outputs such as sufficient force, torque, velocity, acceleration, and flows and so that it has appropriate performance characteristics such as static and dynamic accuracy, resolution, linearity, and dynamic response.

Although there are many types of actuators available, electric motors and especially servo motors are the most widely used robotic actuators. Hydraulic actuators are preferred only for large robots, while pneumatic actuators are preferred for robots with half a degree of freedom, on–off type joints, and placement operations. New actuators, including direct-drive electric motors, electroactive polymer actuators, muscle-wire actuators, and piezoelectric actuators, are mostly used in research and development or as special products for special purposes. More new varieties will be available in the future (Niku, 2011).

3.6.5 MANIPULATOR

The manipulator consists of an end effector that performs the task required by the robot, with the arm, wrist, and hand, which are the joints that provide the robot with the ability to move and work.

The features that affect the manipulation performance such as the working area of the manipulator, the angular displacement of the dexterity, the load-carrying capacity, quickness, and precision are as follows: The type of kinematic structure determines the combined effects of axis drive design and real-time motion control. With the increase in the number of axes of the manipulator, the robot's ability to move also increases, while the control of the robot becomes more difficult and the control algorithm becomes more complex.

The manipulators are called serial manipulators, in which their architecture is modeled as open kinematic chains in which the links are cascaded together with binary joints, according to the kinematic chain of their architecture, and parallel manipulators, where the links are modeled as closed kinematic chains in which the links are articulated to determine the polygonal loops. Also, the kinematic chains of manipulators can be planar or spatial, depending on the field in which they operate. Most of the industrial robotic manipulators are of the serial type, although parallel manipulators have aroused great interest recently and are even used in industrial applications (Ceccarelli, 2004).

The mobility of a manipulator is provided by the presence of joints. The articulation between two cascading connections can be accomplished via a prismatic or rotary joint. In an open kinematic chain, each prismatic or rotary joint provides a single degree of freedom (DOF) to the structure. A prismatic joint creates relative translational motion between two connections, whereas a rotary joint creates relative rotational motion between two connections. Rotary joints are generally preferred to prismatic joints for their compactness and reliability. On the other hand, in a closed kinematic chain, the number of DOFs is less than the number of joints due to the constraints created by the loop. The degrees of freedom must be properly distributed throughout the mechanical structure to have sufficient numbers to perform a given task. In a task consisting of positioning and orienting an object at a desired location in three-dimensional space, six DOFs are required, three to position a point on the object and three to orient the object relative to a reference coordinate frame. If more DOFs are available than task variables, the manipulator is said to be redundant from a kinematics point of view (Sciavicco and Siciliano, 2000).

3.6.6 END EFFECTOR

The end effectors are the parts that enable robots to perform physical movements from one place to another to perform their tasks, as well as activities such as holding, grasping, placing, and carrying objects with hands, wrists, and arms. The end effector consists of a hand or arm attached to the last joint of the manipulator and is designed according to the work that the robot will do. In the design of the robot hand, attention should be paid to the fact that the selected material is durable and light so that it can make forward-backward, up-down, and rotational movements, and so that the robot arm design can achieve flexible and fast work. The movements of the end effector are controlled by the controller such as the motor, wheel, etc., and done with its help.

End effectors are designed as a combination of multiple joint connections like the human hand and arm. Due to the serial nature of the connections, their load-carrying capacity and stiffness are much lower compared to those of machines. For this reason, skillful manipulation has been developed to perform tasks that require a high level of accuracy, as positioning accuracy can be reduced. Skillful manipulation is not limited to just holding objects. At the same time, a controlled movement of objects grasped in the palm of the hand with multiple fingers is provided (Angeles, 2014).

3.7 ROBOT KINEMATICS AND DYNAMICS

Robot kinematics examines the movements of the points where the objects are displaced from one point to another during their movement in the working areas as position, velocity, and acceleration, regardless of the forces and torques that cause the movements of the objects in the robotic system. Because robots must move to perform their tasks, robot kinematics is the most defining feature of robot design for the dynamics, stability characteristics, and control of the robot. For this reason, in robot kinematics, the kinematic chains formed by the limbs connected to each other by kinematic pairs connected to the geometric structure of the robot are examined. If all the kinematic pairs used are closed kinematic pairs, they are closed kinematic chains, and if one of the kinematic pairs is open, they are open kinematic chains. Forward kinematics determines the position and orientation of the end effector based on covariates. Inverse kinematics determines covariates that correspond to the position and orientation of a particular end effector.

In robot dynamics, the relations between the mass and inertia properties, the forces and torques related to the motion, which are the factors that cause and change the motion of the robot, are examined. Knowing the forces and torques, one can predict how a mechanism will act. Robot dynamic modeling deals with deriving dynamic equations of robot motion using the Newton-Euler method and the Lagrangian method. In robot dynamics, it is important to find a solution to two types of problems related to the calculation of forward dynamics and reverse dynamics. Forward dynamics allows us to describe the motion of the real physical system in terms of joint accelerations when a set of joint torques assigned to the manipulator is applied. In reverse dynamics, the joint torques required to produce the motion are determined by the joint accelerations, velocities, and positions.

3.8 ROBOT (AUTONOMOUS) TRACTOR

The concept of autonomy describes that all functions performed by the tractor are performed without human intervention and that it can operate on its own to overcome the many uncertainties found in the working environment. In addition to determining algorithms, modeling, and methods for the control of movement in autonomous tractors, necessary hardware and software are developed for obstacle avoidance, localization, and map creation. For the implementation of road planning and navigation for this purpose, it is necessary to accurately estimate the position of the vehicle and to perceive the environment sensitively during the movement of the vehicle (Özgüven, 2018).

Various techniques are used in the navigation of autonomous vehicles. The first of these is the control of remotely controlled unmanned vehicles by an operator. Although it can be controlled by various communication infrastructures, the most common type of control is wireless control with remote control. The second is unmanned autonomous vehicles that operate independently of human control, in which the vehicle controls itself. Various research is carried out to determine the location of autonomous unmanned vehicles and to map the places and types of objects in the vicinity. These algorithms are also called simultaneous positioning and mapping (SLAM) algorithms in general (Kavak, 2008). Assuming that SLAM algorithms consist of three main layers, the first of these layers can be considered as reading the information about the objects in the environment at the time k from the sensor and arranging this data; the second layer is matching the observations made with the objects on the map; and the third layer is updating the predicted map at the time k-1 by combining these observations (Castellanos et al., 2006).

The operation of an autonomous tractor can be divided into tasks and behaviors. The task is to teach the tractor the work that needs to be done, such as driving the road, cultivating the soil, and planting seeds. The way of performing the task is also called behavior. In order to increase business success, it will be relatively easy for multiple tools working in the same area to be aware of each other and what others are doing, and to be able to share the same task at the same time with multiple tools.

Many researchers working in the field of robotics think that behavior-based robotics applications are the most appropriate way to develop truly autonomous vehicles. In this way, a definition of autonomous tractor behavior can be expressed as "sensible long-term behavior, unattended, in a semi-natural environment, while carrying out a useful task" (Blackmore and Griepentrog, 2006):

- First is the identification of this sensible behavior that is independent of the device. It is the definition of a set of behavioral modes (as defined by humans) that respond sensibly to a predefined set of stimuli in the form of an expert system.

- The second is the ability to perform its duties alone during extended processes. It should be capable of returning to headquarters and restocking when needed to refuel.
- Third, safety behaviors are important. The functional modes of the machine should make it safe for others as well as itself. But when subsystems are out of use, they must be capable of failing safely. Multiple levels of system backup must be designed into the tool to avoid catastrophic errors.
- Fourth, since the vehicle interacts with a complex semi-natural environment (in horticulture, agriculture, parkland, and forest uses), it uses advanced sensing and control systems to behave correctly.

Noguchi et al. (2002) developed a robot tractor. This tractor is a conventional 56 kW MD77 Kubota tractor modified for use as an autonomous tractor. Table 3.1 shows the list of robot controllable items by a PC. The connected internal controller is in the tractor cab along with control actuators for these functions. Internal communication is based on a serial RS232C and communication between the PC and the internal controller is via the controller area network (CAN) bus. The advanced navigation system mainly consists of a mission planner and autonomous operation. It enables the user to control the robot in terms of a hitch function, a PTO, an engine speed set, a transmission, etc., as well as a steering angle during autonomous operation. This control is based on the posture information from the RTK-GPS and the IMU in reference to a navigation map (Noguchi, 2013).

All stages of field operations including tillage, planting, spraying, and harvesting can be automated when robot tractors have a travel path. Moreover, it is possible to fully automate the entire operation sequence because the robot tractor can drive out by itself from the machine hangar, travel along the farm road to the field, complete the necessary operation, and then return to the hangar by itself. The driving accuracy of the tractor robot is ± 5 cm, which is more accurate than a human operator (Noguchi, 2013).

Moorehead et al. (2012) developed a multi-autonomous tractor that autonomously mowed and sprayed lawns in an orange orchard of 1,300 hectares. In the garden, where orange trees are planted in blocks in areas called pumping zones, there are many lagoons and wild areas, roads, and canals as seen in Figure 3.8. Fixed obstacles such as telephone poles and irrigation pump stations are accompanied by mobile equipment such as people, vehicles, orange-picking boxes, and ladders. Moveable obstacles cannot be placed on the orchard map, which makes them more challenging to deal with. These objects can either be detected by sensors on the tractor or operational applications can be instituted that ensure no such objects exist in the orchard while autonomous tractors are operating.

Each tractor developed has a perception system to detect unexpected obstacles in the orchard, a laser scanner and color cameras registered with RTK-GPS for localization to accurately follow

---

TABLE 3.1

Controllable Maneuvers of a Robot (Autonomous) Tractor

- Steering
- Transmission change (eight for every two subtransmissions)
- Switch of forward and backward movements
- Switch of power take-off
- Hitch functions
- Engine speed set (two sets: Manual and maximum)
- Engine stop
- Brake

Source: Noguchi (2013).

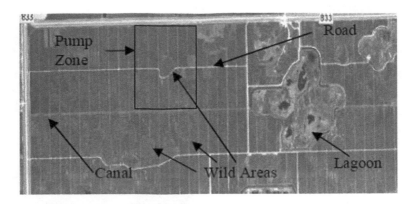

FIGURE 3.8 Satellite image of the orchard (Moorehead et al., 2012).

the planned path, and the ability to control the tractor functions such as the PTO and propulsion with an onboard computer. When a problem is encountered during autonomous operation, it can communicate with the center. Two separate communication links are used for communication. The first is the 900 MHz link, which enables low-bandwidth communication for critical data and signaling messages. If this communication link drops for any reason, the autonomous tractor stops. The second is a 2.4 GHz network that provides the necessary bandwidth for the transmission of images and video but has poor vegetation penetration. The developed software decides whether to continue driving, slow down, or stop the tractor by receiving the information from the sensors. In the tractor detection system, a sequence routing algorithm, which also uses a tree classifier that combines 3D lidar and camera data, is used to distinguish trees and tall weeds. This algorithm uses the computed treemap to find the lateral deviation in the row relative to the original planned path that will prevent the tractor from driving toward trees on either side. This deviation is continuously calculated and applied to the original planned path to create a new path for the tractor to follow, as shown in Figure 3.9.

3.9 ROBOT TRACTOR EQUIPMENT

Various equipment and systems are used to determine the positions of robot (autonomous) tractors, to create the desired route correctly, and to map the surrounding obstacles and objects correctly. The information from the sensors must be of a quality that will enable the autonomous tractor to move safely. Since the autonomous tractor works in the open field and in a wide variety of environments, some sensors may not receive information or errors may occur in the incoming information. For this reason, it is preferred to process the data from different sensors together and they are used together by transforming the unique advantages of different types of sensors used into a more comprehensive detection. In this way, information from the sensors provides more detailed information about the position of the tractor during movement, its surroundings, and surrounding objects. In order to process this obtained information and turn it into usable useful information, advanced decision mechanisms using applications such as image, audio, and video processing algorithms, artificial neural networks, machine learning, and statistical data analysis are used, and the autonomous tractor is operated successfully. Equipment used in autonomous tractors is listed below (Özgüven, 2018).

3.9.1 RADAR SENSORS

Radar is an important sensor used in autonomous tractors. It is used to determine the distance, height, direction, and speed of the radio waves by calculating the return times of the reflection from

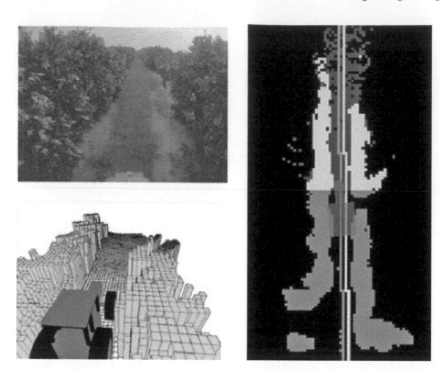

FIGURE 3.9 Queue routing examples. Top left: Tree row; lower left: Green tree cells computed treemap. Right image: Original path as white line and adjusted row orientation path as pink line (Moorehead et al., 2012).

the objects and the reflection from the nearby objects. Radar sensors used in autonomous tractors detect every object in the area covered by the detection beam at 250 m in all weather conditions and calculate the speed and distance of the object. In this way, collision avoidance is ensured.

3.9.2 Laser Scanners

A laser scanner measures by sending laser wavelength light to the object. Although it has various configuration structures, the most used is the one with the elements that emit and receive the beam in a single device. The most widely used laser type in autonomous vehicles is the LMS 200 laser produced by the SICK company, known as a 2D laser range finder, which measures both the distance of the object to the laser and the direction of the object relative to the laser in terms of angle (Kavak, 2008). These laser scanners (Figure 3.10) emit a pulsed rotating laser beam at 75 Hz over 180°, and the distance for each point is calculated at 1° intervals. Although the distance depends on the reflectivity of the detected object, this value varies between 30 m and 150 m. Although the resolution is nominally 10 mm, the statistical error reaches 40 mm at the upper limits of the range (Blackmore and Griepentrog, 2006).

3.9.3 Lidar

A lidar sensor is used to determine the obstacles that the tractor encounters during its movement. Lidar determines the distance of an object or a surface using laser beams. It works similarly to how radar technology works. The difference is that instead of radio waves, laser pulses hit the surrounding objects and the distance value is calculated using the reflection time. Three-dimensional (3D) point information of the area measured with lidar can be obtained in a very short time, at the desired frequency, and with high accuracy.

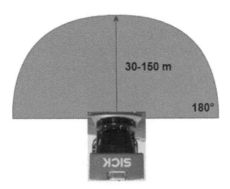

FIGURE 3.10 Laser scanner used for forward proximity sensing.

The laser beams reflected by the rotating mirror on the lidar can scan the environment at various angles depending on the characteristics of the lidar. These angles are read with the help of the encoder on the motor that rotates the mirror, and it is known at which angles the rays are sent. With the help of this angle and time information, the distance of the objects in the surrounding environment and their positions to the lidar can also be calculated (Karaahmetoğlu, 2011). The HDL-64E model lidar sensor produced by Velodyne Lidar can be given as an example of a lidar sensor that is produced and continuously developed by different companies. This sensor is a system with 64 lasers providing 2.2 million data points per second, 360° horizontal and 26.9° vertical viewing angles, and a very high data rate. With a range of 120 meters, this sensor with its innovative laser array allows more environmental observation for the most demanding sensing applications, high-resolution 3D mapping applications (Anonymous, 2020a).

3.9.4 GPS/Inertial Navigation System (INS)

GPS is used to determine the coordinates of the autonomous tractor's position in the world. However, GPS can only provide this information to the user in open areas. For this reason, GPS must be integrated with the inertial navigation system (INS) to provide continuous navigation data in case of signal interruption and weakening. By using the gyroscope and accelerometers in the INS used with GPS, the position, linear movements, and speed of the tractor can be measured continuously.

The ANS-530K inertial navigation system developed by ASELSAN is an integrated position and direction determination system developed for land vehicles. It continuously provides linear acceleration, linear and angular velocity, position, and orientation information to the system it is on. ANS-530K consists of an inertial measurement unit, system processor unit, power supply, embedded GPS receiver, and chassis. It uses a 12-channel P(Y) coded military GPS receiver as an embedded GPS receiver. It also has the ability to work with an external GPS receiver. ANS-530K can provide integrated (Inertial + GPS + Odometer), inertial-only, and GPS-only navigation solutions at the same time. Thanks to its odometer support, it provides high-precision location and orientation even in the absence of GPS. It minimizes the maintenance and repair needs thanks to the long average time between failures and the in-device testing capability (Anonymous, 2020b).

3.9.5 Ultrasonic Sensor

An ultrasonic sensor is used to assist warning systems and as a parking aid in autonomous tractors. They are sensors that detect the distance depending on the return time of an ultrasonic sound wave without contacting an object. The piezoelectric transducers in them convert the electrical energy into sound waves and send them to the object. When AC electricity is supplied to piezoelectric transducers, they produce sound waves with frequencies between 20 kHz and 500 kHz. The smooth linear progression of the waves at these frequencies and the high energies allow them to be easily reflected by the objects they come into contact with. Piezoelectric crystals are used in the sensing part. When a load or sound intensity is applied to the piezoelectric crystals, the sound waves reflected from the object cause the piezoelectric crystals to generate voltage in proportion to the distance, due to the generation of voltage.

3.9.6 Cameras

Although a camera is sensitive to adverse weather conditions and lighting changes, it provides detailed information about the environment by processing complex images taken with its technology and high resolution. This information includes the classification of objects in the environment of autonomous tractors, determination of texture, color, and contrast information, and real-time 3D images. The use of large amounts of data during the processing of the acquired images makes the algorithm complex and computationally intensive.

3.10 AGRICULTURAL ROBOTS

Agriculture is a necessary and strategic activity for the continuity of life, as well as having an important economic return. The aim of agricultural production is to increase productivity and product quality in agriculture, use minimum inputs, ensure food safety, protect natural resources and the environment, and provide economic, sustainable, and productive management. The agricultural sector is adversely affected by economic, social, structural, and climatic problems. Some of these problems are as follows: Global market fluctuations, economic crises, drought, tornadoes and floods resulting from climate change, diseases, the emergence of alternative uses of agricultural products such as biofuels, misuse of agricultural lands such as mining activities, reduction of natural resources such as water, migration of the young population from the villages to the cities, and the increase in the elderly population in the villages (Özgüven, 2020).

As a result of the rapid developments in information technology following the mechanization, automation, and control technologies during the development period of agricultural production, today, intelligent machines and production systems that control machines have begun to take over traditional production methods. Information technology consists of hardware, algorithms, and software developed for the management of the collection, processing, storage, transfer, and use of information processes. The implementation of present knowledge and experiences in agriculture together with machine learning, deep learning, artificial intelligence, modeling, and simulation applications enabled the development of real-time and automated expert systems, autonomous tractors or agricultural machines, and agricultural robotics applications (Ozguven, 2018). These advances provide some of the impetus for agricultural robots as a technology enabling food production. The trend is for agricultural industries to move away from urban centers to areas with less manpower to participate in the increasingly challenging tasks of feeding the growing population. With the aim of achieving integrated autonomous farming systems, agricultural robots not only provide the means to perform necessary agricultural tasks with less manpower but also enable them to exert greater control over crops, improving food quality and reducing dependency on factors (Oetomo et al., 2009). In addition, agricultural robots are a very important tool that will increase farmer welfare by increasing productivity and product quality in agricultural production, reducing production costs and manpower in many laborious agricultural jobs (Özgüven et al., 2016).

Agricultural robots are cheaper, faster, and safer than existing traditional agricultural production systems. However, significant technical challenges stand in the way of the large-scale use of robotics in agriculture. The main obstacle is the failure to develop systems adapted to agricultural conditions. Because many automatic and robotic elements used in industry and smart cities cannot be used in agriculture. For example, many processors used in industry cannot withstand a greenhouse environment due to humidity, temperature, and corrosive factors and will degrade in a short time. In addition, the connectivity problem of robotic systems in large fields at long distances is also a serious problem. These facts are major challenges. Of course, more advanced systems such as processors used in the oil and aerospace industries or military wireless systems can be used in agriculture. However, this is directly related to the increased cost of agriculture and it is of primary importance to reduce the cost of agricultural production. Despite these challenges, robotics in agriculture is an inevitable trend and will be increasingly advanced and used more and more (Albiero, 2019).

In the last two decades, specialized sensors (machine vision, GPS, RTK, laser-based devices, and inertial devices), actuators (hydraulic cylinders, linear and rotational electric motors), and electronic equipment (embedded computers, industrial PC and PLC) have been used by many autonomous vehicles, especially enabling the integration of agricultural robots. If these semi-autonomous/autonomous systems are equipped with the appropriate equipment (agricultural implements or equipment), they provide accurate positioning and guidance in the work area where precision farming tasks are performed (Emmi et al., 2014). Autonomous agricultural robots are the alternative to tractors in the fields today. Cultivation operations such as planting seeds, spraying, fertilizing, and harvesting could be carried out by fleets of autonomous agricultural robots in the future. The agricultural robot must have some basic capabilities and the ability to support multiple applications. Among its core capabilities, a navigation system is required for safe and autonomous navigation (Biber et al., 2012).

The number of commercialized agricultural robots has been limited due to difficulties such as the inability to structure agricultural cultivation and production areas of agricultural robots, the complexity of the studies, the diversity of work areas and applications, and the fact that the studies in this field have generally remained in academic and research dimensions. However, the learning and product development process continues. The new control and automation techniques, smart sensors, electronic hardware, and processors that emerged with the development of information technologies, being affordable and easily accessible, ensure that the added value in the developed products and systems is high. The fact that the importance of healthy and quality agricultural products was better understood during the pandemic process, the incentives and supports brought to those working in these areas, the increase of livestock enterprises established especially in greenhouses and closed areas, and the improvement of infrastructures have managed to attract the attention of companies, SMEs, and academic studies working in this field. When the benefits of agricultural robots, such as efficiency, quality increase, and cost reduction, and the successes achieved in agricultural robot studies are evaluated, it is thought that these studies will become widespread in the near future, and the potential to turn into commercial products is high.

Agricultural robots have been developed and continue to be developed for the realization of many applications of crop and animal production. With the widespread use of techniques such as artificial intelligence, machine learning, deep learning, image processing, and machine vision in location-based and autonomous robot applications, the diversity of agricultural robots has increased. Many agricultural robots such as tractors, sowing, planting, spraying, fertilizing, harvesting, pruning, grafting, soil analysis, irrigation success tracking, disease, pest, and weed control, plant health tracking, object recognition, mapping, milking, feeding, feed pushing, barn cleaning, and calf feeding robots have been developed.

3.11 EXAMPLES OF AGRICULTURAL ROBOTS

Before giving examples of agricultural robots, it is useful to give a brief explanation to the reader about this section. There is no commercial relationship with the companies of the robots selected in this section. In addition, if there is another robot competing with the robot belonging to the selected company, it does not indicate that the robot given in this section is superior to the other robot. Agricultural robots in this section have been chosen to show what kind of work agricultural robots do in general terms. Considering those who work or are interested in animal production, or want to learn new things about it, robots used in animal production are given in this section.

3.11.1 MILKING ROBOT

A milking robot consists of some hardware that performs the milking process and computer software that controls these hardware parts. A milking robot consists of a milking chamber, teat sensors, and a robot arm or arms (Figure 3.11). A manger is placed to encourage cows to enter the milking parlor. The milking chamber can be placed in the middle of the barn, or it can be positioned in a different place in the barn. Ultrasonic sensors detect the location of the teats. The sensors first locate the cow's right front teat, then base the other three teats off this teat. In addition, special camera systems can monitor all movements of the animal during milking (Demir and Öztürk, 2010; Türkyılmaz, 2005). After each cow is milked, the milking robot is washed and cleaned. This washing is very important for cow health and milk quality. If an animal has mastitis, it is prevented from contaminating another animal with this washing and the milk with mastitis is separated and destroyed.

3.11.2 FEEDING ROBOT

The automatic feeding robot seen in Figure 3.12 automatically carries out the process of taking the feed rations from the feed silos and warehouses and delivering them to the animals. While pushing the feeds so that there is always enough fresh feed in the feeder path, it determines whether more feed is needed by controlling the amount of feed with the feed height sensors. In this way, it is ensured that the feed is given to the animals fresh.

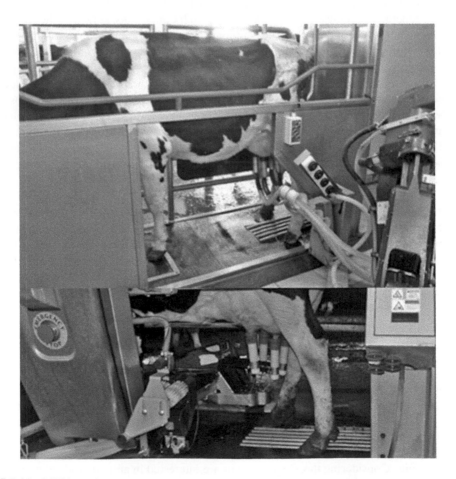

FIGURE 3.11   Milking robot.

FIGURE 3.12 Feeding robot.

FIGURE 3.13 Calf feeding robot.

3.11.3 CALF FEEDING ROBOT

The calf foods used in feeding the calves are in the central feeding unit. When the calves enter the feeding unit, the system recognizes the calf with the identification information on it and freshly prepares the necessary food for the calf's feeding. The system works fully automatically with computer control (Figure 3.13) (Yıldız and Özgüven, 2018). Calves are weaned slowly by gradually reducing the amount of milk. Slowly reducing the amount of milk allows the calves to automatically increase their roughage intake. The weaning process is further improved with optional concentrated feeding. In this way, calves learn to eat more concentrated feed faster and consume less milk.

3.11.4 BARN CLEANING ROBOT

The barn cleaning robot cleans the barn by following the programmed routes in the barn using its advanced software, navigation technology, two ultrasonic sensors, and a gyroscope. The routes and schedule of the cleaning program can be adjusted to the daily rhythm of the farm and the cows. The robot also performs the manure collection, manure discharge, water filling, and water spraying processes required for cleaning on its own, and when its task is completed, it goes to the charging station and charges up (Figure 3.14).

3.11.5 MECHANICAL WEED CONTROL ROBOT

Machleb et al. (2021) designed a new weeding tool for agricultural robots to perform in-row mechanical weed control in sugar beets. In the study, a conventional finger weeder was modified and equipped with an electric motor (Figure 3.15). This allowed the rotational motion of the finger weeders independent of the forward travel speed of the tool carrier. The new tool was tested in field trials with the bi-spectral camera.

The setup of the mechanical weed controller is shown in Figure 3.16. The hoeing system had two cameras (Figure 3.16a). The first camera was the Garford camera for general row alignment of the hoe to perform interrow hoeing. The second camera was the bi-spectral camera (Figure 3.16b) positioned above a sugar beet row and responsible for motorized finger weeders and in-row treatments. A controller adjusted the speed of the motorized finger weeders by performing two different setups. At the location of a sugar beet plant, the rotational speed was equal to the driving speed of the tractor. The rotational speed between two sugar beet plants was either increased by 40% or decreased by 40%. In-row weed control efficiency of this new system in the two-year trial ranged from 87% to 91% and from 91% to 94%, respectively. The sugar beet yields were not adversely affected by the mechanical treatments compared to conventional herbicide application.

3.11.6 SPRAYING ROBOT

Baltazar et al. (2021) developed a smart and novel electric sprayer that can be mounted on a robot (Figure 3.17). The sprayer has a crop perception system that calculates the leaf density based on a

FIGURE 3.14 Barn cleaning robot.

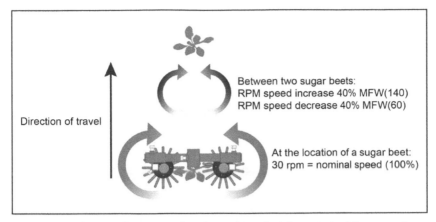

(a) Diagram showing the rotation direction and speed of the motorized finger weeders.

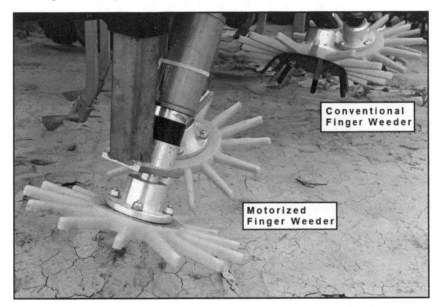

(b) The motorized finger weeders (front) and conventional finger weeders (back).

FIGURE 3.15   Normal and motorized finger weeders and their principles of use (Macleb et al., 2021).

support vector machine classifier using image histograms (local binary pattern, vegetation index, average, and hue). This density is then used as a reference value to feed a controller that determines the sprayer's airflow, water ratio, and water density. This perception system has been tested with a created data set. The results of the leaf density classifier display an accuracy score ranging from 80% to 85%. Testing proves that the solution has the potential to improve spraying accuracy and precision.

The dataset consists of 475 images of 640 × 480 collected from vineyards. Data were collected in video format and saved to a ROSBAG file. Then, images were extracted from one of the lenses from the recorded file and sampled every five frames to reduce the correlation between images and avoid annotating similar images. Three ROIs were selected to feed the independent controllers of the three sprayer drums. The application identifies three ROIs (Figure 3.18) and then asks for leaf density in each ROI. The value is added to four classes: 0% leaves, 33% leaves, 66% leaves, or 100% leaves. When there are few leaves or small amounts that do not exceed half of the total area, it is

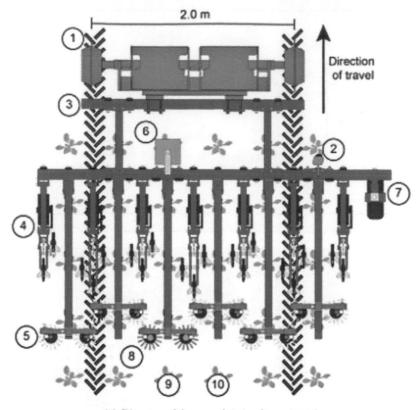

(a) Diagram of the complete implement used.

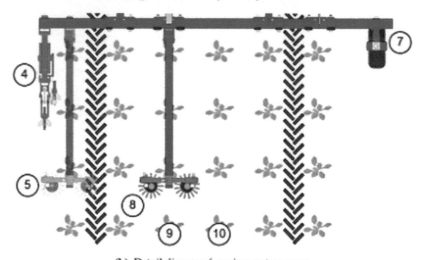

(b) Detail diagram focusing on two rows.

FIGURE 3.16 The setup of the hoe used in the experiments. The image shows one plot width (six sugar beet rows). (1) Garford Robocrop, (2) Garford Robocrop camera for interrow weeding, (3) Argus hoe frame, (4) parallelograms with goosefoot sweeps, (5) conventional finger weeders, (6) bi-spectral camera for intrarow weeding (color blue), (7) odometry wheel, (8) motorized finger weeders (color blue), (9) and (10) the two center sugar beet rows of each plot that were treated with the motorized finger weeders and used for harvesting (Macleb et al., 2021).

FIGURE 3.17   Sprayer mounted on the robot (Baltazar et al., 2021).

considered 33%. If a large part of the area is covered with leaves but not completely covered, it is considered 66%, and when the entire area is filled with leaves, it is considered 100% (Figure 3.19). This value is then saved in a text file with the name of the image.

3.11.7   CUCUMBER HARVESTING ROBOT

Van Henten et al. (2002) developed an autonomous robot for harvesting cucumbers in the greenhouse. The cucumber harvesting robot was tested in a greenhouse and it was determined that the robot was capable of picking cucumbers with 80% success without human intervention. On average, it takes 45 seconds for the robot to select a cucumber. In the Netherlands, where the robot was developed, cucumbers are grown in greenhouses with a uniform arrangement. A 2-ha greenhouse requires four harvesting robots and a docking station during peak season. Based on this information, the design features of the harvesting robot were defined. The basic requirement is that a single harvesting process can take no more than ten seconds. In a standard 2-ha greenhouse, there are corridors each approximately 60 m long perpendicular to the main road and the main road along the center of the greenhouse. There are 100 rows of cucumber plants on either side of the main road. A corridor is located between rows of plants called a path. The distance between the two rows on both sides of the corridor is 0.9 m. The distance between the stalks of cucumbers in the row is 0.35 m. In total, there are approximately 180 stems in a row. Then the density is 3.6 sap/m². In corridors, floor heating pipes are used as transport rails during crop maintenance and harvesting. During manual harvesting, fruits are collected in a cart in a crate. Full crates are replaced by empty crates on the main road. The harvested cucumbers are then transported to the landfill, where they are sorted, packaged, and stored until sale.

The autonomous cucumber robot consists of an autonomous vehicle, manipulator, robot arm tip, two computer vision systems for sensing and 3D imaging of fruit and environment, and a control scheme that generates collision-free motion for the manipulator during harvest (Figure 3.20). The manipulator has seven degrees of freedom. This is enough for the harvest task. The tip of the robot

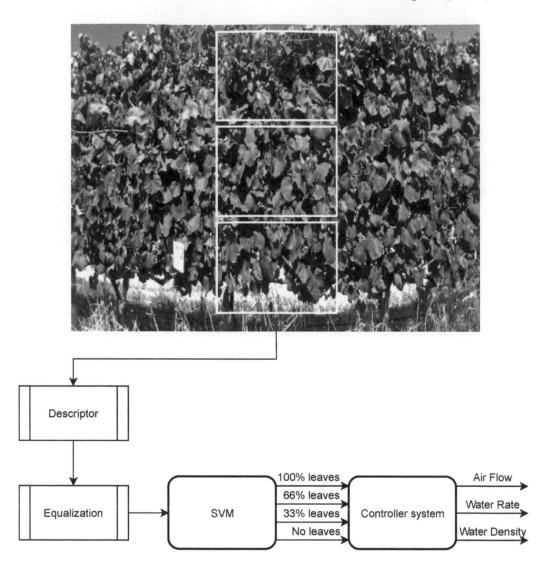

FIGURE 3.18   Software architecture (Baltazar et al., 2021).

arm is designed to hold the soft fruit without loss of quality, and the thermal cutting device inside prevents the spread of viruses throughout the greenhouse.

The computer vision system can detect more than 95% of the cucumbers in the greenhouse. Detection of cucumber fruit using computer vision is shown in Figure 3.21. The maturity of cucumbers is determined using geometric models. A motion planner based on a search algorithm ensures eye-hand coordination without collision.

3.11.8   PADDY PLANTING ROBOT

Nagasaka et al. (2009) developed an autonomous rice planting robot. The rice planting robot is a modified six-row rice planter manufactured by the Iseki Corporation (Matsuyama, Japan). The parts modified for autonomous operation are: Attaching a geared DC motor to the steering system, connecting an absolute rotary encoder to the steering axle to sense the steering angle, installing a servomotor to actuate the hydrostatic transmission arm, and a motor output, powered by the tie rod ring on the hydrostatic transmission arm (Figure 3.22).

FIGURE 3.19 Classes representation, from top to bottom and from left to right: 0%, 33%, 66%, and 100% leaves (Baltazar et al., 2021).

FIGURE 3.20 Cucumber harvester robot (Van Henten et al., 2003).

The autonomous paddy planting robot is guided by GPS location data with a tilt correction feature during straight driving and INS direction data during turning. The GPS antenna is installed at a height of 1.5 m and 0.4 m in front of the vehicle's front axle. The actuator control command and data communication protocols conform to the controller area network (CAN) bus. The planting pattern is controlled by a vehicle microprocessor. A vehicle microprocessor detects the planter lift and lower and start and stop functions when the vehicle is started manually. A host sends the switch function to the vehicle microprocessor to control the sewing pattern during autonomous operation.

FIGURE 3.21 Determination of cucumber. The cucumber shown with the dashed line illustrates that in one harvesting position the fruit may be hidden behind a leaf whereas it can be clearly visible from the next position (Van Henten et al., 2003).

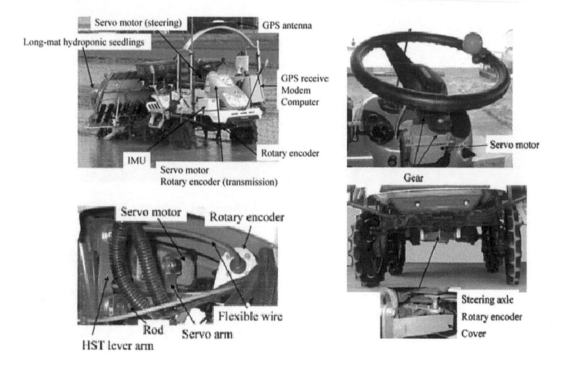

FIGURE 3.22 Improved autonomous paddy planting robot (Nagasaka et al., 2009).

3.11.9 BoniRob Multipurpose Agriculture and Weed Control Robot

The BoniRob™ agricultural robot (Figure 3.23) was developed as a joint project by the University of Osnabrueck, the Robert Bosch company, Deepfield Robotics, and the machine builder Amazonen-Werker. While working in the field, this robot can determine plant phenotype, determine soil compaction, and control weeds among row crops. With the built-in application modules of the robot, its mechanical structure can be changed and customized for various functions for different field operations. Desired mechanical, electrical, and logical interfaces can be integrated into the robot platform. There are wired or wireless communication interfaces to communicate with different vehicles such as monitoring stations or robots and drones. Multiple BoniRob applications can be used on the robot platform. It also allows the driver to control the robot for defined states depending on the situation.

BoniRob is a flexible four-wheel platform that can be steered individually. The wheels are driven by hub motors. By moving the arms, the track width can be adjusted from 0.75 m to 2 m and the chassis clearance from 0.4 m to 0.8 m. BoniRob is equipped with four motor controllers,

FIGURE 3.23   BoniRob robot (Chebrolu et al., 2017).

a motor and hydraulic interface, a navigation control unit, and application control units. Except for motor controllers connected via CAN bus, different control units are connected via ethernet and communicate using TCP/IP. For navigation, it uses a 3D MEMS lidar (Nippon Signal, FX6) that measures distance and reflectivity using infrared pulsed laser light at up to 16 fps with a flight time of approximately 30 picoseconds and an accuracy of 1 cm, an inertial sensor unit (XSens), and optionally, an RTK-GPS. There is also a remote control (Logitech Rumblepad) and a WLAN access point with a dedicated outdoor antenna. All navigation sensors are connected to a fanless onboard computer (Intel Atom) designed to operate in target ambient conditions (Biber et al., 2012). Multi-sensor fusion, including complex sensor systems such as spectral imaging and 3D time-of-flight cameras, and an RTK-DGPS system are used for site-specific individual plant phenotyping. The criteria looked at include, for example, plant number, plant height, stem thickness, and biomass, as well as information on water supply and indications for possible plant diseases (Ruckelshausen et al., 2009).

Phenotyping, penetrometer, and precision spray applications demonstrate the flexibility of the BoniRob platform (version 2). Another BoniRob-App has been developed for mechanical weed control. This application is integrated as part of a more complex environment, including web-based communication, a server, a web client, and a remote human. The application has an actuator for the mechanical processing of weeds. In addition, synchronously triggered cameras and lighting units at different wavelengths that can capture high-contrast images of plants are used. Cameras and lighting units are mounted in a shaded area under the robot. A camera-based solution has been implemented for chemical weed control. The camera is used to detect green areas. The spray nozzles are controlled so that herbicides are applied only to the areas where the plants are present. This technology has the potential for a large reduction in pesticides applied compared to applications with homogeneous norms (Bangert et al., 2013).

REFERENCES

Albiero, D., 2019. Agricultural Robotics: A Promising Challenge. *Current Agriculture Research Journal*, 7(1), 1–3.

Angeles, J., 2014. *Fundamentals of Robotic Mechanical Systems Theory, Methods, and Algorithms*. 4th Edition. Springer International Publishing, Cham. ISBN: 978-3-319-01850-8.

Anonymous, 2007. Robotics. Appin Knowledge Solutions. Infinity Science Press LLC. ISBN: 978-1-934015-02-5.

Anonymous, 2020a. http://www.velodynelidar.com/hdl-64e.html. [Accessed 05.01.2020].

Anonymous, 2020b. https://www.aselsan.com.tr/tr/cozumlerimiz/aviyonik-ve-seyrusefer-sistemler/ataletsel -seyrusefer-sistemleri/ans530k-kara-ataletsel-navigasyon-sistemi. [Accessed 05.01.2020].

Antonelli, G., 2006. *Underwater Robots Motion and Force Control of Vehicle-Manipulator Systems.* 2nd Edition. Editors: Bruno Siciliano, B., Khatib, O. and Groen, F. Springer-Verlag, Berlin Heidelberg. ISBN: 13 978-3-540-31752-4.

Baltazar, A.R., Santos, F.N.D., Moreira, A.P., Valente, A. and Cunha, J.B., 2021. Smarter Robotic Sprayer System for Precision Agriculture. *Electronics*, 10, 2061. https://doi.org/10.3390/electronics10172061.

Bangert, W., Kielhorn, A., Rahe, F., Albert, A., Biber, P., Grzonka, S., Haug, S., Michaels, A., Mentrup, D., Hänsel, M., Kinski, D., Möller, K., Ruckelshausen, A., Scholz, C., Sellmann, F., Strothmann, W. and Trautz, D., 2013. Field-robot-based agriculture: RemoteFarming. 1 and BoniRob-Apps. VDI. Agricultural Engineering 2013, 439–446.

Biber, P., Weiss, U., Dorna, M. and Albert, A., 2012. Navigation System of The Autonomous Agricultural Robot Bonirob. In Workshop on Agricultural Robotics: Enabling Safe, Efficient, and Affordable Robots for Food Production, Vilamoura, Portugal.

Blackmore, B.S. and Griepentrog, H.W., 2006. Section 4.3 Autonomous Vehicles and Robotics, pp. 204–215 of Chapter 4 Mechatronics and Applications, in CIGR Handbook of Agricultural Engineering Volume VI Information Technology. Edited by CIGR-The International Commission of Agricultural Engineering; Volume Editor, Axel Munack. St. Joseph, Michigan, USA: ASABE. Copyright American Society of Agricultural Engineers.

Bottin, M., Cocuzza, S., Comand, N. and Doria, A., 2020. Modeling and Identification of an Industrial Robot with a Selective Modal Approach. *Applied Sciences*, 10, 4619. https://doi.org/10.3390/app10134619.

Castellanos, J.A., Neira, J. and Tardos, J.D., 2006. Lecture Notes on SLAM and Consensus in Data Association, Department of Informatics and System Engineering. Univ. Zaragoza.

Ceccarelli, M., 2004. Fundamentals of Mechanics of Robotic Manipulation. Springer Science+Business Media, Dordrecht. ISBN: 978-90-481-6516-2.

Chebrolu, N., Lottes, P., Schaefer, A., Winterhalter, W., Burgard, W. and Stachniss, C., 2017. Agricultural robot dataset for plant classification, localization and mapping on sugar beet fields. *The International Journal of Robotics Research*, 36, 1045–1052.

Demir, B. ve Öztürk, İ., 2010. Robotlu Sağım Sistemleri. *Alınteri*, 19(B), 21–27. ISSN:1307-3311. (Turkish).

Dudek, G. and Jenkin, M., 2010. *Computational Principles of Mobile Robotics.* 2nd Edition. Cambridge University Press, New York. ISBN: 978-0-521-87157-0.

Emmi, L., Gonzalez-de-Soto, M., Pajares, G. and Gonzalez-de-Santos, P., 2014. New Trends in Robotics for Agriculture: Integration and Assessment of a Real Fleet of Robots. *Hindawi Publishing Corporation the Scientific World Journal.* 2014, Article ID 404059, 21 pages. http://doi.org/10.1155/2014/404059.

Fathi, K., van de Venn, H.W. and Honegger, M., 2021. Predictive Maintenance: An Autoencoder Anomaly-Based Approach for a 3 DoF Delta Robot. *Sensors*, 21, 6979. https://doi.org/10.3390/s21216979.

Gray, J.O. and Davis, S.T., 2013. Robotics in the Food Industry: An Introduction. *Robotics and Automation in the Food Industry Current and Future Technologies.* (Editor: Caldwell, D.G.) Woodhead Publishing Limited, Cambridge, 21–35. ISBN: 978-1-84569-801-0.

Gupta, A.K. Arora, S.K. and Westcott, J.R., 2017. Industrial Automation and Robotics. Mercury Learning and Information. ISBN: 978-1-938549-30-4.

Iqbal, J., Xu, R., Halloran, H. and Li, C., 2020. Development of a Multi-Purpose Autonomous Differential Drive Mobile Robot for Plant Phenotyping and Soil Sensing. *Electronics*, 9, 1550. https://doi.org/10.3390/electronics9091550.

Jeong, H., Choi, K., Park, S.J., Park, C.H., Choi, H.R. and Kim, U., 2021. Rugged and Compact Three-Axis Force/Torque Sensor for Wearable Robots. *Sensors*, 21, 2770. https://doi.org/10.3390/s21082770.

Jordan, J., 2016. *Robots.* The Massachusetts Institute of Technology Press, Cambridge. ISBN: 9780262529501.

Karaahmetoğlu, R., 2011. GCDC 2011 Yarı Otonom Kooperatif Adaptif Sürüş Yarış Aracının Tasarımı. İstanbul Teknik Üniversitesi Fen Bilimleri Enstitüsü Mekatronik Mühendisliği Anabilim Dalı Yüksek Lisans Tezi. İstanbul. (Turkish).

Kavak, D., 2008. İnsansız Kara Araçları Navigasyonunda Genişletilmiş Kalman (GKF) ve Sıkıştırılmış Genişletilmiş Kalman Filtre (SGKF) Tabanlı Slam Yöntemlerinin Geliştirilmesi ve Karşılaştırılması. İstanbul Teknik Üniversitesi Fen Bilimleri Enstitüsü Elektrik Mühendisliği Kontrol ve Otomasyon Anabilim Dalı Yüksek Lisans Tezi. İstanbul. (Turkish).

Kurdila, A.J. and Ben-Tzvi, P., 2020. *Dynamics and Control of Robotic Systems.* John Wiley & Sons Ltd., Hoboken. 9781119524830.

Kyriakopoulos, K.J. and Loizou, S.G., 2006. Section 2.4 Robotics: Fundamentals and Prospects, pp. 93–107. Chapter 2 Hardware, in CIGR Handbook of Agricultural Engineering Volume VI Information

Technology. Edited by CIGR-The International Commission of Agricultural Engineering; Volume Editor, Axel Munack. St. Joseph, Michigan, USA: ASABE. (Çevirmenler: Demircioğlu, P. ve Böğrekci, İ.; Çeviri Editörleri: Tarhan, S. ve Ozguven, M.M.).

Machleb, J., Peteinatos, G.G., Sökefeld, M. and Gerhards, R., 2021. Sensor-Based Intrarow Mechanical Weed Control in Sugar Beets with Motorized Finger Weeders. *Agronomy*, 11, 1517. https://doi.org/10.3390/agronomy11081517.

Moorehead, S.J., Wellington, C.K., Gilmore, B.J. and Vallespi, C., 2012. Automating Orchards: A System of Autonomous Tractors for Orchard Maintenance. http://www.cs.cmu.edu/~mbergerm/agrobotics2012/04Moorehead.pdf.

Moreno, J.C., Bueno, L. and Pons, J.L., 2008. Wearable Robot Technologies. *Wearable Robots: Biomechatronic Exoskeletons*. Editor: Pons, J.L. John Wiley & Sons Ltd., Chichester, 165–200. ISBN: 978-0-470-51294-4.

Nagasaka, Y., Saito, H., Tamaki, K., Seki, M., Kobayashi, K., and Taniwaki, K., 2009. An Autonomous Rice transplanter Guided by Global Positioning System and Inertial Measurement Unit. *Journal of Field Robotics*, 26(6–7), 537–548. https://doi.org/10.1002/rob.20294.

Nenchev, D.N., Konno, A. and Tsujita, T., 2019. Humanoid Robots Modeling and Control. Butterworth-Heinemann is an imprint of Elsevier. ISBN: 978-0-12-804560-2.

Niku, S.B., 2011. *An Introduction to Robotics: Analysis, Control, Applications*. 2nd Edition. John Wiley & Sons, Inc., Hoboken. ISBN: 978-0-470-60446-5.

Noguchi, N., Kise, M., Ishii, K. and Terao, H., 2002. Field Automation using Robot Tractor. Proceedings of Automation Technology for Off-Road Equipment, pp. 239–245.

Noguchi, N., 2013. Agricultural Vehicle Robot. *Agricultural Automation Fundamentals and Practices*. (Editors: Zhang, Q. and Pierce, F.J.) CRC Press Taylor & Francis Group LLC, Boca Raton, 15–39.

Nonami, K., Kendoul, F., Suzuki, S., Wang, W. and Nakazawa, D., 2010. *Autonomous Flying Robots Unmanned Aerial Vehicles and Micro Aerial Vehicles*. Springer, Tokyo. ISBN: 978-4-431-53855-4.

Oetomo, D., Billingsley, J. and Reid, J.F., 2009. Editorial: Agricultural Robotics. *Journal of Field Robotics*, 26(6–7), 501–503. https://doi.org/10.1002/rob.20302.

Ozguven, M.M., 2018. The Newest Agricultural Technologies. *Current Investigations in Agriculture and Current Research*, 5(1), 573–580. https://doi.org/10.32474/CIACR.2018.05.000201.

Özgüven, M.M., Tan, M., Közkurt, C., Yardım, M.H., Özsoy, M. and Sabancı, E., 2016. Çok Amaçlı Tarım Robotunun Geliştirilmesi. *GOÜ, Ziraat Fakültesi Dergisi*, 33 (Ek sayı), 108–116. (Turkish).

Özgüven, M.M., 2018. Hassas Tarım. Akfon Yayınları, Ankara. ISBN: 978-605-68762-4-0. (Turkish).

Özgüven, M.M., 2019a. Teknoloji Kavramları ve Farkları. International Erciyes Agriculture, Animal & Food Sciences Conference 24–27 April 2019-Erciyes University – Kayseri, Turkiye. (Turkish).

Özgüven, M.M., 2019b. Tarım Robotlarının Sürdürülebilir Tarıma Katkıları. 3. Uluslararası UNİDOKAP Karadeniz Sempozyumu. 21–23 Haziran 2019, Tokat. Sempozyum Kitabı. S.354–367. ISBN: 978-605-80568-1-7. (Turkish).

Özgüven, M.M., 2020. Tarımda Dijital Dönüşüm ve Akıllı Makineler. Yeni Türkiye Dergisi. Tarım Politikaları Özel Sayısı. Sayı 114, Cilt 2. S.105–132. (Turkish).

Özgüven, M.M., Beyaz, A., Ormanoğlu, N., Aktaş, T., Emekci, M., Ferizli, A.G., Çilingir, İ. ve Çolak, A., 2020. Hasat Sonrası Ürünlerin Korunmasına Yönelik Mekanizasyon Otomasyon Ve Mücadele Teknikleri. Türkiye Ziraat Mühendisliği IX. Teknik Kongresi. Ocak 2020, Ankara. Bildiriler Kitabı-1, s.301–324. (Turkish).

Perez, T., Bawden, O., Kulk, J., Russell, R., McCool, C., English, A. and Dayoub, F., 2018. Overview of Mechatronic Design for a Weed-Management Robotic System. *Robotics and Mechatronics for Agriculture*. (Editors: Zhang, D. and Wei, B.) CRC Press Taylor & Francis Group LLC, Boca Raton, 23–49. ISBN: 978-1-1387-0240-0.

Pons, J.L., Ceres, R. and Calderon, L., 2008. Introduction to Wearable Robotics. *Wearable Robots: Biomechatronic Exoskeletons*. (Editor: Pons, J.L.) John Wiley & Sons Ltd., Chichester, 1–16. ISBN: 978-0-470-51294-4.

Ruckelshausen A, Biber P, Dorna M, Gremmes H, Klose R, Linz A, Rahe, R., Resch, R., Thiel, M., Trautz, D. and Weiss, U., 2009. BoniRob–an Autonomous Field Robot Platform for Individual Plant Phenotyping. *Precision Agriculture*, 9(841), 1.

Sciavicco, L. and Siciliano, B., 2000. *Modelling and Control of Robot Manipulators*. 2nd Edition. Springer, London. ISBN: 1-85233-221-2S.

SoftGripping, 2022. Soft Gripping. https://soft-gripping.com/. [Accessed 18.05.2022].

Türkyılmaz, M.K., 2005. Süt Sığırcılık İşletmelerinde Sağım Robotu Kullanımı. *Erciyes Üniv Vet Fak Derg*, 2(1), 61–64. (Turkish).

Van Henten, E.J., Hemming, J., Van Tuijl, B.A.J., Kornet, J.G., Meuleman, J., Bontsema, J. and Van Os, E.A., 2002. An Autonomous Robot for Harvesting Cucumbers in Greenhouses. *Autonomous Robots*, 13, 241–258. Kluwer Academic Publishers. Manufactured in The Netherlands.

Van Henten, E.J., Van Tuijl, B.A.J., Hemming, J., Kornet, J.G., Bontsema, J. and Van Os, E.A., 2003. Field Test of an Autonomous Cucumber Picking Robot. *Biosystems Engineering*, 86(3), 305–313. https://doi.org/10.1016/j.biosystemseng.2003.08.002.

4 Use of Unmanned Aerial Vehicles in Agriculture

4.1 WHAT IS AN UNMANNED AERIAL VEHICLE?

Unmanned aerial vehicles (UAVs) are vehicles that can operate remotely or autonomously with their own power system, and which can be loaded and removed from the main body according to the place of use (Figure 4.1). There are two types of unmanned aerial vehicles that can fly autonomously on a spontaneous flight plan: Unmanned aerial vehicles (UAVs) and remotely controlled drones. Although these vehicle names are commonly used interchangeably, the term unmanned aerial vehicles is a general title for all aircraft used as unmanned, whether the aircraft is remotely controlled, autonomous, or not (Özgüven, 2018). The characteristics of the UAV are large, operating in outdoor environments, uncertain and dynamic working environments, with three-dimensional movement, which means that the planning area must be four-dimensional. It can be classified as the presence of moving obstacles and environmental forces that affect movement, such as winds and changing weather conditions, and differential constraints on movement (Sebbane, 2016).

A typical UAV system consists of aircraft, one or more ground control stations and/or mission planning and control stations, payload, and data link. In addition, many systems include launch and recovery subsystems, aircraft carriers, and other ground handling and maintenance equipment (Fahlstrom and Gleason, 2012). A very simple general-use UAV system is shown in Figure 4.2. The fact that UAVs are more complex and have more parts than drone systems increase the cost of UAV system installation. Drone systems, on the other hand, can be used immediately without the need for any other expense, just by purchasing the drone with its apparatus. Drones are preferred to be used in agricultural applications due to their lower purchasing cost, ease of use, and capabilities (Özgüven, 2018).

UAVs can be classified in various ways (Akyürek et al., 2012):

1. By size, altitude, flight time, and payload capacity: Micro, mini, small, tactical, operative, or strategic;
2. By payload type: Armed UAVs or unarmed UAVs;
3. By fuel type: Internal combustion engine or electric-powered UAVs;
4. According to the flight method: Fixed wing or rotary wing;
5. By command type: Autopilot or remote controlled;
6. According to the intended use: Fake/target, reconnaissance surveillance, attack, or logistic support;
7. According to the take-off and landing method: Taking off/ejected from the ramp, taking off from the runway, dropped from the aircraft, hand-launched, landing on the fuselage, or landing with a parachute.

UAVs are classified into three classes and six different groups (micro, mini, small, tactical, operative, and strategic) according to their size and other basic features. According to the classification method in question, the main distinguishing features and the main models in each group are given in Table 4.1. While the basic criterion for UAV classification in Turkey is altitude, weight is taken as a basis in NATO and EU countries (Akyürek et al., 2012).

DOI: 10.1201/b23229-4

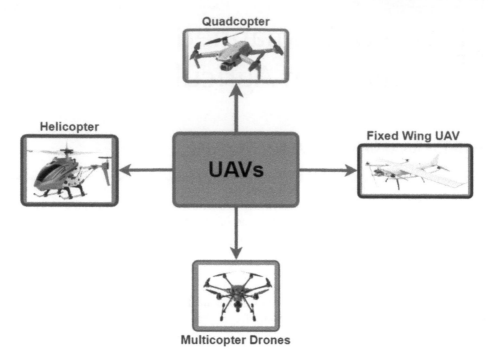

FIGURE 4.1   Different types of UAVs.

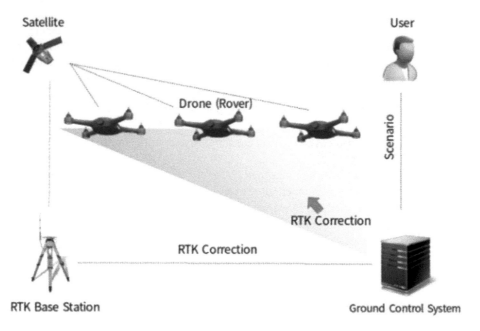

FIGURE 4.2   General UAV system (Moon et al., 2021).

Key attributes that enable UAV operation are (Clarke 2014):
1. Ability to go and return to the target position within the operation area;
2. A set of controls and maneuverability over the UAV attitude, direction, and speed of movement;
3. Remote data streams to maintain timely awareness of movement, attitude, and location;

TABLE 4.1

Classification of UAVs According to Their Size and Other Basic Features by Developing the Table

Class	Category	Operating altitude (feet)	Range radius (km)	Airtime (hour)	Example systems
Class 1 (< 150 kg)	Micro < 2 kg	AGL* + 200	5	1	Black Widow, MicroStar, Microbat, FanCopter, QuattroCopter, M05quito, Hornet, Mite, Arı
	Mini 2–20 kg	AGL + 3,000	25	< 2	ScanEagle, Skylark, DH3, Mikado, Aladin, Tracker, DragonEye, Raven, Pointer II, Carolo C40/P50, 5korpio, R-Max and R-50, RoboCopter, YH- 3005L, Bayraktar, Efe, Gözcü
	Small > 20 kg	AGL + 5,000	50	3–6	Hermes 90, Scorpi 6/30, Luna, SilverFox, EyeView, Firebird, R-Max Agri/Photo, Hornet, Raven, Phantom, GoldenEye 100, Flyrt, Neptune
Class 2 (150–600 kg)	Tactics	AGL + 10,000	200	6–10	Sperwer, İview 250, Watchkeeper, Hunter B, Mücke, Aerostar, Sniper, Falco, Armor X7, Smart UAV, UCAR, Eagle Eye+, Alice, Extender, Shadow 200/400, Taktik (ODTÜ), Çaldıran, Karayel
Class 3 (> 600 kg)	Operative (MALE)	AGL + 45,000	Unlimited	24–48	Reaper, Hermes 900, Skyforce, Hermes 1500, Heron TP, MQ-l Predator, Predator-IT, Eagle1/2, Darkstar, E-Hunter, Dominator, Anka
	Strategic (HALE)	AGL + 65,000	Unlimited	24–48	Global Hawk, Raptor, Condor, Theseus, Hellos, Predator B/C, Libellule, EuroHawk, Mercator, SensorCraft, Global Observer, Pathfinder Plus
	Offensive attack	AGL + 65,000	Unlimited	> 48	Pegasus

Sources: Anonymous (2012); Akyürek et al. (2012).
*AGL: above ground level; MALE: medium altitude long endurance; HALE: high altitude long endurance.

4. Sufficient power to sustain movement, perform the controls, and run sensors and data streams for the period of the flight;
5. Situational awareness through tracking operational space;
6. Ability to avoid collisions and navigate through obstacles;
7. Robustness to withstand various dangerous situations, such as bird strikes, strokes of lightning, turbulence, and wind shear;
8. Ability to fly in all atmospheric conditions.

Two important features that emerged with the development of new hardware and software in UAVs are "smart" behavior development and "autonomous" operation. The feature of being smart includes the UAV's being aware of its surroundings during the flight and deciding on changes such as changing the course of action or speed according to predetermined situations and applying the decision. For its autonomous feature, detection, evaluation, decision-making, control and fault detection, and safe flight capability are developed. Although the autonomous feature has various application levels, it is basically the operation of the UAV without human intervention.

An intelligent and autonomous UAV can be characterized as follows (Sebbane, 2016):

1. Being smaller, lighter, faster, and more maneuverable and maintainable;
2. Sensing the presence of wind and characterizing its effect on UAV performance, stability, and control;
3. Having an independent path planning function;
4. Having advanced guide technologies;
5. Quickly evaluating the information from multi-sensor information;
6. Operating the protection system, providing flight envelope protection, and adapting flight controls when some degradation in performance and control is expected;
7. Activating and managing the protection system and providing the autopilot with feedback on system status;
8. Changing the flight envelope of the UAV using the flight control system to avoid situations where the flight is potentially uncontrollable;
9. Adapting the control system to maintain safe flight with a reduced flight envelope;
10. Automatically creating a flight plan that optimizes multiple objectives for a predefined mission target;
11. Optimizing altitude transitions through weather;
12. Offering actionable intelligence instead of raw information with increased system-sensor automation;
13. Increasing mission performance with situational awareness and weather sensing;
14. Using advanced airborne detect and avoid technologies;
15. Identifying and tracking time-critical targets;
16. Creating a mission plan with uncertain information that meets different and possibly conflicting decision objectives;
17. Applying validation methods to these algorithms.

UAVs can have different levels of autonomy (Wang and Liu, 2012; Anonymous, 2013):

1. Humans make all the decisions;
2. The computer calculates a complete set of alternatives;
3. The computer selects several alternatives;
4. The computer suggests an alternative;
5. The computer executes the suggestion by approving it;
6. The human can veto the computer's decision within the time frame;
7. Runs the computer and then reports to the human;
8. The computer only reports when prompted;
9. Computer only;
10. The computer ignores the human.

The differences between UAVs and drones are listed here (Anonymous, 2017):

1. *Remote Wi-Fi control*: There are versions of drones controlled by remote Wi-Fi. UAVs, on the other hand, do not use Wi-Fi technology because they are mostly military-grade unmanned vehicles;
2. *Control via satellite*: Since drones are for hobby use, there are no such advanced control systems. On the other hand, a UAV can be controlled via satellite in important missions (search and rescue, reconnaissance, attack);
3. *Range*: A drone's range is as long as its battery is sufficient, which is not much. A UAV, on the other hand, has as high a range as possible;

4. *Speed*: Since a drone is used for operations such as taking pictures, its speed is medium. UAVs have high speeds;
5. *Weights*: A drone has a structure as light as possible. It is as durable as its weight. The size of a UAV is large. Its motors and battery are also more powerful than those of drones.

4.2 DRONE SYSTEMS

Due to the rapid spread of commercial UAVs as one of the most innovative technologies in the world and being a new sector with rapidly changing rules, legal regulatory laws and regulations have been enacted around the world. Civilian applications of UAVs have increased significantly in recent years due to the greater availability and smaller size of sensors, GPS, inertial measurement units, and other equipment. This technology is used by UAVs for infrastructure maintenance, agriculture, mining, emergency response, cargo delivery, etc. with applications ranging from sub-meter real sizes and autonomous operation to inspection, mapping, surveying, and transportation (Sebbane, 2018). As drone use increases, countries include drones in their aviation regulatory frameworks, and these regulations are constantly being reassessed. Various regulatory standards are applied for all types of UAVs in the world. National drone regulations generally tend to have the following few elements (Um, 2019):

1. Flight-restricted areas based on drone weight, flight altitude, population density, etc.;
2. UAV pilot license in case of professional use;
3. Drone recording in case of professional use;
4. Radio wave regulation;
5. Insurance in case of professional use.

Among the compact UAVs that can be used for various missions, rotary-wing vehicles that combine the ability of vertical flying vehicles to hang in the air with the long-range advantages of horizontal flying vehicles attract attention with their mobility and adaptability to various missions. Since aircraft that combine vertical and horizontal flight are suitable in terms of flight stability, energy efficiency, and controllability, the design of vehicles with double rotors and four rotors is preferred (Çetinsoy et al., 2009). These vehicles are called drone systems and although it is not a generally accepted design, various designs can be made depending on the required technical features (Tan et al., 2015). One of the most preferred applications of drone systems is the quadrotor drone system shown in Figure 4.3. Quadrotor, as the name suggests, is a general name given to a drone system with four independent rotors. The most important advantage of the quadrotor is its high

FIGURE 4.3 Quadrotor drone system.

maneuverability. This superiority gives the quadrotor the ability to take off and land vertically in dangerous and narrow spaces. The quadrotor's four rotors, which give it maneuverability, cause high power consumption, so it cannot fulfill a long-term flight mission. By increasing the number of rotors, the load-carrying capacity of the device also increases. Six-engine hexacopters and eight-engine octocopters can be given as examples of the different forms of quadrotors obtained by increasing the number of rotors (Merç and Bayılmış, 2011).

The quadrotor drone seen in Figure 4.3 is a DJI Phantom 3 Advanced and has a 12-megapixel camera that shoots 1,080 P video and records up to 60 FPS. In addition, its camera has a 94-degree field of view and an f/2.8 lens. It can instantly send an image in 720p HD format to a smartphone or tablet within a range of about 2 km. There is a GPS positioning system on the drone, and it can determine its position by scanning the ground level with the help of ultrasonic sensors. With the autopilot feature, it can start the engine with one touch and glide at a pre-set height. When GPS is active, it can reach the position it took off from with the return home button. When the battery level drops to a low level or the connection with the remote control is interrupted for any reason, it can return to the take-off position with the failsafe feature and land (DJI, 2017).

4.3 COMPONENTS OF DRONES

Although drones can be produced in different ways according to their usage areas, their basic components are listed here (Figure 4.4) (Akyüz, 2013; Johnson, 2015; Baichtal, 2016; Szabó et al., 2018):

1. *Propellers*: Propellers used by mechanically connecting to motors for spatial movements of drones are generally made of carbon fiber material. For example, the propellers of a quadcopter typically consist of two standard and two pusher propellers rotating in opposite directions.
2. *Motors*: Although DC or AC motors might be used, electric motors are used most, either brushed or brushless direct current motors. So as to do the same amount of work on all rotors, the same kind of motor is used. Compared to brushed motors, brushless motors are more widely used owing to their advantages such as quiet operation, long life, much more efficiency, less wearing of parts, no electrical noise, no regular maintenance, and the ability to run at higher speed and high torque in a lower voltage range.
3. *Electronic speed controllers (ESCs)*: ESCs convert DC to AC for brushless motors and also trigger the motors' power supply. One is used for each engine. ESCs' firmware might

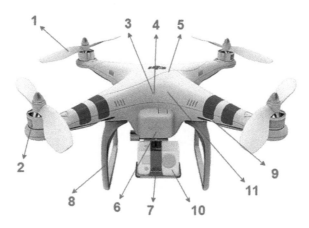

FIGURE 4.4 Components of the drone.

be changed to create different motor behaviors. For instance, ESCs are often configured to slow down the motor rather than stop abruptly.

4. *Flight controller*: The flight controller controls the entire electromechanical system of the drone. It assists manual flight with certain autonomous functions. For instance, many flight controllers have an accelerometer sensor that keeps the drone level. The flight controller provides movement and stabilization in the desirable direction by changing the speed of the motors according to the data from the sensors. Thus, the drone can stabilize itself even if the engines are given different pushers.

5. *Airframe*: The airframe consists of components such as including motor booms alongside an enclosure or platform for housing the electronics. The drone ought to be light and thin enough to lift off and tough enough not to break in a minor accident. Carbon fiber, plastic, wood, and aluminum alloy materials are generally preferred in the construction of the main body.

6. *Battery pack*: Usually a LiPoly battery, the drone's battery pack keeps the propellers turning while also powering whatever electronics are onboard.

7. *Gimbal*: The gimbal is a rotating platform on which a camera is placed. Servomotors allow the operator to turn and angle the camera during flight.

8. *Landing struts*: Landing struts are used to prevent damage to the camera or other protuberance under the drone. Drones without cameras, on the other hand, do not have landing struts, and the drone lands with its entire airframe.

9. *Front indicator*: Especially the front side of the drone must be known by the operator. For this, different colored lights, LEDs, reflective materials, or colored balls are used.

10. *Video camera*: Cameras with different resolutions are used that send images to a tablet with radio waves. HD cameras are often used for mapping, surveying, and image capture, such as multispectral cameras, hyperspectral cameras, thermal cameras, laser scanners, and synthetic aperture radars.

11. *Receiver*: Commands given by the pilot to move the drone are converted by the receiver into the flight controller instructions. A five-channel communication module is sufficient for propeller acceleration control and control mode (Tx/Rx) for yaw, roll, and pitch angles.

4.4   DRONE OPERATION

There are four rotors placed at the ends of the plus-shaped skeleton structure of the quadrotor drone system. While the opposite propellers one and three (front and rear) rotate counterclockwise in the same direction, propellers two and four (right and left) rotate clockwise in the same direction. All rotors exert lift by generating downward airflow. All movements of the quadrotor drone system are provided by changing the speeds of the four rotors. There are four basic movements (Figure 4.5). These are (Altaş, 2017):

1. *Vertical movement (throttle)*: Elevation is achieved by equally increasing the speeds of the propellers two and four rotating clockwise and the propellers numbered one and three rotating counterclockwise. Landing is achieved by reducing the speed of these two pairs of propellers at the same rate. It also remains constant in the air when the ratio of the velocities is made equal to the gravitational force.

2. *Pitch movement*: Forward or backward movement is provided in opposite speed changes of the propellers one and three (front and rear) rotating in the same direction.

3. *Roll movement*: As in the pitch movement, the propellers two and four (right and left) rotating in the same direction provide movement to the right or left in opposite speed changes.

4. *Yaw movement*: By changing the relative speeds of propellers one and three (front and rear) and propellers two and four (right and left), the drone makes a yaw movement.

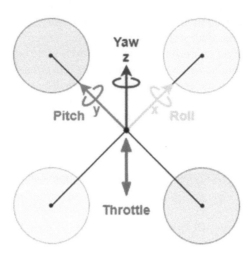

FIGURE 4.5   Basic movement patterns of the quadrotor drone system.

4.5   ISSUES TO CONSIDER IN DRONE DESIGN

Due to the ease of use and capabilities of drones, they can be used for different tasks in a wide variety of business lines. Therefore, they can be manufactured with different designs according to the need. If a drone is to be designed for commercial purposes, customer-oriented requests and prices are considered. If a drone is to be purchased, first, the suitability of the drone for the needs and its price are checked. The technical features to consider in drone design can be listed as follows:

1. *Working environment*: Drone's engine and electronic components and connections can be used in rainy weather if there is no protection. When used in potentially dangerous areas such as birds, protection should be considered during design for this purpose.
2. *Size*: The size of the drone should be chosen in accordance with the task to be carried out, in a size that will allow the necessary equipment and useful load to be carried.
3. *Flight time*: The flight time varies according to the characteristics of the battery, which is the power unit used in the drone design. Multiple batteries can be used. Work continues the battery in order to increase the flight time. In addition, there are studies on wired power supplies for this purpose.
4. *Payload capacity*: It varies according to the weight of vehicles such as machines or weapons that can be placed on the drone.
5. *Altitude and range*: It varies according to the amount of connection with the control consisting of radio frequencies, the size and characteristics of the battery, and the strength of the material from which the drone is made.
6. *Mode of operation*: Autonomous or controlled working status.
7. *Speed*: Depending on the selected engines, it varies according to the weight and design of the drone.

In addition to the necessary technical features suitable for the nature of the work to be done, the most important issue affecting the design is the budget. Apparatus, tools, and devices should be selected depending on the budget. First, a suitable size and a suitable platform should be selected for the work to be done. Carbon fiber materials, which are both robust and light, are preferred as platform materials. The main determinant of the design is the engines chosen after the platform. Engines are selected according to the desired payload. In addition, the selected engine has an effect on the basic features of the drone such as speed, altitude, and range. Brushed or brushless direct current motors are most used as motors. After that, the engines, drivers, and propellers suitable for the engines are selected. One ESC should be used for each motor. The ESC controls each of the motors separately, providing the amount of voltage that will keep the drone in balance during flight, and they are the devices that trigger the power supplies of the motors. Direction, spacing, and size should be considered when choosing a propeller.

Electronic components and control tools should be selected in accordance with the selected power source. LiPo battery is generally preferred as a power source. The power supply is the most important drone part that affects the flight time depending on the weight of the drone and the equipment selected. As a remote control, a suitable model should be selected from the RC controls of different companies. Moving the drone forward and backward and left to right, moving the camera mechanism up and down, moving the gimbal mechanism left to right and up and down, and motor step changes are used by defining these channels on the remote control. In commercial drones, camera functions are also controlled remotely via the tablet placed on the remote control, and the captured images can be saved to the memory card on the drone and also to the tablet.

The Pixhawk module is widely used as a flight controller for autonomous flights. Pixhawk is an autopilot module based on open source Stm32-based software base called ArduPilo. There is a Se100 M8N GPS module to determine location in the system. A gimbal mechanism is used to connect the camera and camera to the drone which can take photos and videos on the drone. Finally, the landing gear is mounted so that the drone is not damaged while landing. After all the connection procedures are completed in the laboratory environment, the system is calibrated, and the flight tests of the drone are carried out after the calibration processes.

4.6 CAMERAS AND SENSORS USED WITH DRONES

4.6.1 OPTICAL CAMERAS

Most low-cost UAV mapping research using visible band sensors focuses on the ability to produce high-resolution digital orthophotos and digital surface models. Some of the important parameters to be considered during the design phase are sensor type and resolution, pixel size, frame rate, shutter speed, focal length to be used, and weight of the camera-lens system. While the most common types of optical sensors are CCD and CMOS DSLR cameras, mirrorless cameras are becoming increasingly popular, especially due to their small weight (Georgopoulos et al., 2016). Table 4.2 summarizes the optical cameras used with the UAV.

TABLE 4.2
Optical Cameras Used with UAVs

Manufacturer and model	Sensor type resolution (MPx)	Format type	Sensor size (mm²)	Pixel pitch (μm)	Weight (kg)	Frame rate (fps)	Max shutter speed (s⁻¹)
Canon EOS 5DS	CMOS 51	FF	36.0×24.0	4.1	0.930	5.0	8,000
Sony Alpha 7R II	CMOS 42	FF MILC	35.9×24.0	4.5	0.625	5.0	8,000
Pentax 645D	CCD 40	FF	44.0×33.0	6.1	1.480	1.1	4,000
Nikon D750	CMOS 24	FF	35.9×24.0	6.0	0.750	6.5	4,000
Nikon D7200	CMOS 24	SF	23.5×15.6	3.9	0.675	6.0	8,000
Sony Alpha a6300	CMOS 24	SF MILC	23.5×15.6	3.9	0.404	11.0	4,000
Pentax K-3 II	CMOS 24	SF	23.5×15.6	3.9	0.800	8.3	8,000
Canon EOS 7D Mark II	CMOS 20	SF	22.3×14.9	4.1	0.910	10.0	8,000
Panasonic Lumix DMC-GX8	CMOS 20	SF MILC	17.3×13.0	3.3	0.487	10.0	8,000
Ricoh GXR A16	CMOS 16	SF	23.6×15.7	4.8	0.550	2.5	3,200

Source: Georgopoulos et al. (2016).

Numerous images can be captured during a drone flight. These images can be evaluated with the eyes of experts, as well as with the developed image processing software, and comments and evaluations can be performed about the images. In addition, artificial intelligence–based software that can perform real-time and automatic evaluation has been improved recently, new methods and models have been improved to obtain better results, and development studies are continuing rapidly to apply this software in new areas. There are plenty of successful studies on this subject, especially with the deep learning method (Ozguven and Yanar, 2022).

4.6.2 MULTISPECTRAL CAMERAS

Multispectral sensors were first used on satellites, then platforms and manned aircraft. More recently, it has emerged as a very popular and cost-effective RS technology consisting of small-sized and lightweight sensors that can be integrated into UAVs. The ability of hyperspectral sensors to measure hundreds of bands adds complexity given the huge amount of data acquired. Therefore, ensuring the usefulness of multispectral data depends on both calibration and corrective tasks occurring in the pre- and post-flight phases (Adão et al., 2017). Multispectral sensors seen in Table 4.3 combine high-resolution RGB cameras with four, five, or six different spectral bands. As a result, it provides high spatial resolution suitable for beam tuning and extraction of geometric parameters. These cameras offer discrete multispectral sensors equipped with world-class interference filters and provide high-precision spectral measurements that can replace ground-based spectral reference measurements in the future (Szabó et al., 2018).

Multispectral cameras are used widely in agriculture to obtain information about plant growth, soil, and water properties. Plants reflect especially in the near-infrared (NIR) region. The NDVI (NDVI = (NIR − R)/(NIR + R)) values obtained by proportioning the NIR band with the red band give information about green vegetation. Thus, when the NDVI value approaches 1, the plant is healthy, while the NDVI value approaches 0 if the plant is weak or stressed. The band values vary as to the characteristics of the developed sensor. For instance, the multispectral bands of the Landsat satellite are as stated here: 0.45–0.52 µm (blue), 0.52–0.60 µm (green), 0.63–0.69 µm (red), and 0.76–0.90 µm (NIR) (Ozguven and Yanar, 2022).

4.6.3 HYPERSPECTRAL CAMERAS

Multispectral sensors collect information from only some spectral bands, whereas hyperspectral sensors can detect hundreds of very narrow spectral channels. Therefore, hyperspectral sensors (Table 4.4) can obtain more detailed information compared to multispectral cameras. This is

TABLE 4.3
Multispectral Cameras Used with UAVs

Manufacturer and model	Resolution (MPx)	Size (mm)	Pixel size (µm)	Weight (kg)	Number of spectral bands	Spectral range (µm)
Tertracam MiniMCA-6	1.3	131 × 78 × 88	5.2 × 5.2	0.7	6	450–1,000
Tetracam ADC micro	3.2	75 × 59 × 33	3.2 × 3.2	0.9	6	520–920
Quest Innovations Condor-5 UAV-285	7	150 × 130 × 177	6.45 × 6.45	1.4	5	400–1,000
Parrot Sequoia	1.2	59 × 41 × 28	3.75 × 3.75	0.72	4	550–810
Parrot Sequoia		120 × 66 × 46		0.18	5	475–840

Sources: Colomina and Molina (2014); Szabó et al. (2018).

TABLE 4.4

Hyperspectral Cameras Used with UAVs

Manufacturer and model	Resolution (MPx)	Size (mm²)	Pixel size (lm)	Weight (kg)	Spectral range (nm)	Spectral bands and resolution (nm)
Rikola Ltd. Hyperspectral Camera	CMOS	5.6 × 5.6	5.5	0.6	500–900	40/10
Headwall Photonics Micro-Hyperspec X-series NIR	InGaAs	9.6 × 9.6	30	1.025	900–1,700	62/12.9

Source: Colomina and Molina (2014).

TABLE 4.5

Thermal Cameras Used with UAVs

Manufacturer and model	Resolution (MPx)	Sensor size (mm²)	Pixel pitch (µm)	Weight (kg)	Spectral range (µm)	Thermal sensitivity (mK)
FLIR Vue Pro 640	640 × 512	10.8 × 8.7	17	< 0.115	7.5–13.5	50
FLIR Vue Pro 336	336 × 256	5.7 × 4.4	17	< 0.115	7.5–13.5	50
Thermoteknix Miricle 307K	640 × 512	16.0 × 12.0	25	< 0.170	8.0–12.0	50
Thermoteknix Miricle 110K	384 × 288	9.6 × 7.2	25	< 0.170	8.0–12.0	50/70
Workswell WIRIS 640	640 × 512	16.0 × 12.8	25	< 0.400	7.5–13.5	30/50
Workswell WIRIS 336	336 × 256	8.4 × 6.4	25	< 0.400	7.5–13.5	30/50

Source: Georgopoulos et al. (2016).

because each of the recorded pixels covers the entire spectrum. However, while there are special multispectral sensors designed for use in UAVs, it is not easy to find a hyperspectral sensor that can be used directly in UAVs. Also, the integration of these cameras into a UAV is complicated because the captured frames do not overlap. Therefore, much more attention and care must be given to the acquisition of the image and its post-processing (Horstrand et al., 2019).

Hyperspectral image sensing has the ability to resolve several hundred spectral bands in the region from visible light to short-wave infrared and may make it possible to ensure more phyto-biological information by analysis of continuous spectral properties, compared with multispectral analysis. Hyperspectral analysis may provide information on productivity and stresses of plants, biochemical and mineral components in living plants and soils, classification of species, soil types, and parts of plants (Omasa et al., 2006).

4.6.4 THERMAL CAMERAS

UAV-based thermal imagers have been available for decades (Table 4.5). However, it is still generally used for single image analysis, only in oblique view format, or for visual evaluations of video sequences. Photogrammetric workflows allow the processing of thermal images like RGB data. However, current thermal cameras contain sensors with low spatial resolution, typically uncooled, and coated with VOx (vanadium oxide). Therefore, significant geometric lens aberrations are present, and typically low contrast requires an adapted workflow (Boesch, 2017). Unlike optical image solutions, thermal imaging requires special geographic reference procedures. Basically, it is difficult to find natural ground control points in lower-resolution thermal images. Therefore, there is a need for artificial control points using predominantly aluminum as the material. Aluminum ground control points show a sharp boundary in the thermal image and automated identification algorithms are available for them (Szabó et al., 2018). Most of these algorithms rely on processing thermal data with motion formation photogrammetry (SfM) to create orthophotos (Maes et al., 2017). A general three-step framework for processing thermal images with UAV data is presented here (Turner et al., 2014):

1. Image preprocessing, which is the removal of blurry images and conversion to 16-bit TIFF files where all images have the same dynamic scale range to ensure that a temperature value corresponds to the same digital number value in all images;
2. Image alignment where initial estimates of image position are derived from the built-in GPS log file and the time log of each image;
3. Spatial image co-registration to RGB (or other) orthophotos which is performed by manually adding ground control points with known positions from real-time kinematics (RTK) GPS or processed RGB images.

Thermal infrared imaging (a passive spectral imaging method) is effective for the early diagnosis of plant stresses along with the measurement of surface temperatures of soils and plants (Hashimoto et al., 1984; Omasa and Aiga, 1987; Omasa, 1990). Image analysis of the energy balance on the canopy and the leaf provides phytobiological information on stomatal response and evapotranspiration (Omasa and Croxdale, 1992; Jones, 1999; Omasa, 2002). Although low-cost thermal cameras are widely available today for UAVs, spatial resolution is quite limited. Also, thermal camera lenses have a significant radial distortion (Boesch, 2017).

4.6.5 Light Detection and Ranging (Lidar)

Lidar sensors determine the distance of an object or a surface using laser beams. It works similarly to how radar technology works. The difference is that instead of radio waves, laser pulses hit the surrounding objects and the distance value is calculated using the reflection time. 3-D point information of the area measured with lidar can be obtained in a very short time, at the desired frequency, and with high accuracy (Özgüven, 2018). Lidar works as an active sensor like radar. Unlike radar, lidar sends artificially produced laser signals in the visible and near-infrared regions to the earth, instead of microwaves (Çelik, 2017). The range is determined by the delay in the travel and return of the light waves to the target. The nanosecond pulses used in this pulsed lidar generally have high instantaneous peak power. Therefore, centimeter resolution can be achieved in single pulses over a wide aperture window (Royo and Ballesta-Garcia, 2019).

Lidar sensors can be grouped into two groups based on the platforms they are installed on: Airborne lidar sensors (ALS) and terrestrial lidar sensors (TLS). Airborne scanning consists of lidar sensors, INS, and an onboard computer, while ground scanning systems do not have INS. In both types of lidar sensors, the beams emanating from the laser source are not continuous but are periodically directed to the area to be scanned by hitting the reflector. Thus, 20,000 to 150,000 points per second of the area to be scanned from an aircraft or a tripod on land are scanned. When a one-unit signal reaches any ground, it is reflected again from that surface, first reaching the reflector and then the receiving sensor. Meanwhile, the time difference between the departure and the return of the signal is measured by the internal computer (Anonymous, 2019). Lidar sensors used with the UAV are shown in Table 4.6.

4.6.6 Synthetic Aperture Radar (SAR)

Synthetic aperture radar (SAR) technology is a method used to obtain higher-resolution images in the flight direction of the platform with a smaller antenna length. By moving the radar antenna along the desired aperture, it takes measurements at certain time intervals and collects these data

TABLE 4.6
Lidar Sensors Used with UAVs

Manufacturer and model	Scanning pattern	Range (m)	Weight (kg)	Angular res. (deg)	FOV (deg)	Laser class and λ (nm)	Frequency (kp/s)	Application
ibeo Automotive Systems IBEO LUX	Four scanning parallel lines	200	1	(H) 0.125 (V) 0.8	(H) 110 (V) 3.2	Class A 905	22	Automotive
Velodyne HDL-32E	32 laser/detector pairs	100	2	(H) – (V) 1,33	(H) 360 (V) 41	Class A 905	700	Terrestrial mobile mapping
RIEGL VQ-820-GU	One scanning line	≥ 1,000	–	25.5 (H) 0.01 (V) N/A	(H) 60 (V) N/A	Class 3B 532	200	Hydrography

Source: Colomina and Molina (2014).

TABLE 4.7
SAR Sensors Used with UAVs

Manufacturer and model	Spectral bands	Weight (kg)	Transmitted power (W)	Resolution (m)
IMSAR NanoSAR B	X and Ku	1.58	1	Between 0.3 and 5
Fraunhofer FHR MIRANDA	W	N/A	0.1	0.15
NASA JPL UAVSAR	L	200	2,000	2
SELEX Galileo PicoSAR	X	10	–	1
OrbitSAR UAV-SAR	X and P	30	30 (X), 50 (P)	0.5 (X), 1.5 (P)
SARVANT UAS-System SAR	X and P	30	30 (X), 50 (P)	0.5 (X), 1.5 (P)

Source: Szabó et al. (2018).

simultaneously to form a synthetic aperture. Thus, a large artificial opening equal to the real physical opening is created (Irak, 2009). SAR technology is a traditional method implemented by satellite systems. Although it is not fully applied to UAVs, studies working toward this result are ongoing. The main problem with the concept is that this type of research is mainly affected by various weather conditions (Szabó et al., 2018). Table 4.7 summarizes some of the most common synthetic aperture radars used for UAV mapping.

4.7 USE OF DRONES IN AGRICULTURE

Increasing the productivity and product quality in agricultural production depends on the good follow-up of the development process of the plants and the implementation of the necessary processes at the most appropriate time. It provides farmers with the opportunity to plan agricultural activities by creating high-resolution images and 3-D images with sensors and cameras placed on drone systems, which have a simple technical structure and easy use. With drone systems, studies are carried out for applications such as product development monitoring, plant species separation, crop yield determination, automatic harvesting, drought, disease, agricultural pests, damage detection, fruit, vegetable, and soil moisture classification, field management, organization of agricultural activities, and agricultural insurance (Tan et al., 2015).

4.7.1 PRECISION AGRICULTURE

By using drones in precision agriculture and collecting data from infrared sensors and image compositing files, fields can be mapped, crop yields estimated, plant health evaluated, weeds or diseased plants identified, plant growth recorded, and hydration levels measured. All this information is then analyzed and used to make the best plant management decisions. Fertilizer and pesticide recipes can then be selectively applied to individual plants using drones in an improved site plan. It is used to spray the various formulas prescribed by spraying in a specific way, in order, exactly only to the plants that need it. Successful UAV precision farming programs exist in Canada, Brazil, and Sweden, with the prominent model being in Japan. In 1983, the Japanese Ministry of Agriculture commissioned Yamaha to develop an unmanned aerial system for agriculture. The Yamaha RMAX unmanned helicopter has been spraying crops since 1991. Today, RMAX is spraying 40% of Japan's rice fields, approximately 2.5 million acres. With an impeccable safety record, productivity gains are estimated to increase by as much as 30% (Kilby and Kilby, 2016). There are five effective uses of drone systems in precision agriculture. These are (Grassi, 2014):

1. *Monitoring plant status*: Farmers can examine their developing plants more quickly and effectively with drones with normalized difference vegetation index (NDVI) or near-infrared (NIR) sensors.
2. *Monitoring irrigation systems*: Large enterprises can monitor irrigation systems for the supply of water needed after some crops such as corn, which are spread over large areas, reach certain sizes.
3. *Weed identification*: Weed maps are created by processing NDVI sensor data and postflight images. This way, growers can easily distinguish high-density weed areas growing alongside healthy plants.
4. *Variable rate applications*: Instead of variable rate application maps made with ground-based or satellite images, variable rate maps are prepared quickly and practically by using NDVI sensors in drone systems. In this way, it is possible to increase the yield by reducing the costs of pesticides and fertilizers.
5. *Herd management and monitoring*: It is possible to monitor the quantities and activity levels of free-raised sheep and cattle from above with drones.

4.7.2 EXAMINING THE GROWTH STATUS OF PLANTS

Plant height measurements are made to examine critical aspects of plant physiology, genetics, and environmental conditions during plant growth. This process is difficult and time-consuming to do manually and repetitively. Measurements can be made in a short time in large areas with 3-D point clouds created from drone images by using motion occurrence photogrammetry (SfM) techniques for the measurement of plant height. With Pix4D image processing and digitization software, which is a commercial product, a 3-D color point cloud, digital elevation model, digital terrain model, 3-D model, and classified end product analysis can be made.

Malambo et al. (2018) obtained high-resolution images in the visible spectrum collected over 12 weekly dates from April (planting) to July (harvest) 2016 on 288 maize (*Zea mays* L.) and 460 sorghum (*Sorghum bicolor* L.) plots using a DJI Phantom 3 Professional UAV. In the study, a methodology was developed to estimate plant height data from point clouds obtained with a drone and created using the multi-time 3-D plant modeling SfM technique, and the results were evaluated. The study compared terrestrial lidar (TLS) and SfM point clouds on two dates to assess the ability of SfM point clouds to accurately capture ground surfaces and crop canopies, both of which are critical for plant height estimation. Extended plant height comparisons were implemented between SfM plant height (the 90th, 95th, and 99th percentiles and maximum height) per plot and field plant height measurements at six dates throughout the growing season to test the repeatability and consistency of SfM estimates. High correlations were observed between SfM and TLS data (R^2 = 0.88–0.97, RMSE = 0.01–0.02 m and R^2 = 0.60–0.77, RMSE = 0.12–0.16 m for ground surface and canopy comparison, respectively). Extended height comparisons also indicated strong correlations (R^2 = 0.42–0.91, RMSE = 0.11–0.19 m for maize and R^2 = 0.61–0.85, RMSE = 0.12–0.24 m for sorghum). The researchers reported that the accuracy of SfM plant height estimations fluctuated throughout the growing period, likely influenced by the changing reflectance regime due to plant growth, reducing the laborious manual height measurement of their study results. In addition, the researchers noted that drones and SfM show a potential way to improve plant research programs; 3-D point clouds and orthomosaics are created using the Pix4DMapper software, which processes all the images collected for a given day and uses a standard three-step semi-automatic processing workflow (Figure 4.6).

Pagliai et al. (2022) used different digital tools, namely MA, MLS, and UAV, to generate 3-D point clouds of test vines (*Vitis vinifera* L.) to evaluate three different phenological stages of the canopy size parameters such as thickness, height, and volume. The investigators reported that the results of the study showed a good correlation between all tools in terms of detecting intra-field variability and canopy size parameters. Pix4Dmapper Pro software was used for 3-D point cloud reconstruction. Figure 4.7 shows a 3-D point cloud reconstruction of vineyard rows and vines. The

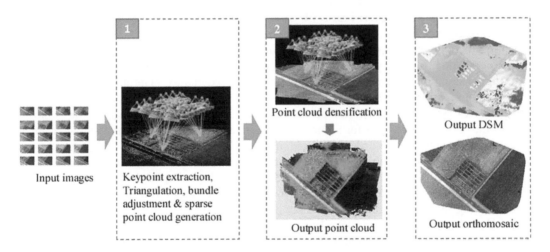

FIGURE 4.6 Pix4D rendering workflow: (1) Key point extraction, triangulation, beam tuning, and initial processing of sparse point cloud generation, (2) point cloud condensation, (3) digital surface model (DSM) and orthomosaic generation (Malambo et al., 2018).

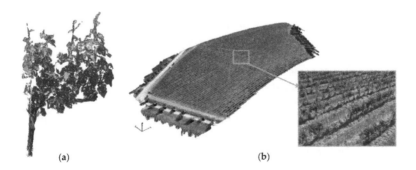

FIGURE 4.7 3-D point cloud reconstruction of vineyard rows and vines. (a) MA vine point cloud processed with Pix4DMapper and cleaned with CloudCompare, (b) UAV vineyard point cloud processed with Pix4DMapper (Pagliai et al., 2022).

UAV and MA 3-D point clouds were processed by an algorithm that was coded in MATLAB. The 3-D point cloud of a vineyard row section, in which the x, y, and z axes align with the vineyard row, canopy width, and vertical axis, was processed through a series of spatial manipulations taking into account the local soil gradient. Then the canopy density, height, and thickness were calculated. The present code used the best numerical descriptors as numerical descriptors. Thus, the code reads the processed UAV and MA 3-D point clouds and gives as results the main canopy size parameters such as thickness, height, and volume. The working algorithm for UAV and MA 3-D point clouds is given in Figure 4.8.

Quick and accurate monitoring of plant height (PH), canopy area (CC), and leaf nitrogen concentration (LNC) is essential for the precise management of irrigation and fertilization. Lu et al. (2021) conducted the study to estimate summer maize PH by choosing the optimal percentile height of the point cloud, extracting CC from images by using the point cloud method, and determining if the combination of PH and CC with visible vegetation index (VI) could improve the prediction accuracy of LNC. At the end of the study, the 99.9 growth height of the point cloud was found to be the optimal value for estimating maize PH. It was seen that the point cloud method is better than the maximum likelihood classification method for maize CC estimation. Newly developed VIs integrating pH, CC, and visible VIs have been reported to significantly improve the LNC prediction accuracy of maize. Normalized redness intensity (NRI) had a potential for estimating LNC (R2 1/4 0.474) compared with the green, red ratio VI, green, red VI, and atmospherically resistant VI and NRICCH exhibited the highest correlation with maize LNC (R2 1/4 0.716).

Counting the number of plants grown in a certain area in plant production, that is, obtaining the plant population, may vary according to the effect of various factors during and before the season. Early season plant numbers are often the result of seed quality, planting performance, and the response of seedlings to soil and air at emergence. Mid- to late-season plant numbers are often the result of weather, soil, fertilizer, pest pressure, and other management practices. Plant numbers are one of the most common ways for farmers to evaluate crop growth conditions and management practices throughout the season and can be used to make management decisions, such as determining whether replanting is necessary. However, the traditional method for early season plant counting is manual inspection, which is time-consuming, laborious, and spatially limited in scope. In recent years, UAV-based RS has been used to provide low-altitude, high-spatial-resolution images to aid decision-making in crop counting and in many different applications in agriculture. Pang et al. (2020) designed a system that uses deep learning and geometric descriptive information to determine early season maize plant numbers from relatively low spatial resolution (10 to 25 mm) UAV images covering an area of 10 to 25 hectares (Figure 4.9). The researchers reported that instead of detecting individual crops in a row, they processed the entire row at once, which significantly reduced their requirement for clarity of the crops. In addition, the researchers stated that their newly developed MaxArea Mask Scoring R-CNN algorithm can separate the rows of plants in each patch image regardless of field conditions, and the system was tested on data collected in two different areas in different years, and the accuracy of the estimated output speed reached 95.8%. In addition, the newly developed MaxArea Mask Scoring R-CNN algorithm is able to segment crop rows in each patch image regardless of terrain conditions, and the robustness of our plan was tested on data collected in two different areas in different years. It has been reported that the accuracy of the estimated emergence rate reaches 95.8% and the system has the potential for real-time applications in the future due to its high processing speed.

Accurate mapping of cropland is an important prerequisite for precision agriculture, as it aids in field management, yield estimation, and environmental management. Optical imaging with sensors mounted on UAVs is today a cost-effective option for capturing images covering cultivated areas. However, only visual inspection of such images can lead to both difficult and inaccurate assessments, especially identifying plants and rows in one step. Therefore, the development of an architecture that can extract individual plants and plantation rows simultaneously from UAV images is an important need to support the management of agricultural systems. For this purpose, Osco et al.

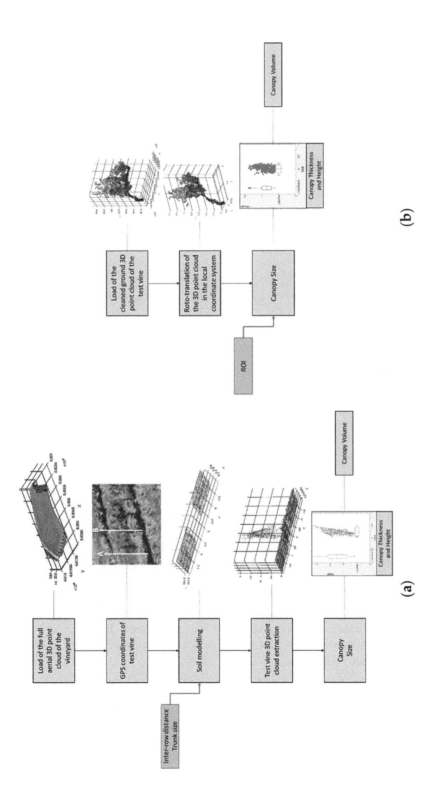

FIGURE 4.8 Algorithm processing scheme. (a) Processing workflow for the aerial 3-D point cloud, (b) processing workflow for the ground 3-D point cloud (Pagliai et al., 2022).

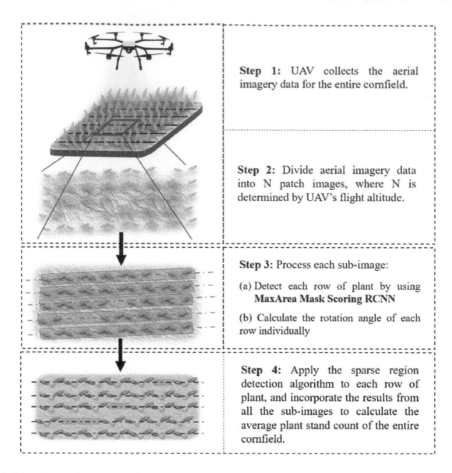

Step 1: UAV collects the aerial imagery data for the entire cornfield.

Step 2: Divide aerial imagery data into N patch images, where N is determined by UAV's flight altitude.

Step 3: Process each sub-image:

(a) Detect each row of plant by using **MaxArea Mask Scoring RCNN**

(b) Calculate the rotation angle of each row individually

Step 4: Apply the sparse region detection algorithm to each row of plant, and incorporate the results from all the sub-images to calculate the average plant stand count of the entire cornfield.

FIGURE 4.9 System for determining the number of early season maize plants (Pang et al., 2020).

(2021) proposed a new deep learning method based on a convolutional neural network (CNN) that simultaneously detects and locates plantation rows while counting plants, considering high-density planting configurations. To prove the robustness of the proposed approach, the experimental setup was evaluated in a corn field (*Zea mays* L.) with different growth stages (Figure 4.10) and in a citrus orchard (*Citrus sinensis*) (Figure 4.11). Both datasets characterize different plant density scenarios at different locations, different types of crops, and different sensors and dates. It can be concluded that the CNN method results outperform the results from other deep networks (HRNet, Faster R-CNN, and RetinaNet) evaluated with the same task and dataset and that the proposed method can be used to count and position plants and plant rows in UAV images from different plant species. It has also been reported that it can contribute to the sustainable management of agricultural systems by applying it to decision-making models.

Plants reflect especially in the near-infrared region. The results obtained by comparing the NDVI values or the near-infrared band with the red band give information about green vegetation, as well as identify areas where the vegetation is weak or without vegetation. In addition, the closer the plant index is to 1, the stronger the plant; the closer to 0, it shows that the vegetation disappears; and when it is negative, it shows that the areas are absolutely plantless (Düzgün, 2010). Figure 4.12 shows a map showing the use of NDVI images in agriculture. The plants in the dark green area photosynthesize with high efficiency, while the plants in the green area do photosynthesis at medium activity. The figure shows two management zones by (high–medium) vine viability for the grower to decide on fertilization and varying yields, together with water condition maps.

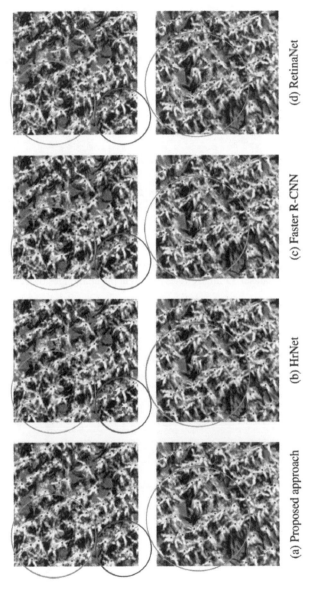

FIGURE 4.10 Comparison of object detection methods: (a) proposed approach, (b) HRNet, (c) Faster R-CNN, and (d) RetinaNet on a dataset of higher growth stage (mature with cobs) corn plants (orange and blue circles highlight usual and challenging detections, respectively) (Osco et al., 2021).

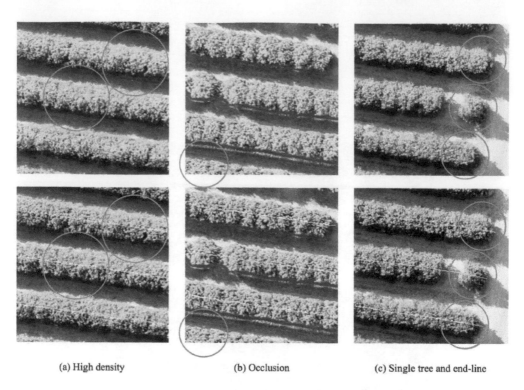

(a) High density   (b) Occlusion   (c) Single tree and end-line

FIGURE 4.11   Plant and plantation row determinations examples of (a) high density, (b) occlusion, and (c) single tree and end-line in the citrus dataset. Orange circles highlight the challenges overcome by the approach in each scene (Osco et al., 2021).

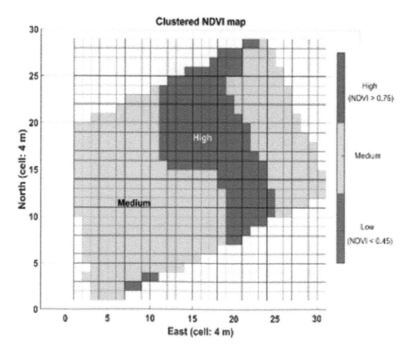

FIGURE 4.12   NDVI map of an agricultural area (Saiz-Rubio and Rovira-Más, 2020).

4.7.3 CLASSIFICATION OF PLANTS

Separation of plant species is essential for a wide variety of agricultural applications, especially when information is needed during the production period. In general, RS can provide such information with high accuracy, but very high spatial resolution data is required in small fields. Böhler et al. (2020) present a study involving spectral and textural features obtained from near-infrared (NIR) red-green-blue (NIR-RGB) band datasets, captured using a UAV, and an imaging spectroscopy (IS) dataset of 2 m spatial resolution and 173 spectral bands between 399 nm and 2,431 nm obtained by the Airborne Prism EXperiment (APEX) (Figure 4.13). In the study, it was reported that these datasets were used both alone and in combination, were analyzed with a random forest-based method, and different band reduction methods based on feature factor loading were analyzed to distinguish plants.

At the end of the study, the most accurate crop separation results were obtained using both the IS dataset and the two combined datasets with an average accuracy of > 92%. In addition, it was concluded that in the case of a reduced number of IS features (i.e., wavelengths), the accuracy can be compensated by using additional NIR-RGB texture features (average accuracy > 90%).

4.7.4 DETERMINATION OF PLANT PHENOTYPES

Near RS approaches using drones are used for high-throughput in-field phenotyping in the context of plant breeding and biological research. Data on canopy cover (CC) and canopy height (CH) and their temporal changes during the growing season provide information on plant growth and performance. In study by Borra-Serrano et al. (2020), sigmoid models were fitted to multi-temporal CC and CH data obtained using RGB images captured with a drone for a large set of soybean genotypes. It was stated that the Gompertz and Beta functions were used to fit the CC and CH data,

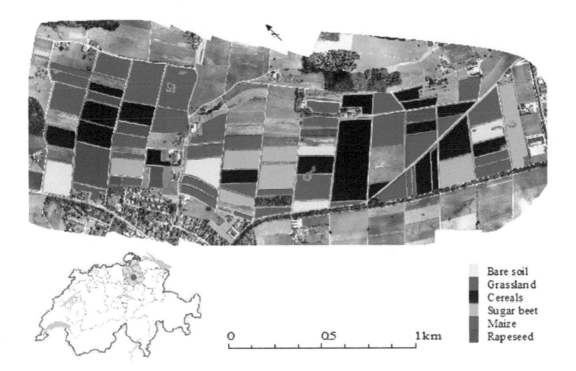

FIGURE 4.13 The study area where the plants are classified by superimposing the datasets obtained with the RGB camera and IS attached to the UAV (Böhler et al., 2020).

respectively; overall, 90.4% fits for CC and 99.4% fits for CH reached an adjusted $R^2 > 0.70$, demonstrating good performance of the models chosen (Figure 4.14). In the study, parameters such as maximum absolute growth rate, early vigor, maximum height, and senescence were calculated for a collection of soybean genotypes using these growth curves, and this information was also used to predict seed yield and maturity (R8 stage) (adjusted $R^2 = 0.51$ and 0.82). Combinations of parameter values were also tested to identify genotypes with interesting traits. It has been determined to result in biointerpretable parameters that provide information for relevant properties.

4.7.5 Detection of Plant Diseases and Pests

Plant diseases cause economically important income losses in agricultural production all over the world (Savary and Willocquet, 2014; Avelino et al., 2015). To reduce crop loss, plant disease severity must be determined accurately and rapidly. Therefore, determining the outbreak, severity, and progression of diseases in a timely and accurate manner is of great importance for effective integrated disease management (Bock et al., 2010). There is an immediate need to develop faster and more practical methods, which could reduce human errors in the identification of plant diseases, their severity, and progress, especially in large production areas (Altas et al., 2018). In the event of a disease, plants exhibit visual signs in the shape of colorful spots with different shapes and sizes according to the type of disease and in the shape of lines seen on stems and different sections or organs of the plants. These symptoms alter color, shape, and size while the disease progresses. With image processing methods, colored objects might be distinguished, and the severity of plant diseases might

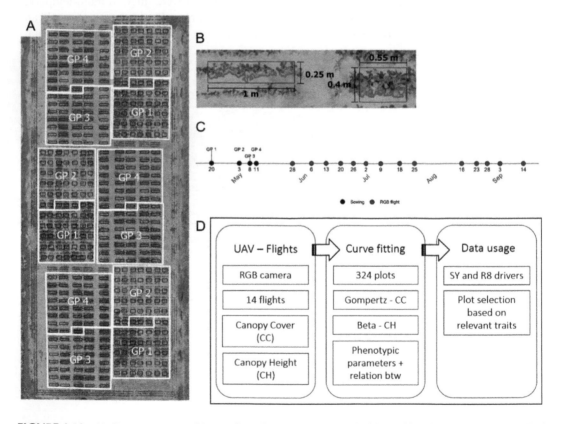

FIGURE 4.14 (a) Group locations, (b) zooming of two graphs marked with a red box in (a) with mask applied to images to select pixels corresponding to vegetation, (c) timeline of planting dates and UAV flights, and (d) workflow followed in the study (Borra-Serrano et al., 2020).

be determined. Besides image processing methods, expert systems might be improved to allow instant disease diagnosis with machine learning methods (Ozguven, 2020). Recently, the potential use of image processing and machine learning methods for disease detection in whole plants and/ or different plant parts (leaves, stems, fruit, and such) has been comprehensively studied by many researchers (Ozguven and Adem, 2019). Monitoring pests and diseases is an extremely important activity to increase productivity in agriculture. RS combined with machine learning techniques opens new possibilities for tracking and identifying characteristics such as identifying diseases, pests, water, and nutritional stress (Calou et al., 2020).

Altas et al. (2018) used a drone image processing technique to determine the level of sugar beet leaf spot disease (*Cercospora beticola* Sacc.). Images taken with a DJI Phantom 3 Advanced brand drone, taken from an area of approximately 200 m² where the disease is seen intensely in a local farmer's field where sugar beet is grown, were processed using the Image Processing Toolbox module of the R2014a version MATLAB program. Leaf spot disease in sugar beets is detected by observations made by plant protection experts. It has been determined whether there is a sugar beet leaf spot disease and in case of disease, the stage of the disease. Observations made may vary from expert to expert. In the study, the images taken using the camera attached to the drone were processed with the developed image processing algorithms, and the diseased areas were detected. The expert evaluation and image processing application for a sample image (m image) from the study are given in Figure 4.15. As a result of the local field observation evaluation, the disease severity was

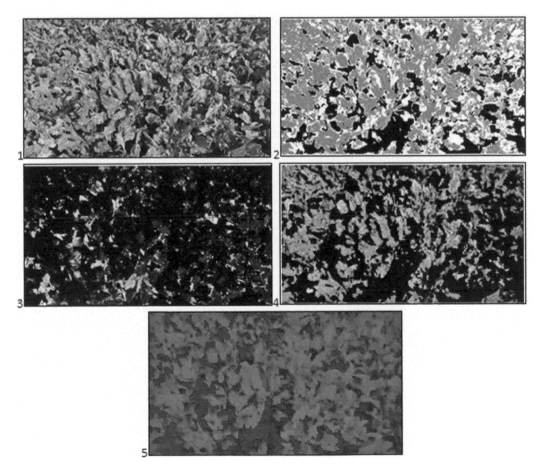

FIGURE 4.15   M image: (1) original image, (2) pixel labeling, (3) brown segment, (4) green segment, (5) contrast enhancement (Altas et al., 2018).

found to be 50–55%. As a result of image processing, the disease severity was calculated as 51% in the area of the m image. Since it is almost impossible to read the value of 51% by expert observation (50–55%), it is understood that the closest range of numbers containing these values was chosen by the expert.

Keeping fire blight under control in pear-growing areas depends on regular visual inspections of orchards for this disease. However, visual tracking is labor-intensive and time-consuming. Existing field investigations with spectral sensors mounted on UAVs make it possible to monitor larger areas in a short time. Unlike traditional RS platforms such as manned aircraft and satellites, the use of UAVs offers greater flexibility and an extremely high level of detail. Schoofs et al. (2020) mapped a heavily infected pear orchard using a UAV platform carrying a hyperspectral COSI-cam camera. From the 440 trees within the scope of the study, 24 reference "infected" trees were selected because they showed visible disease symptoms throughout the whole season and 23 trees of class 0 were selected as reference "healthy" trees. These 47 trees formed the training set for subsequent tree-based modeling (TBM). Standardized difference vegetation index (SDVI) and TBM were used in modeling. All pixels with reflectance values at 611 nm above 0.08 were considered infected. The remaining pixels were categorized as healthy (0) if reflectance at 784 nm was equal to or higher than 0.62. A training accuracy of 85% (kappa = 70%) was obtained by using this TBM model for dividing the dataset into infected and healthy trees. The SDVI combining the two selected spectral bands (611 and 784 nm) was calculated for all pixels of the trees under investigation (around 650 pixels per tree, 440 trees) (Figure 4.16). The researchers reported that the ratio of infected pixels to healthy pixels was used to represent the probability of infection of each tree in the garden (Figure 4.17) and that red in the figure indicates a high probability, while green corresponds to a low probability of infection.

Tetila et al. (2020) developed a computer vision system for the acquisition of soybean pests by UAV images and their identification and classification using deep learning (Figure 4.18). Researchers

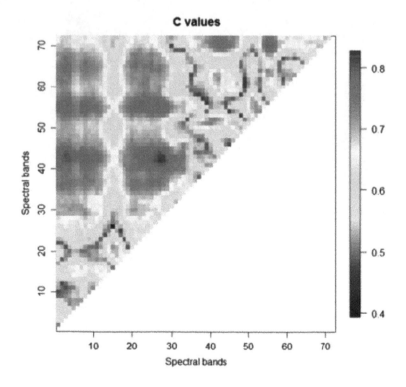

FIGURE 4.16 Logistic regression results in all possible two-band combinations (SDVI) for 47 selected reference trees (Schoofs et al., 2020).

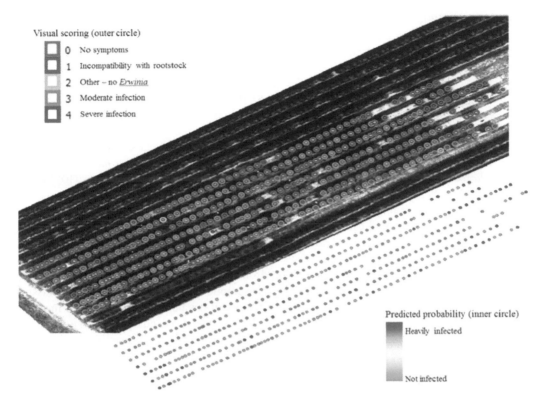

FIGURE 4.17   For each tree, probability of infection calculated from COSI-cam images by visual scoring (outer circle) (Schoofs et al., 2020).

FIGURE 4.18   System developed to detect soybean pests with UAV images using deep learning (Tetila et al., 2020).

used a dataset of 5,000 images captured under real growing conditions and evaluated the performance of five deep learning architectures Inception-v3, Resnet-50, VGG-16, VGG-19, and Xception for the classification of soybean pest images. The experimental results showed that the deep learning architectures trained with fine-tuning can lead to higher classification rates in comparison to other approaches, reaching accuracies of 93.82%. Deep learning architectures outperformed traditional feature extraction methods, such as SIFT and SURF with the bag-of-visual-words approach, the semi-supervised learning method OPFSEMImst, and supervised learning methods used to classify images, such as SVM, K-NN, and random forest.

The proposed approach uses a set of 5,000 images divided into 13 classes: (1) *Acrididae*, (2) *Anticarsia gemmatalis*, (3) *Coccinellidae*, (4) *Diabrotica speciosa*, (5) *Edessa meditabunda*, (6) *Euschistus heros* adult, (7) *Euschistus heros* nymph, (8) *Gastropoda*, (9) *Lagria villosa*, (10) *Nezara viridula* adult, (11) *Nezara viridula* nymph, (12) *Spodoptera* spp., and (13) without the presence of pests. Figure 4.19 shows examples of superpixel images from the data set. It has been reported that the images were collected in real field conditions providing various lighting conditions such as sun reflection, cloud cover, shadow, size and positioning of objects, overlap, background variations, and mating and development stages.

Calou et al. (2020) used high-spatial-resolution drone images to monitor the extent of Yellow Sigatoka damage in a banana plant, following basic assumptions about the identification, classification, quantification, and estimation of phenotypic factors. It has been reported that the monthly flights were conducted on a commercial banana plantation using a drone equipped with a 16-megapixel RGB camera; five classification algorithms were used to identify and quantify damage, and field assessments were made according to traditional methodology. The researchers found that the SVM algorithm achieved the best performance (99.28% overall accuracy and 97.13 Kappa index), followed by the ANN and minimum distance algorithms. In quantifying the disease, the SVM algorithm was more effective than other algorithms compared to the traditional methodology used to estimate the extent of Yellow Sigatoka, demonstrating that the tools used for monitoring leaf spots can be handled with RS, machine learning, and high-spatial-resolution RGB images. Figure 4.20 shows the damage levels of Yellow Sigatoka pests on banana leaves.

4.7.6 Detection of Weeds

Correct detection and control of weeds in agricultural areas is a necessary procedure to increase plant yield and prevent herbicide pollution. Identifying weeds and precision spraying only weeds are the ideal solution for weed eradication. For this purpose, superior features of imaging systems are used. With the advent of UAVs, the ability to acquire images of the entire agricultural area at very high spatial resolution and at low cost becomes possible, and the resulting input data meets high standards for weed localization and weed management. The similarity of weeds to plants can sometimes be a problem in the automatic detection of weeds. Color information alone is not sufficient to distinguish between plants and weeds. Multispectral cameras provide luminance images with high spectral resolution and spectral reflectance is estimated in several narrow spectral bands for detection. Supervised and unsupervised learning methods can be used to solve this problem automatically.

Bah et al. (2018) propose a novel fully automatic learning method using CNNs with an unsupervised training dataset collection for weed detection from UAV images (Figure 4.21). The proposed method has three main stages (Figure 4.22). The first step is to automatically detect crop rows and use them to identify inter-row weeds. The second step is to use inter-row weeds to generate the training dataset. The third step is to perform CNNs on this dataset to create a model that can detect crops and weeds in the images. As a result of the study, the difference was obtained with an accuracy of 1.5% in the spinach field and 6% in the bean area (Figure 4.23).

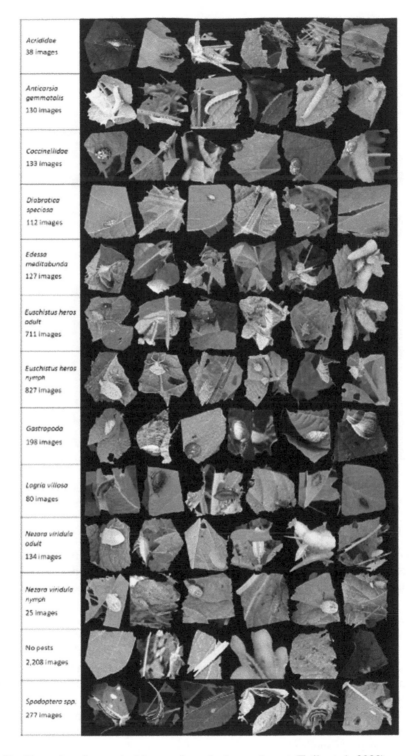

FIGURE 4.19 Examples of superpixel images from the image dataset (Tetila et al., 2020).

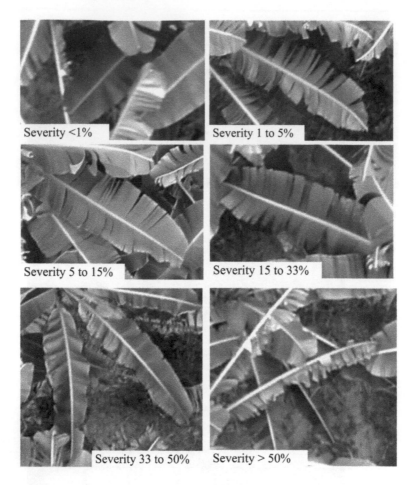

FIGURE 4.20 Damage levels of Yellow Sigatoka on banana leaves (Calou et al., 2020).

4.7.7 AGRICULTURAL SPRAYING

Low-volume UAV applications using small and very small diameter droplets in agricultural spraying have started to be widely used due to their ease of use and high efficiency. Wang et al. (2020) conducted a study to compare the drag potential of three different volume median diameters of 100, 150, and 200 μm from a commercial quadcopter equipped with centrifugal nozzles exposed to different wind speeds in real field conditions to determine whether there is pesticide drift during spraying with UAV applications (Figure 4.24). The results indicated that the relationship between rotation speed and $Dv_{0.5}$ agrees with the negative power function. Field tests found that the deposition at 12 m downwind direction decreased by an order of magnitude compared with the average deposition within the field zone. The deposition of almost all the treatments at 50 m downwind is lower than the detection limits of 0.0002 μL.

Koç (2017) designed and manufactured a hexacopter drone for pesticide application (Figure 4.25). It has been reported that the latest developments in UAV technologies made multi-rotor UAVs suitable for precision pesticide applications as these vehicles do not damage the crop due to field traffic, can operate freely on sloping terrain, and have data storage capabilities, the developed drone consists of a five-liter tank volume, a 222 W battery, a special roof made of aluminum material, a camera, GPS and related control units, and the laboratory and field trials of the drone have been carried out successfully.

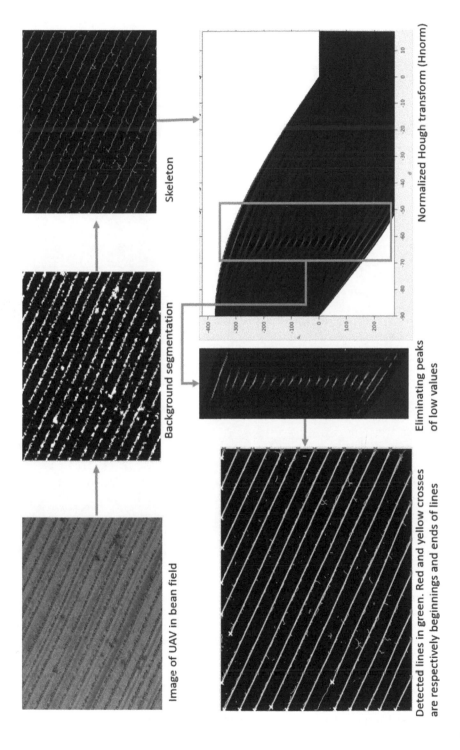

Image of UAV in bean field

Background segmentation

Skeleton

Normalized Hough transform (Hnorm)

Eliminating peaks of low values

Detected lines in green. Red and yellow crosses are respectively beginnings and ends of lines

FIGURE 4.21 Flowchart of crop line detection method (Bah et al., 2018).

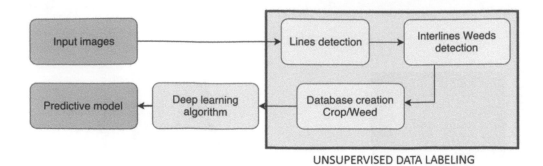

UNSUPERVISED DATA LABELING

FIGURE 4.22   Flowchart of the proposed method (Bah et al., 2018).

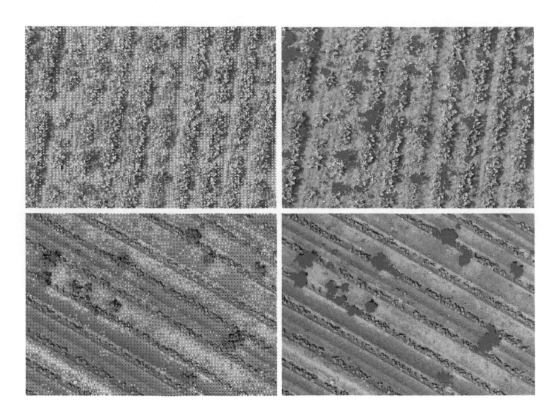

FIGURE 4.23   Examples of UAV image classification with models created by unsupervised data in a spinach (top) and bean (bottom) field. On the left are examples obtained using a floating window without crop line and background information. Blue, red, and white dots mean that the plants are identified as crop, weed, and uncertain decision, respectively. On the right in red are the weeds detected after crop line and background information has been applied (Bah et al., 2018).

Drone components and their weights are listed in Table 4.8. The total weight of the drone with all its components, including pesticide, was 8,720 g.

The flight path was determined using the Mission Planner program (Figure 4.26), three different gas levels (55%, 70%, and 85%) in the field, and three different drone speeds (3.4 kmh^{-1}, 6.3 kmh^{-1}, and 9.62 kmh^{-1}) for field tests. Food coloring paint and water-sensitive papers were used to control

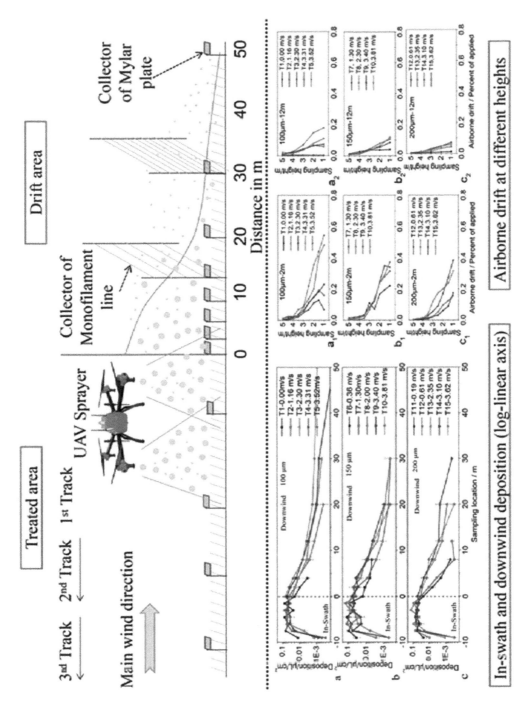

FIGURE 4.24 Results of the study of determination of drift in agricultural spraying with UAV (Wang et al., 2020).

FIGURE 4.25   View of the developed drone (Koç, 2017).

TABLE 4.8
The Drone Components and Their Weights

Components	Component name	Weight (g)
Controller	APM flight controller set APM 2.6 6 M/H GPS + OSD + radio telemetry, etc.	124
Connectors	20× male + female 4 mm banana plug + 2× XT60 connectors	17
Frame	Aluminum chassis, 6 arms 56 cm + 2 landing skids + motor coupling + vibration absorber	580
Tank	5 L prismatic tank 170 × 185 × 160 mm	210
Pump	3 bar 12 V 3 A pump 0.4 l/m + pipes and nozzles	182
Batteries	3 s 11 V 1 Ah 30 c +1.2 Ah 30 c battery	70
Cargo	5 Lt pesticide ~5 kg	5,000
Diğer	Screws, connectors, jumpers, plastic cable ties, adhesive tapes	50
Motors	6× SunnySky X4108S 480 KV brushless DC motor	678
ESC	6× BLHeli 30 A brushless ESC with LED light support 2–6 s for RC multicopter	156
Propeller	6× entire carbon fiber 1555 15 × 5.5 CW/CCW propeller	113
Battery	Tunelsan 5 s 18.5 V 12 Ah 35 C battery	1,540
Total weight:		8,720

Source: Koç (2017).

the spray areas in the field, and conical hole nozzles with a diameter of 0.5 mm were used in the experiments. The cost of the developed hexacopter drone was approximately $1,000. This amount was about one-tenth of the cost of a similar drone sold commercially (such as DJI). Aluminum was used as the frame material in this prototype, as it was easy to use, lightweight, and relatively inexpensive.

4.7.8   ARTIFICIAL POLLINATION

Pollination is the first step for a plant's fruit production; without pollination, fruit formation will not occur. The lack of pollination in the desired way causes low yield in fruit production. Pollination occurs by pollen being carried from the male flower to the mature female organ. Bees, insects, birds, or wind are required for natural pollination to occur. However, other conditions must be

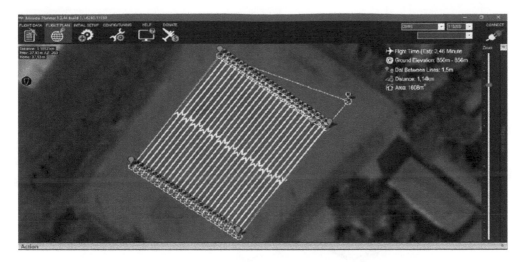

FIGURE 4.26 The marker path followed by the drone during the experiments (Koç, 2017).

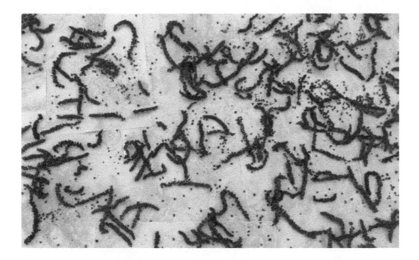

FIGURE 4.27 Pollen taken out of male walnut flowers.

provided for natural pollination. Failure to meet these conditions creates problems in pollination. Some of these problems are the absence of pollinator flowers in the environment, the short natural pollination period, male and female flowers that do not have simultaneous phenological periods required for pollination, etc. Therefore, artificial pollination applications are the best way to solve these problems. Artificial pollination by drone is the most successful form of artificial pollination. In Figure 4.27, a cross-section of the processes of taking out pollen from male flowers required for the artificial pollination of female flowers in walnut trees is presented. Figure 4.28 shows an artificial pollinating machine used with a drone. With these artificial pollinating machines, effective pollination can be done in large gardens in a short time. Figure 4.29 shows the artificial pollination of walnut trees by drone.

4.7.9 VARIABLE RATE FERTILIZATION

Fertilizer is considered one of the most important inputs of agricultural production. Increasing productivity in agriculture requires the use of more input per unit area, and therefore fertilizer and

FIGURE 4.28   Artificial pollination machine used with drone.

FIGURE 4.29   Artificial pollination of walnut trees by drone.

fertilization, which play an important role in plant nutrition, come to the fore. Incorrect and excessive use of fertilizers causes soil and water pollution. Harmful results in terms of human health can be encountered when chemicals such as nitrates and phosphates that do not dissolve easily in water mix with surface and underground waters (Zengin, 2008). According to the results of soil analysis made with mixed sampling taken from different parts of the land, as a result of the homogeneous fertilization applied, fertilizer will fall more than needed in some parts of the land and less than needed in some parts of the land. This will cause the fertilizers to accumulate or wash out in the soil in areas where fertilizer is given above the need, and it will cause low yield in areas where fertilizer is given below the need. This situation will also lead to the deterioration of the proportional balance between the nutrients (Karaman et al., 2012).

VRT technology is required for site-specific fertilizer application. VRT is used not only to minimize the amount of fertilizer used and manure pollution, but also to increase work efficiency and reduce production costs. However, the use of the technology has been limited to existing ground machinery, and studies of UAV-based variable rate fertilization have rarely been reported. Song et al. (2021) developed a variable rate fertilizer control system (VRFCS) based on a recipe map for a UAV-based granular fertilizer spreader (GFS) (Figure 4.30). The VRFCS basically consists of a flight controller and a spread controller. The control software is designed to perform variable rate fertilization by modulating the measuring apparatus with a slotted roller according to the real-time coordinates of the GFS (Figure 4.31). Laboratory and field experiments have been reported to

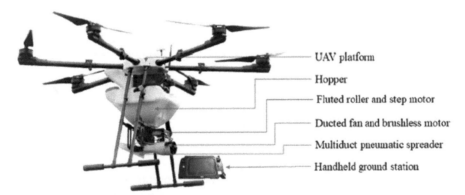

FIGURE 4.30   Key components of UAV-based granular fertilizer spreader (Song et al., 2021).

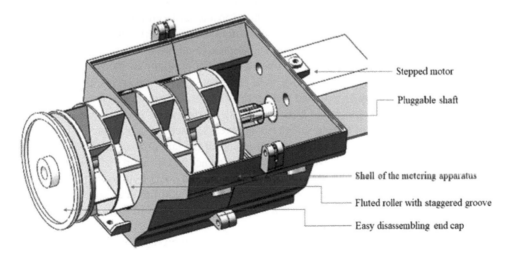

FIGURE 4.31   Optimized measuring apparatus (Song et al., 2021).

evaluate the response characteristics and fertilization accuracy of the control system. The results showed a certain relationship between the target discharge rate and the rotational speed of the fluted roller motor, which is almost linear for the operational range. The control system can respond quickly (0.1 s) and accurately to changes in the target discharge rates, whether in a fixed rate or a variable rate pattern. The error in the discharge rate between the average actual values and the target values in the fixed rate pattern was 6.05%. During the accuracy test of the variable rate pattern, errors in rotational speed and the actual application amount were approximately 4.50% and 6.64%, respectively. The field experiment based on the prescription map showed an error of less than 6.07% and a high efficiency (average of 0.72 ha/h). This study demonstrated the possibility and adaptability of variable rate fertilization by UAV-based GFSs using the developed control system.

4.7.10 Yield Estimate

Yield and quality in plant production may vary depending on various factors. Factors affecting yield and quality are listed as water, topography, nutrient availability, weeds, pests, genetics, and weather. Various studies are carried out on increasing productivity and product quality in agriculture, using minimum inputs, ensuring food safety, and protecting natural resources and the environment. Productivity increase can be achieved by the development of high-yield quality seed material, the use of machinery in agriculture, and especially the use of precision agriculture technologies. Yield mapping in precision agriculture is used to determine how much yield is obtained from which point of the field during harvest, to determine the variability in the field, and to map the data showing this variability by digitizing the data related to the product performance in a given year. By examining the yield maps, the regions with low yield and the factors limiting the yield in these regions can be determined. The management decisions to be taken by considering these factors in the production to be made in the following year will ensure an increase in efficiency. Yield mapping can be done during harvest; harvested product weight, harvested area, and average yield can be monitored instantly. These data can be transferred to computers and further analyses can be made with special software (Özgüven, 2018). Early estimation of yield by using UAVs is an important activity that can contribute to the early introduction of farmers' products to the market, marketing their products to be sold at their true value and increasing their income.

Yu et al. (2020) show that crop yield prediction plays an important role in agricultural development and management decision-making and that simulating crop growth at field scales requires high-resolution RS data and a viable model. However, RS data based on satellite sensors were affected by atmospheric interference, limiting its use in precision agriculture. The researchers reported that plant height data were applied to the crop model to improve yield estimation and optimize agricultural water management; plant height data were regularly monitored by a UAV and ground measurements at 60 plots and two-year sugarcane. In addition, the researchers stated that a SWAP-PH model was created considering the plant height simulation, the performance of the developed model was analyzed under the effects of different observation errors, plant group sizes, and development stages on crop growth and yield estimation, and the plant height estimation results from two types of observation platforms were compared. At the end of the study, the results reveal that the crop growth simulation at field scales is improved by incorporating plant height data into the model. Also, incorporating plant height measurements at the late growth stage with an absolute error of 1–2 cm and plant group size over 50 can achieve an acceptable crop yield estimation result. In addition, the assimilation of UAV-derived measurements can contribute to better yield estimation results than ground-measured plant height data. Finally, an additional irrigation and drainage strategy was proposed in July and August to maximize sugarcane yield in the study area.

Grasslands and pastures provide a wide range of ecosystem services worldwide. To protect these ecosystems and to better plan studies, it is necessary to develop highly sensitive models that consider the spatial nature of the ecosystem's structure, processes, and functions. Pecina et al. (2021) estimated the above-ground biomass at very high spatial resolution at nine study sites. The study used a combination of UAV-derived datasets to generate vegetation indices and micro-topographic models, and a random forest algorithm was used to generate above-ground biomass maps and evaluate the contribution of each predictive variable. The researchers reported that the developed model successfully predicted biomass with very high accuracy, the structural heterogeneity of the grassland was evaluated using UAV-derived datasets and vegetation indices, and when the results were subsequently related to management history in each study area, it was determined that continuous, monospecific grazing management tends to simplify grassland structure. These results also indicate that UAV-based research can serve as reliable grassland monitoring tools and help develop site-specific management strategies.

UAVs equipped with hyperspectral cameras can provide high spatial and temporal resolution remote sensing data on the rice canopy and provide possibilities for flowering monitoring. Wang et al. (2021) conducted two consecutive years of rice field experiments to investigate the performance of fluorescence spectral information in improving the accuracy of VIs-based models for yield predictions. Figure 4.32 shows the study area. In the study, fluorescence ratio reflection and fluorescence difference reflection and their first derivative reflections were defined and their correlations with rice yield were evaluated. It was determined that on the seventh day of rice flowering, fluorescence spectral information showed the highest correlation with yield. The researchers reported that incorporating fluorescence spectral information at the flowering stage into traditional VIs-based yield estimation models helped improve rice yield estimation accuracy. The graphic summary of the study is given in Figure 4.33.

4.7.11 Determination of Soil Fertility

Healthy plant production depends on the availability of plant nutrients in sufficient and balanced levels in the soil. It is not possible to say how long the production of cultivated plants will continue without adding nutrients to the soil. Soil fertility determination techniques are of great importance in determining which plant nutrients should be applied to agricultural soils, how much, when, and by which method. For this, it is necessary to develop an appropriate and economical method for the determination of insufficient nutrients in the soil (Karaman et al., 2012). According to the soil analysis results, maps showing the soil characteristics of the field from which soil samples were taken are prepared. A separate map is created for each feature of the soil, such as plant nutrient element, pH level, organic matter content, and soil compaction. Land diseases such as low soil pH and soil compaction, which restrict the development of the predetermined plant, are attempted to be eliminated in agricultural production. In cases where it cannot be eliminated, it is possible to reduce the amount of input applied during agricultural applications with variable rate application. By reducing the amount of input that cannot be used by the plant, unnecessary input usage is prevented, contributing to economic production, and preventing these inputs from harming the environment (Özgüven, 2018). Various monitoring tasks in agricultural areas can be performed quickly and successfully using drones. With the sensors and imaging technologies developed in recent years, information about some soil properties can be collected, and then high-resolution soil maps can be created.

Recently, drones have been used to measure several different soil properties and soil fertility maps have been prepared by processing ortho-images obtained during their flight over the fields. The focus of working with the drone in this way is to evaluate soil properties that are directly related to productivity. Data obtained using drone images or sensors in drones can be effectively superimposed with maps prepared by ground analysis. For example, soil survey maps show various ripples, hills, troughs, etc., and topographic maps showing features and crop yield maps can be superimposed. Soil maps showing erosion, acidity, salinity/alkalinity, water stagnation, moisture deficiency, and diseases can be overlaid with drone images showing normalized vegetation cover index (NDVI), green normalized vegetative index of difference (GNDVI), and leaf chlorophyll (Franzen and Kitchen, 2011).

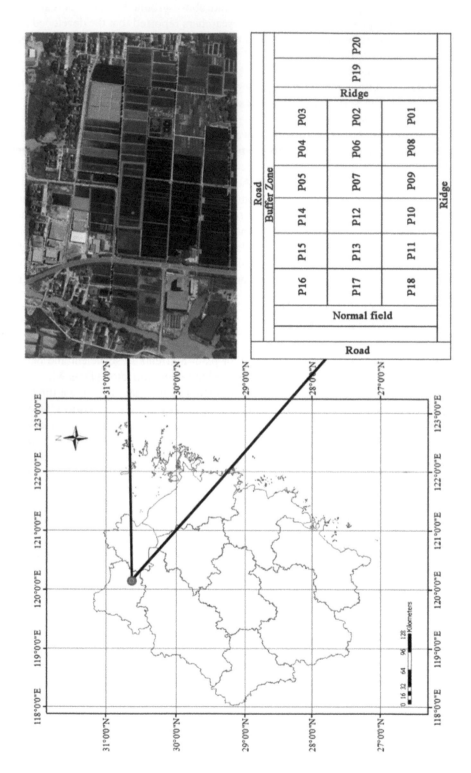

FIGURE 4.32   General location of the study area (Wang et al., 2021).

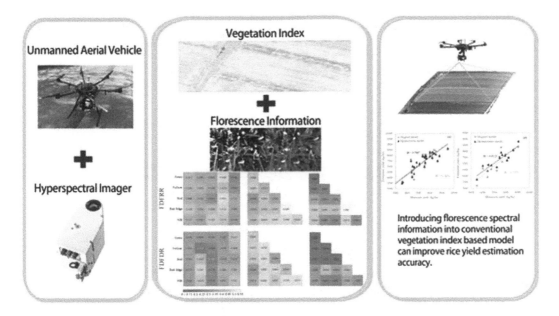

FIGURE 4.33 The graphic summary of the study (Wang et al., 2021).

Ge et al. (2021) collected 70 soil samples (0–10 cm) and UAV-based hyperspectral imaging data with 4 cm resolution from agricultural lands (2.5 × 104 m²) in order to determine the soil moisture content in arid regions. In the study, four estimation strategies were tested. These strategies are: Original image (strategy I), first- and second-order derivative methods (strategy II), fractional order derivative (FOD) technique (strategy III), and optimal fractional order combined with optimal multi-band indices (strategy IV). The results showed that FOD technology can extract information effectively (absolute maximum correlation coefficient 0.768). As a result of the comparison, strategy IV yielded the best estimates for SMC from the tested methods (R^2_{val} = 0.921, RMSEP = 1.943, and RPD = 2.736). As a result, the combination of FOD technology and optimal multi-band indices produced a highly accurate model within the XGBoost algorithm for SMC estimation. Researchers reported that this research provides a promising data mining approach for UAV-based hyperspectral imaging data. It can be seen in Figure 4.34 that the distribution of SMC was uneven. A higher SMC was determined in the eastern part of the farmland and a lower SMC in some places. This result indicated that the spatial distribution of SMC was variable.

Soil erosion in agricultural areas continues to be a global social problem. The movement of soils is often accompanied by nitrogen and phosphorus, which are very important for plant growth, thus reducing plant yield and water quality. Although in-site sediment and nutrient movement is measured at a small area scale and by field edge monitoring, this approach cannot determine spatial distributions. Menzies Pluer et al. (2020), pairing soil sampling with UAV data, have presented a new low-cost approach to mapping the spatial distribution of soil properties and nutrient concentrations within a farm area. In this study, the UAV data are used to generate a digital terrain model and then map in-site topographic variation and erosional flow paths. Topographic variation is subdivided into landform elements (flat, shoulder, backslope, footslope) that capture in-site heterogeneity and have the potential to spread the soil sample over wider spatial dimensions (Figure 4.35). The results of the study show that water content and organic matter control factors of plant yield, as represented by NDVI; both parameters show significant differences in water content and organic matter between landform elements with increases in downslope. Upsloping landform elements contain more sand content (9–20%) and had lower NDVI values than downsloping elements. Complementing these findings, significant differences in organic matter, soluble nitrogen, and soluble reactive phosphorus occurred along erosional flow paths. As a result of soil properties analysis, NDVI values showed a

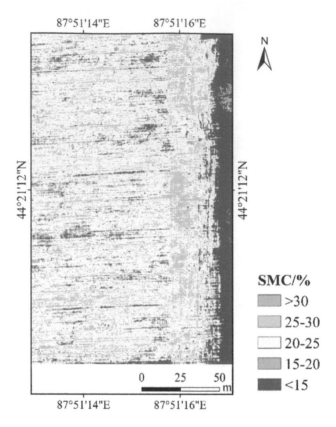

FIGURE 4.34 Soil moisture content (SMC) mapping under the optimal estimation model strategy (Ge et al., 2021).

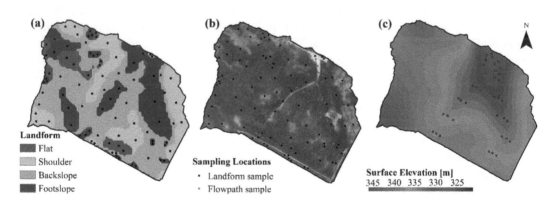

FIGURE 4.35 Landform element classification: (a) Orthomosaic spatial distribution of overlaid landform element sampling locations, (b) all overlaid sampling sites, and (c) flow path sampling locations placed on topographic map of the study basin (Menzies Pluer et al., 2020).

positive correlation with water content (0.05), organic matter (0.15), silt (0.36), and clay (0.17) content, while soluble nitrogen (–0.47) and phosphorus (–0, 30) showed a negative correlation with their concentrations. In addition to discussing the challenges and opportunities to expand the research presented, a simple proof-of-concept hydrological model was used to illustrate the potential role of hydrological connectivity and the variable source area as a driver of in-site nutrient movement. Researchers concluded that the results of the study have important implications for agricultural production, including the combination of experimental results showing water content and organic matter as control factors on agricultural yield, the role of hydrological connection and climate predictions of increased storm intensity, new time series soil sampling, and erosion and runoff data generated by UAV. They emphasized that changes in soil properties and nutrient concentrations in individual farm areas can advance the understanding of changes.

4.7.12 Use in Irrigation Applications

Irrigated agricultural areas constitute 47% of total agricultural land. Moisture status of the soil, seasonal fluctuations, precipitation-use efficiency, availability of stored water, and the irrigation methodology adopted are important parameters that affect the crop growing systems preferred by the farmers, the grain types that dominate an agricultural region, and the yield targets envisaged by the farmers. Although different irrigation methods are used during plant production, irrigation needs to be observed 24 hours a day in all methods. In this context, images obtained using predetermined flight paths on periodic flights with drones with high-resolution sensors can help farmers. With their ability to take accurate images using visual, NIR, and IR sensors, drones tend to effectively assess and map water bodies, mark irrigation channels, determine crop water status, and analyze drought or flood effects on a large farm or a village. Such data is acquired by drones in a few minutes of flight over the fields. Based on this, management blocks for irrigation can be created. Adopting such water-blocking and precision irrigation techniques helps improve water use and maximize plant productivity (Krishna, 2018).

There is a need for effective and efficient methods for mapping agricultural underground drainage systems. Visible color (VIS-C), multispectral (MS), and thermal infrared (TIR) images obtained with UAVs can be used to determine the locations of drainage pipes. Allred et al. (2020) conducted aerial surveys of 29 farmland sites to evaluate the potential of this technology for mapping buried drainage pipes using a UAV with VIS-C, MS, and TIR cameras. The results of the study show that VIS-C images detected at least some drainage lines in 48% (14 out of 29) areas, MS images detected drainage lines in 59% (17 out of 29) of areas, and TIR images detected drainage lines in 69% (20 out of 29) of areas, and visible, multispectral, and thermal infrared images were found to be useful for drainage mapping. At the end of the study, the researchers concluded that although TIR performed best overall (Figure 4.36), it was found that VIS-C or MS was more effective than TIR in

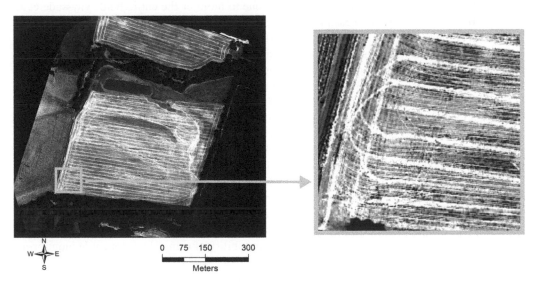

FIGURE 4.36 Orthomosaic from TIR images. The enlarged interior shows the turning lines connecting the linear features, which is indicative of field operations and does not show the dump lines (Allred et al., 2020).

mapping subsurface drainage systems; therefore VIS-C, MS, and TIR images were combined in use for increased success (Figure 4.37). It was emphasized that the timing of UAV investigations after recent rainfall can sometimes have a significant impact on the drainage pipe detection results, linear features representing drainage lines and farm field operations can be confused with each other, and often both can be determined from aerial images in the field.

Vegetation index maps show spatial variations of crop water status. Little is known about the ability of vegetation indices to indicate foliage water stress in vegetables under drip systems using under-irrigation in tropical sub-humid areas. Mwinuka et al. (2021) conducted a study to evaluate the feasibility of mobile-phone-based thermal and UAV-based multispectral imaging to evaluate the irrigation performance of African eggplant. In the study, a random block design (RBD) was applied, with sublands irrigated at 100% (I100), 80% (I80), and 60% (I60) of the crop water requirements calculated using drip irrigation (Figure 4.38). Leaf moisture content was monitored at different soil moisture conditions during the early, vegetative, and fully vegetative stages, and thermal images were obtained throughout the irrigation cycle for all days after irrigation to record canopy temperature at different soil moisture levels (Figure 4.39). At the end of the study, the plant water stress index (CWSI) obtained from mobile-phone-based thermal images was found to be sensitive to leaf moisture content (LMC) at I80 and I60 at all vegetative stages. NDVI and optimized soil adjusted crop index (OSAVI) from the UAV correlated with LMC at vegetative and full vegetative stages for all three irrigation treatments. NIR and red and green bands were found to be sensitive in detecting leaf moisture content.

4.7.13 Use in Animal Breeding

By attaching a GPS-mounted bracelet/leash to each animal individually or as a herd, the movements of the animals can be monitored continuously (or at specified time intervals, for example, every five minutes). The data to be obtained in this way is transferred to the GIS database in real time and necessary analyses are made. These analyses will be an indicator of animal/herd movements (recorded speeds at different times of the day and in grazing areas, lying, standing, and walking times, etc.). These data will enable the determination and mapping of the most productive grazing times of the day so that an effective grazing plan can be realized by determining the grazing areas that the animals most prefer and stay in the longest (Kahveci, 2014). However, using active sensors can quickly become expensive when several animals must be monitored simultaneously. Another method is to use a passive sensor attached to the drone to monitor the entire herd. Vayssade et al. (2019) developed a method for processing images taken by a drone to automate the tracking of animal activities (Figure 4.40). In the study, they reported that their activities could be monitored using a combination of thresholding and supervised classification methods that automatically detect goats from images, tested the method on 571 drone images taken in 11 days, and found a sensitivity of 74% for animal detection and 78.3% for activity detection.

Due to the ease of use of UAVs, they can assist in monitoring and management activities in open-field cattle breeding. In addition, deep learning's automatic extraction of useful information by processing images can enable UAVs to be used more effectively. For Barbedo et al. (2019), the aims of their study are as follows: To identify Canchim animals with the highest possible accuracy, to determine the ideal ground sample distance required for this detection, and to determine the most accurate CNN architecture for this detection. The study included 1,853 images containing 8,629 samples of animals and tested 15 different CNN architectures. A total of 900 models were trained (15 CNN architecture × 3 spatial resolution × 2 datasets × 10-fold cross-validation). The results revealed that many CNN architectures are robust enough to reliably detect animals in aerial images even at far from ideal conditions, demonstrating the feasibility of using UAVs for cattle monitoring. Examples of images found in the dataset are shown in Figure 4.41.

Li and Xing (2019) examined the problem of deploying a group of UAVs for monitoring and monitoring animals such as cattle and sheep in a pasture. In the study, it was assumed that all targeted animals were fitted with GPS collars and that the mobility of each targeted animal could not

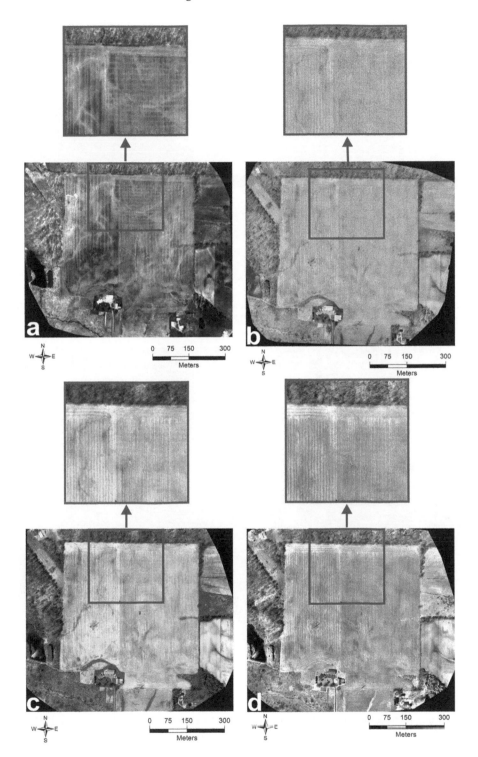

FIGURE 4.37 Fieldwork results with UAV. (a) TIR, (b) VIS-C, (c) false CIR (false color infrared), and (d) MS red edge (Allred et al., 2020).

FIGURE 4.38 Experimental field with irrigation application at 100% (I100), 80% (I80), and 60% (I60) of the calculated plant water requirements (Mwinuka et al., 2021).

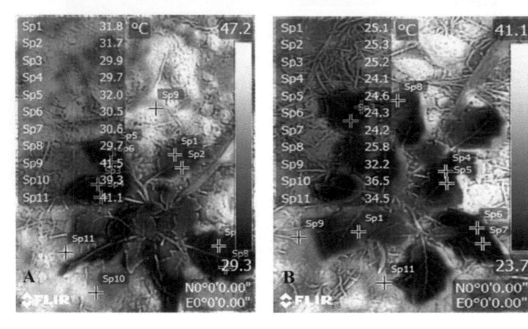

FIGURE 4.39 Thermal image temperature variation at canopy and (a) soil surface under water stress and (b) non-stressed conditions (Mwinuka et al., 2021).

be ignored. In addition, the number of UAVs was assumed to be sufficient to cover the entire pasture and it was aimed to find the optimal distribution of UAVs to minimize the average UAV–animal distance. At the end of the study, it was reported that the procedure for performing sweep coverage by UAVs was presented and it was determined that the initial positions of all targeted animals were achievable by deploying UAVs to obtain sweep coverage on the entire pasture. In addition, it was stated that the deployment of UAVs was determined and updated by clustering flowing K-means according to their starting positions and updated positions from GPS collars, and they reported that their solution proposal can always provide a lower UAV–animal distance compared to a standard K-means clustering algorithm, regardless of the mobility of the targeted animals.

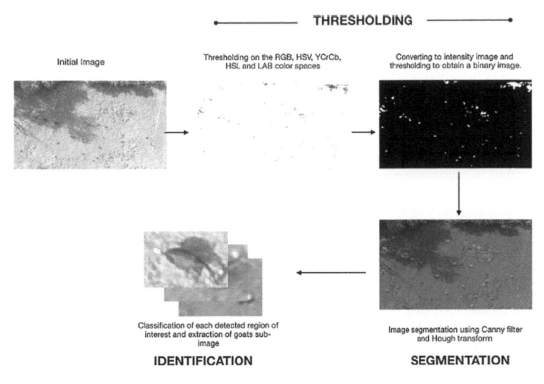

FIGURE 4.40   Schematic flow of the goat detection process (Vayssade et al., 2019).

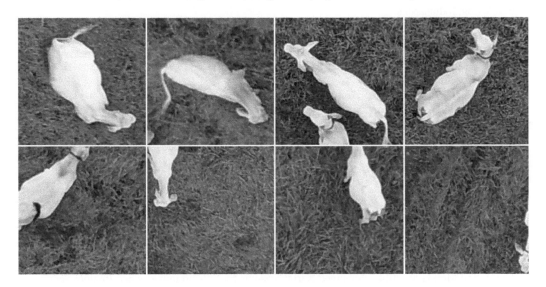

FIGURE 4.41   Examples of images found in the dataset (Barbedo et al., 2019).

REFERENCES

Adão, T., Hruška, J., Pádua, L., Bessa, J., Peres, E., Morais, R. and Sousa, J.J., 2017. Hyperspectral Imaging: A Review on UAV-Based Sensors, Data Processing and Applications for Agriculture and Forestry. *Remote Sensing*, 9(11), 1110. https://doi.org/10.3390/rs9111110.

Akyürek, S., Yılmaz, M.A. ve Taşkıran, M., 2012. İnsansız Hava Araçları Muharebe Alanında ve Terörle Mücadelede Devrimsel Dönüşüm. Rapor No: 53, Bilge Adamlar Stratejik Araştırmalar Merkezi. İstanbul. (Turkish).

Akyüz, S., 2013. Dört Rotorlu İnsansız Hava Aracı (Quadrotor)'un Pd ve Bulanık Kontrolcü Tasarımı ve Benzetim Uygulaması. Ege Üniversitesi Fen Bilimleri Enstitüsü Elektrik-Elektronik Mühendisliği Anabilim Dalı Yüksek Lisans Tezi. İzmir. (Turkish).

Allred, B., Martinez, L., Fessehazion, M.K., Rouse, G., Williamson, T.N., Wishart, D., Koganti, T., Freeland, R., Eash, N., Batschelet, A. and Featheringill, R., 2020. Overall Results and Key Findings on the use of UAV Visible-Color, Multispectral, and Thermal Infrared Imagery to Map Agricultural Drainage Pipes. *Agricultural Water Management*, 232, 106036. https://doi.org/10.1016/j.agwat.2020.106036.

Altaş, Z., 2017. Drone Kullanılarak Görüntü İşleme Tekniği İle Şeker Pancarı Yaprak Lekesi Hastalık (Cercospora beticola) Düzeyinin Belirlenmesi. Gaziosmanpaşa Üniversitesi Fen Bilimleri Enstitüsü Biyosistem Mühendisliği Ana Bilim Dalı Yüksek Lisans Tezi. Tokat. (Turkish).

Altas, Z., Ozguven, M.M. and Yanar, Y., 2018. Determination of Sugar Beet Leaf Spot Disease Level (*Cercospora Beticola* sacc.) with Image Processing Technique by Using Drone. *Current Investigations in Agriculture and Current Research*, 5(3), 621–631. https://doi.org/10.32474/CIACR.2018.05.000214.

Anonymous, 2013. Department of Defense (USA) Unmanned Systems Integrated Roadmap: FY 2013–2038, ref. number 14–S–0553 (129 pages).

Anonymous, 2017. http://keremcep.com/drone-ve-iha-arasindaki-fark-nedir/. [Accessed 15.08.2017].

Anonymous, 2012. Ministry of Defence, "Current Terminology, Doctrine, and Classification", Joint Doctrine Note 2/11 The UK Approach to Unmanned Aircraft Systems (2011): 2–7. http://www.mod.uk/DefenceInternet/MicroSite/DCDC/OurPublications/JDNP/211TheUkApproachToUnmannedAircraftSystems.htm. [Accessed 23.10.2012].

Anonymous, 2019. https://www.elektrikport.com/makale-detay/Lidar-nedir-nasil-calisir/18602#ad-image-0. [Accessed 12.02.2019]. (Turkish).

Avelino, J., Cristancho, M., Georgiou, S., Imbach, P., Aguilar, L. and Bornemann, G., 2015. The Coffee Rust Crises in Colombia and Central America (2008–2013): Impacts, Plausible Causes and Proposed Solutions. *Food Security*, 7(2), 303–321.

Bah, M.D., Hafiane, A. and Canals, R., 2018. Deep Learning with Unsupervised Data Labeling for Weed Detection in Line Crops in UAV Images. *Remote Sensing*, 10, 1690. https://doi.org/10.3390/rs10111690.

Baichtal, J., 2016. *Building Your Own Drones: A Beginners' Guide to Drones, UAVs, and ROVs*. Que Publishing, Indianapolis.

Barbedo, J.G.A., Koenigkan, L.V., Santos, T.T. and Santos, P.M., 2019. A Study on the Detection of Cattle in UAV Images Using Deep Learning. *Sensors*, 19, 5436. https://doi.org/10.3390/s19245436.

Bock, C.H., Poole, G.H., Parker, P.E. and Gottwald, T.R., 2010. Plant Disease Severity Estimated Visually, by Digital Photography and Image Analysis, and by Hyperspectral Imaging. *Critical Reviews in Plant Sciences*, 29(2), 59–107.

Boesch, R., 2017. Thermal Remote Sensing with UAV-based Workflows. *International Archives of the Photogrammetry, Remote Sensing and Spatial Information Sciences*, 42, 41.

Borra-Serrano, I., De Swaef, T., Quataert, P., Aper, J., Saleem, A., Saeys, W., Somers, B., Roldán-Ruiz, I. and Lootens, P., 2020. Closing the Phenotyping Gap: High Resolution UAV Time Series for Soybean Growth Analysis Provides Objective Data from Field Trials. *Remote Sensing*, 12(10), 1644. https://doi.org/10.3390/rs12101644.

Böhler, J.E., Schaepman, M.E. and Kneubühler, M., 2020. Crop Separability from Individual and Combined Airborne Imaging Spectroscopy and UAV Multispectral Data. *Remote Sensing*, 12(8), 1256. https://doi.org/10.3390/rs12081256.

Calou, V.B.C., Teixeira, A.D.S., Moreira, L.C.J., Lima, C.S., De Oliveira, J.B. and De Oliveira, M.R.R., 2020. The use of UAVs in Monitoring Yellow Sigatoka in Banana. *Biosystems Engineering*, 193, 115–125. https://doi.org/10.1016/j.biosystemseng.2020.02.016.

Clarke, R., 2014. Understanding the Drone Epidemic. *Computer Law and Security Review*, 30, 230–246.

Colomina, I. and Molina, P., 2014. Unmanned Aerial Systems for Photogrammetry and Remote Sensing: A Review. *ISPRS Journal of Photogrammetry and Remote Sensing*, 92, 79–97. https://doi.org/10.1016/j.isprsjprs.2014.02.013.

Çelik, H., 2017. Taşkın Modellemede Lidar Verisi İle Performans Analizleri. İstanbul Teknik Üniversitesi Fen Bilimleri Enstitüsü, Geomatik Mühendisliği Anabilim Dalı (Doktora Tezi), İstanbul. (Turkish).

Çetinsoy, E., Sırımoğlu, E., Öner, K.T., Ayken, T., Hançer, C., Ünel, M., Akşit, M.F., Kandemir, İ. ve Gülez, K., 2009. Yeni Bir İnsansız Hava Aracının (SUAVİ) Prototip Üretimi ve Algılayıcı-Eyleyici Entegrasyonu, Sabancı Üniversitesi, Mühendislik ve Doğa Bilimleri Fakültesi, s:7. (Turkish).

DJI, 2017. Phantom 3 Advanced Manual. DJI Innovations. https://dl.djicdn.com/downloads/phantom_3/20170706/Phantom+3+Advanced+User+Manual+v1.8.pdf.

Düzgün, H.Ş., 2010. Uzaktan Algılamaya Giriş Dersi, Ünite 1 – Uzaktan Algılamaya Giriş. Ulusal Açık Ders Malzemeleri Konsorsiyumu. (Turkish).

Fahlstrom, P.G. and Gleason, T.J., 2012. *Introduction to UAV Systems*. John Wiley & Sons, Ltd, Chichester. ISBN 978-1-119-97866-4.

Franzen, D.W. and Kitchen, N.R., 2011. Developing Management Zones to Target Nitrogen Applications. The Site-Specific Management Guidelines Series No: SSMG-5. Potash & Phosphate Institute.

Ge, X., Ding, J., Jin, X., Wang, J., Chen, X., Li, X., Liu, J. and Xie, B., 2021. Estimating Agricultural Soil Moisture Content through UAV-Based Hyperspectral Images in the Arid Region. *Remote Sensing*, 13, 1562. https://doi.org/10.3390/rs13081562.

Georgopoulos, A., Oikonomou, Ch., Adamopoulos, E. and Stathopoulou, E.K., 2016. Evaluating Unmanned Aerial Platforms for Cultural Heritage Large Scale Mapping. *International Archives of the Photogrammetry, Remote Sensing and Spatial Information Sciences*, XLI-B5, 355–362. https://doi.org/10.5194/isprsarchives-XLI-B5-355-2016.

Grassi, M., 2014. 5 Actual Uses for Drones in Precision Agriculture Today. http://dronelife.com/2014/12/30/5-actual-uses-drones-precision-agriculture-today/.

Hashimoto, Y., Ino, T., Kramer, P.J., Naylor, A.W. and Strain, B.R., 1984. Dynamic Analysis of Water Stress of Sunflower Leaves by Means of a Thermal Image Processing System. *Plant Physiology*, 76, 266–269.

Horstrand, P., Guerra, R., Rodriguez, A., Diaz, M., Lopez, S. and Lopez, J.F., 2019. A UAV Platform Based on a Hyperspectral Sensor for Image Capturing and On-board Processing. *IEEE Access*, 7, 66919–66938.

Irak, H., 2009. SAR Sistem ve Teknolojileri. Elektrik Mühendisliği Dergisi, 437. sayı, Aralık 2009, s.86–90. (Turkish).

Jones, H.G., 1999. Use of Thermography for Quantitative Studies of Spatial and Temporal Variation of Stomatal Conductance Over Leaf Surface. *Plant, Cell & Environment*, 22, 1043–1055.

Johnson, M., 2015. Components for Creating an Unmanned Aerial Vehicle. Application Note. http://www.egr.msu.edu/classes/ece480/capstone/spring15/group14/uploads/4/2/0/3/42036453/johnsonappnote.pdf.

Kahveci, M., 2014. Uydularla Konum Belirleme Sistemleri (GNSS)'nin Hassas Tarımda Kullanımı ve Sağladığı Katkılar. *Harita Teknolojileri Elektronik Dergisi Cilt*, 6(2), 35–48. (Turkish).

Karaman, M.R., Brohi, A.R., Müftüoğlu, N.M, Öztaş, T. and Zengin, M., 2012. Sürdürülebilir Toprak Verimliliği. Koyulhisar Ziraat Odası Kültür Yayınları No:1. (Turkish).

Kilby, T. and Kilby, B., 2016. *Make: Getting Started with Drones*. Maker Media, Inc, San Francisco. ISBN: 978-1-457-18330-0.

Koç, C., 2017. Design and Development of a Low-cost UAV for Pesticide Applications. *Journal of Agricultural Faculty of Gaziosmanpasa University*, 34(1), 94–103. https://doi.org/10.13002/jafag4274.

Krishna, K.R., 2018. *Agricultural Drones: A Peaceful Pursuit*. Apple Academic Press, Oakville. ISBN: 978-1-315-19552-0.

Li, X. and Xing, L., 2019. Use of Unmanned Aerial Vehicles for Livestock Monitoring based on Streaming K-Means Clustering. *IFAC-PapersOnLine*, 52(30), 324–329. https://doi.org/10.1016/j.ifacol.2019.12.560.

Lu, J., Cheng, D., Geng, C., Zhang, Z., Xiang, Y. and Hu, T., 2021. Combining Plant Height, Canopy Coverage and Vegetation Index from UAV-Based RGB Images to Estimate Leaf Nitrogen Concentration of Summer Maize. *Biosystems Engineering*, 202, 42–54. https://doi.org/10.1016/j.biosystemseng.2020.11.010.

Maes, W.H., Huete, A.R. and Steppe, K., 2017. Optimizing the Processing of UAV-Based Thermal Imagery. *Remote Sensing*, 9(5), 476. https://doi.org/10.3390/rs9050476.

Malambo, L., Popescu, S.C. Murray, S.C., Putman, E., Pugh, N.A., Horne, D.W., Richardson, G., Sheridan, R., Rooney, W.L., Avant, R., Vidrine, M., McCutchen, B., Baltensperger, D. and Bishop, M., 2018. Multitemporal Field-Based Plant Height Estimation Using 3D Point Clouds Generated from Small Unmanned Aerial Systems High-Resolution Imagery. *International Journal of Applied Earth Observation and Geoinformation*, 64, 31–42. https://doi.org/10.1016/j.jag.2017.08.014.

Menzies Pluer, E.G., Robinson, D.T., Meinen, B.U. and Macrae, M.L., 2020. Pairing Soil Sampling with Very-High Resolution UAV Imagery: An Examination of Drivers of Soil and Nutrient Movement and Agricultural Productivity in Southern Ontario. *Geoderma*, 379, 114630. https://doi.org/10.1016/j.geoderma.2020.114630.

Merç, Y. ve Bayılmış, C., 2011. Dört Rotorlu İnsansız Hava Aracı (Quadrotor) Uygulaması, 6th International Advanced Technologies Symposium (IATS'11). s:18-20, 16–18 May. (Turkish).

Moon, S., Lee, D., Lee, D., Kim, D. and Bang, H., 2021. Energy-Efficient Swarming Flight Formation Transitions Using the Improved Fair Hungarian Algorithm. *Sensors*, 21, 1260. https://doi.org/10.3390/s21041260.

Mwinuka, P.R., Mbilinyi, B.P., Mbungu, W.B., Mourice, S.K., Mahoo, H.F. and Schmitter, P., 2021. The Feasibility of Hand-Held Thermal and UAV-Based Multispectral Imaging for Canopy Water Status Assessment and Yield Prediction of Irrigated African Eggplant (*Solanum aethopicum* L). *Agricultural Water Management*, 245, 106584. https://doi.org/10.1016/j.agwat.2020.106584.

Omasa, K. and Aiga, I., 1987. Environmental Measurement: Image Instrumentation for Evaluating Pollution Effects on Plants. *Systems and Control Encyclopedia*. (Editor: Singh, M.G.) Pergamon Press, Oxford, 1516–1522.

Omasa, K., 1990. Image Instrumentation Methods of Plant Analysis. *Modern Methods of Plant Analysis*. (Editors: Linskens, H.F. and Jackson, J.F.) Springer-Verlag, Berlin, 203–243.

Omasa, K. and Croxdale, J.G., 1992. Image Analysis of Stomatal Movements and Gas Exchange. *Image Analysis in Biology*. (Editor: Häder, D.P.) CRC Press Taylor & Francis Group LLC, Boca Raton, FL, 171–197.

Omasa, K., 2002. Diagnosis of Stomatal Response and Gas Exchange of Trees by Thermal Remote Sensing. *Air Pollution and Plant Biotechnology*. (Editors: Omasa, K., Saji, H., Youssefian, S. and Kondo, N.) Springer-Verlag, Tokyo, 343–359.

Omasa, K., Oki, K. and Suhama, T., 2006. Section 5.2 Remote sensing from satellites and aircraft, pp. 231–244 of Chapter 5 Precision Agriculture, in CIGR Handbook of Agricultural Engineering Volume VI Information Technology. Edited by CIGR-The International Commission of Agricultural Engineering; Volume Editor, Axel Munack. St. Joseph, Michigan, USA: ASABE. Copyright American Society of Agricultural Engineers.

Osco, L.P., De Arruda, M.S., Gonçalves, D.N., Dias, A., Batistoti, J., De Souza, M., Gomes, F.D.G., Ramos, A.P.M., De Castro Jorge, L.A., Liesenberg, V., Li, J., Ma, L., Marcato, J. and Gonçalves, W.N., 2021. A CNN Approach to Simultaneously Count Plants and Detect Plantation-Rows from UAV Imagery. *ISPRS Journal of Photogrammetry and Remote Sensing*, 174, 1–17. https://doi.org/10.1016/j.isprsjprs.2021.01.024.

Ozguven, M.M. and Adem, K., 2019. Automatic Detection and Classification of Leaf Spot Disease in Sugar Beet using Deep Learning Algorithms. *Physica A-Statistical Mechanics and Its Applications*, 535(122537), 1–8. https://doi.org/10.1016/j.physa.2019.122537.

Ozguven, M.M., 2020. Deep Learning Algorithms for Automatic Detection and Classification of Mildew Disease in Cucumber. *Fresenius Environmental Bulletin*, 29(08/2020), 7081–7087.

Ozguven, M.M. and Yanar, Y., 2022. The Technology Uses in the Determination of Sugar Beet Diseases. *Sugar Beet Cultivation, Management and Processing*. (Editors: Misra, V., Santeshwari and Mall, A.K.). Springer, Singapore. 621–642. https://doi.org/10.1007/978-981-19-2730-0_30.

Özgüven, M.M., 2018. Hassas Tarım. Akfon Yayınları, Ankara. ISBN: 978-605-68762-4-0. (Turkish).

Pagliai, A., Ammoniaci, M., Sarri, D., Lisci, R., Perria, R., Vieri, M., D'Arcangelo, M.E.M., Storchi, P. and Kartsiotis, S.-P., 2022. Comparison of Aerial and Ground 3D Point Clouds for Canopy Size Assessment in Precision Viticulture. *Remote Sensing*, 14, 1145. https://doi.org/10.3390/rs14051145.

Pang, Y., Shi, Y., Gao, S., Jiang, F., Veeranampalayam-Sivakumar, A.-N., Thompson, L., Luck, J. and Liu, C., 2020. Improved Crop Row Detection with Deep Neural Network for Early-Season Maize Stand Count in UAV Imagery. *Computers and Electronics in Agriculture*, 178, 105766. https://doi.org/10.1016/j.compag.2020.105766.

Peciña, M.V., Bergamo, T.F., Ward, R.D., Joyce, C.B. and Sepp, K., 2021. A Novel UAV-Based Approach for Biomass Prediction and Grassland Structure Assessment in Coastal Meadows. *Ecological Indicators*, 122, 107227, https://doi.org/10.1016/j.ecolind.2020.107227.

Royo, S. and Ballesta-Garcia, M., 2019. An Overview of Lidar Imaging Systems for Autonomous Vehicles. *Applied Sciences*, 9(19), 4093. https://doi.org/10.3390/app9194093.

Saiz-Rubio, V. and Rovira-Más, F., 2020. From Smart Farming towards Agriculture 5.0: A Review on Crop Data Management. *Agronomy*, 10(2), 207. https://doi.org/10.3390/agronomy10020207.

Savary, S. and Willocquet, L., 2014. Simulation Modeling in Botanical Epidemiology and Crop Loss Analysis. The Plant Health Instructor. 173 p.

Schoofs, H., Delalieux, S., Deckers, T. and Bylemans, D., 2020. Fire Blight Monitoring in Pear Orchards by Unmanned Airborne Vehicles (UAV) Systems Carrying Spectral Sensors. *Agronomy*, 10(5), 615. https://doi.org/10.3390/agronomy10050615.

Sebbane, Y.B., 2016. *Smart Autonomous Aircraft Flight Control and Planning For UAV*. CRC Press Taylor & Francis Group LLC, Boca Raton., ISBN: 978-1-4822-9916-8.

Sebbane, Y.B., 2018. *Intelligent Autonomy of UAVS, Advanced Missions and Future Use*. CRC Press Taylor & Francis Group LLC, Boca Raton., ISBN: 978-1-138-56849-5.

Song, C., Zhou, Z., Zang, Y., Zhao, L., Yang, W., Luo, X., Jiang, R., Ming, R., Zang, Y., Zi, L. and Zhu, Q., 2021. Variable-Rate Control System for UAV-Based Granular Fertilizer Spreader. *Computers and Electronics in Agriculture*, 180, 105832. https://doi.org/10.1016/j.compag.2020.105832.

Szabó, G., Bertalan, L., Barkóczi, N., Kovács, Z., Burai, P. and Lénárt, C., 2018. Zooming on Aerial Survey (Chapter 4). *Small Flying Drones, Applications for Geographic Observation*. (Editors: Casagrande, G., Sik, A. ve Szabó, G.) Springer International Publishing AG, Cham. 91–126 ISBN: 978-3-319-66576-4.

Tan, M., Özgüven, M.M. ve Tarhan, S., 2015. Drone Sistemlerin Hassas Tarımda Kullanımı, 29. Tarımsal Mekanizasyon Kongresi ve Enerji Kongresi, 2–5 Eylül Diyarbakır, S:543–547. (Turkish).

Tetila, E.C., Machado, B.B., Astolfi, G., De Souza Belete, N.A., Amorim, W.P., Roel, A.R. and Pistori, H., 2020. Detection and Classification of Soybean Pests using Deep Learning with UAV Images. *Computers and Electronics in Agriculture*, 179, 105836. https://doi.org/10.1016/j.compag.2020.105836.

Turner, D., Lucieer, A., Malenovský, Z., King, D.H. and Robinson, S.A., 2014. Spatial Co-Registration of Ultra-High Resolution Visible, Multispectral and Thermal Images Acquired with a Micro-UAV over Antarctic Moss Beds. *Remote Sensing*, 6(5), 4003–4024. https://doi.org/10.3390/rs6054003.

Um, J.-S., 2019. *Drones as Cyber-Physical Systems*. Springer Nature Singapore Pte Ltd, Gateway East. ISBN: 978-981-13-3740-6.

Vayssade, J.-A., Arquet, R. and Bonneau, M., 2019. Automatic Activity Tracking of Goats Using Drone Camera. *Computers and Electronics in Agriculture*, 162, 767–772. https://doi.org/10.1016/j.compag.2019.05.021.

Wang, Y. and Liu, J., 2012. Evaluation Methods for the Autonomy of Unmanned Systems. *Chinese Science Bulletin*, 57, 3409–3418.

Wang, G., Han, Y., Li, X., Andaloro, J., Chen, P., Hoffmann, W., Han, X., Chen, S. and Lan, Y., 2020. Field Evaluation of Spray Drift and Environmental Impact Using an Agricultural Unmanned Aerial Vehicle (UAV) Sprayer. *Science of the Total Environment*, 737, 139793. https://doi.org/10.1016/j.scitotenv.2020.139793.

Wang, F., Yao, X., Xie, L., Zheng, J. and Xu, T., 2021. Rice Yield Estimation Based on Vegetation Index and Florescence Spectral Information from UAV Hyperspectral Remote Sensing. *Remote Sensing*, 13, 3390. https://doi.org/10.3390/rs13173390.

Yu, D., Zha, Y., Shi, L., Jin, X., Hu, S., Yang, Q., Huang, K. and Zeng, W., 2020. Improvement of Sugarcane Yield Estimation by Assimilating UAV-Derived Plant Height Observations. *European Journal of Agronomy*, 121, 126159. https://doi.org/10.1016/j.eja.2020.126159.

Zengin, E., 2008. Küreselleşme Sürecinde Tarım'da Sürdürülebilirlik ve Çevre Sorunları. *Alatoo Academic Studies*, 3(2), S:44–54. (Turkish).

5 Agriculture 5.0 and the Internet of Things

5.1 WHAT IS AGRICULTURE 4.0?

The concept of Industry 4.0 or the Fourth Industrial Revolution first emerged in Germany in 2011. With Industry 4.0, it has been demonstrated that it is possible to develop the product first after the marketing strategy, and this strategy has been successful. After Germany, first Europe and then the world have adapted to this change and strategies have been developed that include a flexible and smart production system that can cope with more complexity in the product, process, and environment in tough competitive conditions. Industry 4.0 has made it possible to collect each data in the production environment and analyze these data in detail, resulting in the emergence of more flexible, productive, and efficient business models and the formation of a smart factory system. With the use of the innovations that emerged with the concept of Industry 4.0 in agriculture, the concept of Agriculture 4.0 has started to be mentioned by researchers.

The concept of Industry 4.0 stands for the Fourth Industrial Revolution, a new level of organization and control of the entire value creation chain throughout the life cycle of products. This cycle addresses increasingly individualized customer demands and extends from concept to ordering, development and production, delivery of a product to the end user, and the recycling process, including related services (The German Plattform Industrie 4.0, 2015). The Fourth Industrial Revolution marks a new level "in the organization and control of the entire value creation chain". To leave no doubt about what is involved, this value creation chain is specifically and comprehensively defined from its inception to the services associated with the products. This makes it clear that we are dealing with a fundamental transformation in industrial production methods and not simply a change of any part of these methods (Sendler, 2018). However, the Fourth Industrial Revolution is not just about smart and connected machines and systems. Its scope is much wider. This includes all of the major developments occurring simultaneously in fields ranging from gene sequencing, nanotechnology, biotechnology, artificial intelligence, robotics, the Internet of Things, autonomous vehicles, 3D printing, materials science, energy storage, and renewable energy to quantum computing. What makes the Fourth Industrial Revolution fundamentally different from previous revolutions is the fusion of these technologies and their interactions in the physical, digital, and biological realms. To fully grasp the speed and breadth of this new revolution, it is enough to consider the limitless possibilities: Connecting billions of people with mobile devices creates unprecedented processing power, storage capabilities, and information access (Schwab, 2017).

The goals of Industry 4.0 are to meet customer needs, improve flexibility, optimize decision-making, create better resource productivity and efficiency, as well as create new services, respond to social challenges such as demographic change and work-life balance, and maintain a competitive high-wage economy. To achieve all these goals, three features of Industry 4.0 should be implemented: Horizontal integration through the value network, end-to-end digital integration across the entire value chain, and vertical integration into the manufacturing system (Kagermann et al., 2013).

The Agriculture 1.0 period is the period in which agricultural activities were carried out depending on manpower and animal power. In this period, simple tools such as sickles and shovels were used. With the development of steam engines in the 19th century, the Agriculture 2.0 period began to be widely used in agriculture as well as in every other field. It is a period when labor-intensive agricultural production was made with agricultural machinery, although it was not very developed. In this period, especially in the 1950s, with the use of chemical applications in agriculture, serious

DOI: 10.1201/b23229-5

increases were seen in yield. In the Agriculture 3.0 period in the 2000s, GPS was opened to civilian use, and electronics, software, and sensors began to be used in precision agriculture applications. In the Agriculture 4.0 period, technologies such as artificial intelligence, robotics, the Internet of Things, autonomous vehicles, drones, advanced computers, cloud computing, and big data have been used in agriculture, and agricultural machinery has begun to be made smart. Agriculture 5.0 is the period when personal solutions will come to the fore instead of mass production, and machines, robots, or systems will gain the feature of being smart and autonomous. This period has not started fully, but studies have started for this period (Figure 5.1). Thus, with the use of these technologies in agriculture, more competitive, efficient, and sustainable agricultural practices are possible.

Development has continued since the emergence of precision agriculture technology and this development continues today. While it was focused on the field-specific determination of variability before, systems with sensor-based applications were developed later. Today, agricultural machines are being converted into ISOBUS-compatible smart machines. Agricultural machinery and production or cultivation areas are equipped with sensors, and even more than that, one agricultural machine can communicate with another. Apart from agricultural machines, it is possible to communicate with agricultural robots or drones and agricultural applications can be made in coordination.

5.2   SUPPORTIVE TECHNOLOGIES IN AGRICULTURE 4.0

Technological developments are constantly being experienced and new techniques, technologies, methods, machines, and devices are emerging day by day. The most important goal in Industry 4.0 or Agriculture 4.0 is to increase efficiency and productivity in production. However, achieving this goal is not a standard and ordinary task. The diversity of production areas and the multitude of application areas make this work difficult. For this reason, it is necessary to plan by taking into consideration the existing technologies specific to each production and application.

Current supporting technologies in this area (Figure 5.2) can be listed as follows:

- *Adaptive robotics*: Increasing processing, communication, and control processes with the addition of microprocessors and artificial intelligence capabilities to machines and systems; they are made smarter by gaining the feature of having autonomy and sociality. Thus, adaptable and flexible robots, combined with the use of artificial intelligence, recognize the sub-segments of each part, allowing for easier production of different products (Wittenberg, 2015).

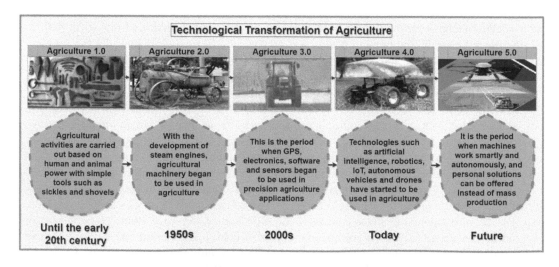

FIGURE 5.1   Technological transformation of agriculture and Agriculture 5.0.

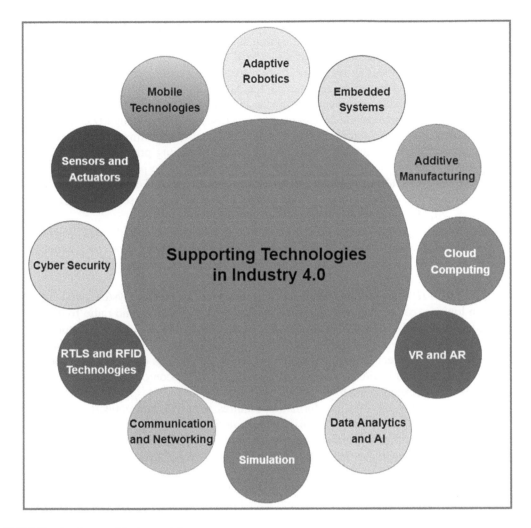

FIGURE 5.2   Scope of Industry 4.0.

- *Embedded systems*: It is a supporting technology for the organization and coordination between the physical infrastructure and computing capabilities of network systems. Physical and digital tools must be integrated and linked with other devices to enable decentralized actions (Bagheri et al., 2015).
- *Additive manufacturing*: Computer-aided design (CAD) and a set of digital features of the product are edited to produce 3D objects of various shapes and sent to industrial machines, and layers of material are added repeatedly until a three-dimensional object emerges from the digital models created. The materials used can be plastics, polymers, ceramics, or metals in the form of liquids, powders, or sheets (Gaub, 2015).
- *Cloud computing*: It is a model that allows access to a common repository of configurable computing resources (computer networks, servers, databases, applications, services, etc.) that can be quickly provisioned and published with minimal management effort or service provider interaction, on demand, anytime, anywhere (Mell and Grance, 2011).
- *Virtual reality (VR) and augmented reality (AR)*: VR and AR applications associate graphical interfaces with the user's view of the current environment. The primary role of graphical user interfaces is that users can directly affect visual representations of items using commands that appear on the screen and interact with these menus referenced by ad

hoc feedback (Salkin et al., 2018). For this reason, equipment such as handheld devices, fixed imaging systems, spatial imaging systems, head-mounted displays, smart glasses, and smart lenses are used for the application (Paelke, 2014).

- *Simulation*: It is a modeling technique that creates an infrastructure for monitoring the properties of the real system by transferring the data of a physical system existing in the real world to a virtual environment. It provides advantages in terms of time, cost, and risk management as it can make the development of processes traceable (Bungartz et al., 2014).
- *Data analytics and artificial intelligence*: A large amount of real-time data generated in R&D, production, operation, and maintenance processes during production is collected from multiple sources (Zhang et al., 2016). This data needs to be processed quickly and may require combining various data sources in diversified formats. For example, data mining techniques should be used where data is collected from various sensors (Obitko and Jirkovský, 2015).
- *Communication and networking*: It can be defined as a link between individually defined physical and distributed systems. In this way, computers and machines can interact by reaching specific goals using communication tools and devices for anyone, anywhere, at any time (IERC, 2011).
- *RTLS and RFID technologies*: Efficient coordination of embedded systems and information logistics can be had with real-time locating systems (RTLS) and radio frequency identification (RFID), providing identification, location detection, and status monitoring of objects and resources within and across the enterprise. Thus, the collection and processing of real-time data collected from production processes and various environmental sources helps integrate organizational functions and enables self-decision of machines and other smart devices (Uckelmann, 2008).
- *Cyber security*: Industry 4.0 transformation requires intensive data collection and processing activities. Therefore, the security of data storage and transfer processes are fundamental concepts for companies. Security should be provided on the basis of personal authority and confidentiality in cloud technologies, machines, robots, and automated systems (Salkin et al., 2018).
- *Sensors and actuators*: Embedded systems consist of a control unit with one or more microcontrollers that monitor sensors and actuators for interaction with the real world. That's why sensors and actuators are the core technology for embedded systems. Sensors handle the processing of the signal, while actuators independently check the current state of production and correct if necessary (Jazdi, 2014).
- *Mobile technologies*: Mobile technologies such as high-quality cameras and microphones enable the reception and processing of large amounts of information connected to the Internet and allow the recording and transmission of information (Salkin et al., 2018).

5.3 WHAT IS THE INTERNET OF THINGS (IOT)?

The Internet of Things (IoT) is a distributed information and communication technology that integrates sensors, computing devices, algorithms, and physical objects known as uniquely identifiable objects (Khan, 2019). In IoT technology, anytime, anywhere, any object can be connected to each other over any network (Guillemin and Friess, 2009). The ultimate goal of IoT is to create new applications and services by connecting everything to the internet in every field (Figure 5.3). It is estimated that billions of IoT devices will connect over the internet and interact with each other, creating a huge amount of data (big data) that needs to be processed with cloud and/or fog computing techniques. IoT technologies are expected to improve quality of life, create new job opportunities, and increase the efficiency of factories, buildings, public infrastructure, and services (Khan and Yuce, 2019).

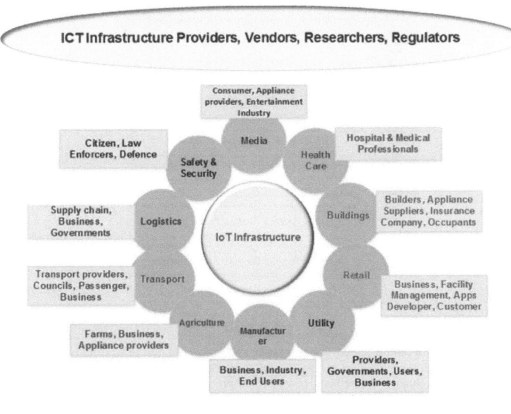

FIGURE 5.3 IoT application areas (Khan, 2019).

While sensors collect data based on their design and architecture, this data can be processed locally on a smart device to a certain extent or sent intermittently to a gateway device for regional processing. This process at the edge of a network is called "edge" or "fog" computing. Data from sensors can also be sent to a cloud-based storage and processing location. Cloud services can be public, private, hybrid, or community cloud depending on the architecture and design of the smart service. The IoT application can also be integrated with a data analytics engine for one-to-one tuning and customization of application output if desired (Seidler, 2019). There is more than one way to describe the IoT, and there is a lot of discussion about it. That's why at the IEEE World Forum on the Internet of Things, a reference model was created by the CISCO company to standardize the concepts and terminology used in IoT and to describe the functionality of IoT at all levels and the interactions between them (Figure 5.4) (CISCO, 2014):

- *Level 1. Physical devices and controllers*: Physical devices and controllers are a wide variety of endpoint devices that send and receive information and are the "Things" in the IoT. IoT devices can be queried or controlled over the internet and are capable of analog-to-digital conversion of signals and data generation as needed. Devices send data intermittently in small units to Level 2 network equipment.
- *Level 2. Connectivity*: The most important function of Level 2 is reliable, timely information transfer. Therefore, the communication and connectivity of devices are kept at this level. At this level, various protocols are implemented, interprotocol translations are made, and switching and routing operations, as well as network security and network analytics operations, are performed.

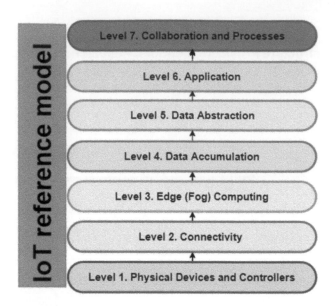

FIGURE 5.4   IoT reference model.

- *Level 3. Edge (fog) computing*: To meet specific IoT service needs, sending data from sensor devices to a central "cloud" for processing and further action can be time-consuming and have high operational overhead. Therefore, at Level 3, calculations such as packet inspection, evaluation of whether data needs to be processed at a higher level, reformatting of data, expansion/decoding of encrypted data, distillation/reduction of data, threshold acquisition, or data evaluation for warning generation are performed. At this stage, information processing is limited and is done on a packet basis. Higher-level processing of information is carried out at Level 4.
- *Level 4. Data accumulation*: Level 4 is the storage level where in-motion, event-driven data from a network is converted into static data for query-based processing by applications in a non-real-time manner as needed. It transforms the database into relational tables through filtering and selective storage. Additionally, this level includes event filtering/sampling, event comparison, event attendance for complex event handling, event-based rule evaluation, event collection, north/south alerts, and event persistence in storage.
- *Level 5. Data abstraction*: This level helps to collect data from multiple devices. It simplifies data access to the application by creating schemas and data views. The basic processes at this level are filtering, selecting, projecting, and reforming data in different formats, normalizing and indexing differences in data form, format, semantics, access protocol, and security of data from different sources.
- *Level 6. Application*: Level 6 is the application level where the intended output is realized by interpreting the information. Software at this level interacts with Level 5 and static data, so it doesn't need to run at network speeds. Applications vary by vertical markets, nature of device data, and business needs. For example, some apps will focus on monitoring device data. Some will focus on controlling devices. Some will merge device and non-device data. Applications can be mobile apps, business intelligence reports, analytics, control apps, etc.
- *Level 7. Collaboration and processes*: At this level, the IoT system and the information it creates are made useful for people and business processes with its communication and collaboration capability. Apps empower people to do the right thing by giving the right data at the right time.

5.4 BIG DATA AND ANALYTICS

Big data describes data that is too large to be managed by traditional databases and processing tools. These big data structures can consist of a combination of structured and unstructured data from various sources such as text, forms, web blogs, comments, video, photos, telemetry, GPS traces, instant messaging chats, and news feed. The problem with these various data structures is that they are very difficult to incorporate or analyze in a traditional structured database. For example, processing large amounts of unstructured data alongside M2M sensor data from thousands or more devices is no easy task. Therefore, to take advantage of industrial IoT (IIoT), data from all sources must be analyzed. Big data types are listed as follows (Gilchrist, 2016):

- *Volume*: The ability to analyze large volumes of data is the main purpose of big data. For example, the larger the repository, the more we can trust its estimates. Analysis of a 500-factor pool is more reliable than a 10-factor pool.
- *Velocity*: The speed with which data enters the system and how long it takes to analyze is related to the speed. Some data, such as M2M sensors, require in-flight or in-memory analysis. Data rate or data flow in an IIoT context requires as much real-time or near-real-time processing and analysis as possible. This restriction places additional pressures on data storage and processing systems.
- *Variety*: Another feature of big data is that it is typically dispersed and comes from various sources such as raw sensor data or web service APIs that do not fit neatly into organized relational structures, thus requiring NoSQL databases. A typical use of big data processing is to extract meaning from unstructured data so that it can be entered as structured data into an application, which requires cleaning up dirty data.
- *Veracity*: The problems with big data arise when we move beyond collecting and storing large amounts of data and analyze data stores using three Vs and think the data is actually correct. Not only can the data be dirty or unreliable, but it can also be completely wrong. For example, let's say you are collecting data from multiple broken sensor sources. If the data is initially wrong, the results will likewise be wrong.
- *Value*: Because not all data is created equal, it is important to decide which data to collect and analyze. In fact, the idea of big data is not to store everything and throw away anything. Data is only valuable if its relationship to business value can be determined. It means nothing to have a big data value set that doesn't produce correlations and trends.
- *Visibility*: Visualizing data is extremely important as it provides a better understanding of trends and correlations. Visualization software facilitates understanding by visualizing data in many forms such as dashboards and spreadsheets or through graphical reports.

Big data analytics is the use of advanced analytical techniques against very large and diverse datasets containing structured, semi-structured, and unstructured data of varying sizes from terabytes to zettabytes from different sources. With the analysis of big data, analysts, researchers, and business users are enabled to make better and faster decisions by using previously unused data. For this purpose, advanced analytical techniques such as text analytics, machine learning, predictive analytics, data mining, statistics, and natural language processing can be used (IBM, 2021). Examples of big data analytics applications include understanding and targeting customer behavior, accepting and optimizing business forms, individual talent, and application improvements, improving healthcare, science, and research, improving machine and device performance, monitoring security and legal performance, and improving traffic flows (Solanki et al., 2019).

Big data has prominent applications in agriculture. For this reason, it is also called big data-driven farming. Big data supports better quality and more informed decisions for both production and business. Some of the contributions of big data to agriculture are listed here (Ahmad and Nabi, 2021):

- Yield/harmful disease/weather forecasts;
- Optimal farming decisions;
- Plant suggestions;
- Intercropping suggestions;
- Market price and profitability analysis;
- Policy recommendations;
- Business/equipment/risk management;
- Efficient agricultural practices;
- Selection of suitable hybrids.

5.5 5G AND ITS USE IN AGRICULTURE

The next generation of cellular technology is focused not only on smartphone internet applications but also on real-time applications related to device-to-device communication, M2M communication, and IIoT (Bangerter et al., 2014; Tehrani et al., 2014; Anonymous, 2016). To meet IIoT requirements, 5G networks are designed to provide a data rate in the 10 Gbps range with an end-to-end latency of less than 1 ms. True 4G networks provide a typical download speed of around 20 Mbps, depending on network types, with a maximum theoretical throughput of 150 Mbps, but these are the highest data rates and do not account for network latency and jitter. 5G networks, on the other hand, allow IoT nodes to be connected directly to 5G base stations, eliminating the need for gateways and other network interfaces that negatively affect network planning and reduce performance in terms of latency and availability. Another important advantage of 5G networks is the ability to interface between a large network of wired or wirelessly connected machines, sensors, and actuators using specific capillary networks (Pereira and Viegas, 2019).

5G exceeds existing 4G and 4G LTE standards by up to 100 times in download and upload speeds. This means that while it would normally take six minutes to download a two-hour movie on 4G, the same download would take less than four seconds on 5G. 5G can connect one million devices per square kilometer and continues to be supported when devices are traveling at very high speeds (about 500 km/h). Therefore, the 5G mobile network is well suited to support smart agriculture by providing wide coverage, low energy consumption, low-cost devices, and high spectrum efficiency.

A fast and reliable internet connection is required for agricultural IoT devices to work. The current generation of mobile networks fails due to poor connectivity in rural areas, and even in areas where high-speed connectivity is available, failures occur due to high demand. A recent report reported that even in the UK, almost 80% of rural areas are outside of the 4G range. In most countries, the current level of network availability in rural areas is insufficient (USDA, 2019). 5G technology makes IoT devices and farm applications much easier to monitor and manage. The areas where the 5G mobile network can be used in agriculture are shown in Figure 5.5. These areas can be counted as IoT sensors, drones, robots, smart devices, real-time monitoring, virtual consultation and predictive maintenance, augmented reality and virtual reality, artificial intelligence–supported robots, data analytics and cloud computing, etc. All these applications take advantage of the most important features of 5G, such as device density, ultra-low latency, ultra-reliability, and security. This makes it possible to work together seamlessly to maximize productivity in agriculture and drastically reduce the resulting cost (Tang et al., 2021).

5.6 IOT APPLICATION EXAMPLES IN AGRICULTURE

The use of IoT in agriculture has become widespread and has begun to replace some traditional methods. The availability of a wide variety of affordable IoT-enabled sensors, the advancement

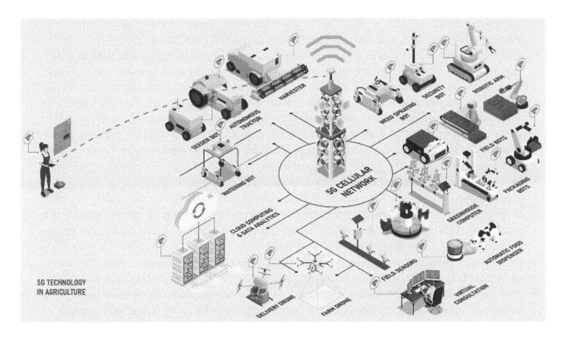

FIGURE 5.5   Use of 5G mobile networks in agriculture (Tang et al., 2021).

of wireless technology, and the economic and easy accessibility of cloud computing have made it possible to make many agricultural applications. The data obtained with the sensors is transferred to the cloud by wireless communication and this information can be accessed from anywhere in the world at any time. In this way, farmers, producers, and agricultural business managers can follow all the activities carried out and take the necessary decisions instantly. In addition, the use of IoT in agriculture is not limited to producers and businesses. In addition to manufacturers and businesses, it serves a wide audience working in the agriculture and food sector such as suppliers, government institutions, cooperatives, vendors, and services. Agricultural productivity can be increased by evaluating data from a wide variety of sensors (climatic data, soil temperature, soil moisture, salinity, pH, etc.) from agricultural fields.

With the smartness of agricultural machinery, maintenance and repair processes of agricultural machinery such as tractors and combines with high purchase prices can be monitored remotely, and defective parts are replaced without losing time by directing the services. For example, the fault of a combine harvester that fails while working in the field can be determined by remote connection. If the detected malfunction is caused by software problems, it is solved by updating the program via remote connection. If it is caused by defective or faulty equipment, a new part of the defective part is supplied by the relevant service and maintenance, or repair operations are carried out without delay. During these processes, tractors or agricultural machines do not experience any loss of operation due to malfunctions or problems. Situations such as the fact that the operators working in the field come to the machine for repair and maintenance without even being aware of these situations have started to be experienced frequently in daily life. In addition, by using the diagnostic test system found in some advanced tractor models, fault codes and error messages received from the electronic brain control unit can be addressed immediately and targeted directly to the problem. In this way, with the efficiency achieved, there are no situations such as lost time for service, unnecessary trips to the service, not finding the problem, or replacing the wrong part.

Following are various applications of IoT-enabled farming systems (Nayak et al., 2020):

- *Monitoring activities*: The most critical task for the farmer today is to monitor, model, and manage environmental parameters that are beyond the estimation of engineers and scientists. Agricultural activities require constant monitoring to get accurate and reliable advice from agricultural engineers. There are many parameters to monitor such as irrigation and water quality, weather, soil and cropland productivity, and farm practices.

- *Crop management*: The agricultural process not only consists of operations such as planting, irrigation, fertilization, spraying, and harvesting from the field but also requires a lot of analytics at every stage of the operations. For example, soil moisture and humid weather conditions are necessary for the germination of seeds. For this reason, a weather forecasting system is required during the germination phase. Similarly, fertilizer is necessary to grow a healthy plant. The amount of fertilizer or pesticide to be applied requires an in-depth analysis for better crop management. Farmers can make decisions by keeping some historical data or with the help of IoT-based management techniques. In real time, information can be accessed or monitored remotely via mobile phones, URLs, etc.
- *Agricultural machinery*: The use of IoT in agricultural machinery can greatly increase crop productivity. Unmanned aerial vehicles and robots operating in autopilot mode are examples of direct applications of IoT in agricultural machinery. Farm equipment manufacturer CLAAS has implemented IoT in some of its equipment. Precision Hawk's UAV sensors can provide information such as wind speed and air pressure. With the use of driverless tractors in tillage and planting applications, it can save farmers time and effort.
- *Precision agriculture*: Precision agriculture deals with the evolution of traditional technology to miniaturize the electronics industry. Precision farming is the management of agriculture-specific information to improve crop production. These include meteorological factors (temperature, humidity, sunlight, wind, and water), smart irrigation details, etc., which are very important for the farmer in planning their activities and making them profitable. The evolution of IoT and embedded systems has led to the development of economically viable systems that are easy to install with minimum power and their use in precision agriculture applications.
- *Disease and pest control*: Farmers previously had difficulties in crop monitoring, disease, and pest prediction. With the use of the IoT, early disease detection has become possible, which may facilitate the taking of some preventive measures. For example, continuous crop monitoring activity can be assigned to an IP camera, and recorded video of the crop can be transmitted daily to a smartphone. These data can be combined with expert advice to select a suitable pesticide. Thus, farmers do not need to visit agronomists in rural areas to get advice on pesticides.
- *Greenhouse applications*: Wireless sensor networks are used in greenhouses to detect environmental conditions (temperature, light, pressure, humidity) and transmit data to the web or a mobile application. This way, the farmers monitor the activity in the greenhouse from afar, and it helps them relax. Farmers can receive the current status of the greenhouse and the forecasts made by data mining instantly and accurately with the notifications coming to their mobile applications.

5.6.1 Digital Agriculture Platforms

With the Digital Agriculture Market (DİTAP) implemented by the Turkish Ministry of Agriculture and Forestry, agricultural products can be supplied directly between the buyer and the seller, without intermediaries. In this way, farmers can find a market for their products, and consumers and sellers can supply the quality product they are looking for. First of all, it is necessary to register to the system at ditap.etarim.gov.tr and sup.etarim.gov.tr and log in to the system. In order for the buyer companies to register to the system, ESBİS or MERSIS records must be registered, while the sellers must be registered to the ÇKS, HBS, AKS, KBS, and SBS systems. Buyers and sellers can create requests and offers or place direct sales advertisements through the special pages reserved for them in the system. The demands created by the buyers can be easily seen by the producers in the region and bids can be made for these demands. More than one manufacturer can receive offers for the created demand and the buyer can evaluate the offer he/she wants. A contract is signed between the buyer and the seller for sales transactions. All details such as quantity, price, production type, cash/ in-kind advance, production place, order date, delivery date, shipping, packaging, etc. are determined in advance in the contract and both parties can trade safely (DİTAP, 2021).

The more real-time data on agricultural production, the more realistic it is to make production decisions. In order to decide on the necessary applications in agricultural production, there is a need for a lot of information such as field area, crop or animal, soil type, fertilizer, yield, irrigation information, machinery, equipment and system information, disease info, feed info, weather conditions, etc. IoT and cloud computing offer the opportunity to instantly measure and process the information necessary for more efficient agricultural applications.

Triantafyllou et al. (2019) developed a smart agriculture monitoring system. The system basically consists of sensing agricultural parameters, transferring the sensed data, evaluating the data, applying the application decision, and delivering the results to the grower through an application. The DIAS platform consists of seven architectural layers as shown in Figure 5.6. At each layer, the necessary technologies are used to ensure the efficient performance and reliable operations of the overall system. The system consists of the following seven layers: The sensor layer, the link layer, the encapsulation layer, the middleware layer, the configuration layer, the management layer, and the application layer. The development of the system and the control of its operation were carried out in an agricultural enterprise that produces saffron.

5.6.2 TELEMETRY SYSTEMS

Telemetry is defined as sensing and measuring information in various remote environments and then transmitting this information to a central or main location. Telemetry is used to monitor and control a transaction at a remote site. In the remote site, a sensor (or sensors) is typically the data source. The output of the sensor(s) is converted into digital data by a small computing device or remote terminal unit (RTU). An RTU is an interface to a modem device that converts digital data into an analog signal that can be transmitted over the air. The radio transmitter then transmits the signal to the main radio receiver. Then the process is reversed. The modem takes the received analog signal and converts it back into a digital form that can be processed by data recovery equipment. In a typical application, the host requests data from the remote site(s). The base station transmits a request to the remote unit instructing it to send its data. The base station returns to receive mode and waits for transmission from the remote site. After sending the remote-control data, it returns to a receive mode pending instructions from the base station. When the base station receives the remote site information, it can send additional instructions to that site or continue to request data from the next remote site. This polling process continues until all remote controls in the system send their data (Kumar, 2004).

Remote monitoring of any mobile machine requires radio technology, internet technology, protocols, and applications. Mobile cellular networks provide both radio and communications for internet services while IoT protocols are under development. A protocol used in industrial automation to connect machine automation to production process control is open platform communications (OPC). The latest version of this technology is OPC unified architecture (OPC UA). Oksanen et al. (2016) examined the suitability of using OPC UA telemetry application on a combine with a yield monitoring system. The study includes both the server-side system on the combine and the client for remote monitoring, and the system's latency was measured. As a result of the study, it was reported that the perceived end-to-end delay over the internet connection was less than 250 ms and this result was sufficient for most telemetry applications in agriculture. The researchers built a laptop (Panasonic Toughbook CF-19) onboard with a CAN bus adapter (NI USB-8473) to integrate data from the CAN bus and RS-232. The software was developed using NI LabVIEW and the main functions of the software are reading CAN bus data, reading RS-232 data, saving data for internal storage, and writing combined data to the CAN bus. In addition to these functions, the developed

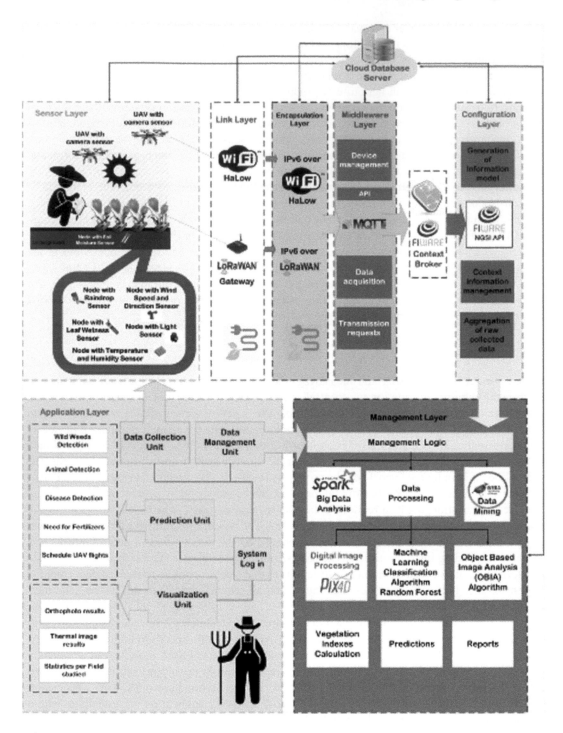

FIGURE 5.6   The DIAS architecture (Triantafyllou et al., 2019).

system provides an interface for the user to enter metadata and monitor the status of the system (Oksanen et al., 2016).

Modern technology and telecommunication systems enable the acquisition of sensitive land data that improves agricultural operations management. Sarri et al. (2017) developed the prototype telemetry system shown in Figure 5.7 for grape producers to monitor the performance of

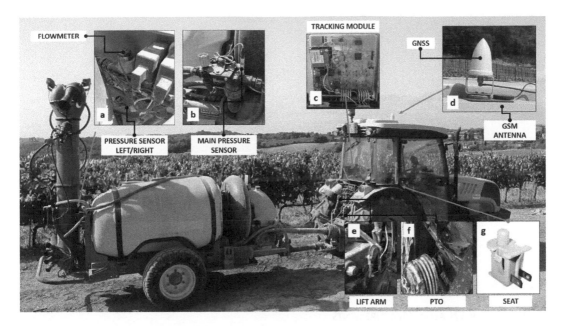

FIGURE 5.7 Prototype telemetry system architecture. (a) and (b) sensors on the sprayer, (c) onboard monitoring module, (d) communication devices placed on the tractor roof, (e), (f), and (g) sensors mounted on the tractor to monitor the driving conditions (Sarri et al., 2017).

their spraying operations in real time and obtain useful data. The prototype telemetry system consists of a monitoring module for data collection, a server for remote monitoring and data storage, a GSM/GPRS/GPS module for data transmission, and a GNSS for localization of the sprayer. Data available are latitude, longitude, sprayer speed, three-point linkage condition, power take-off, operating side of sprayhead (left and right), operator presence, pressure values at centrifugal pump and sprayhead, and flow rate. Developed by researchers, this telemetry system has been tested in different vineyards to evaluate whether all components are working properly. The results showed that the spray pressure and flow rate measured by the sensors of the telemetry system were like the theoretical values defined for the regulation of the sprayer. The estimated value of the application rate, which was a number derived from the provided forward speed, was also like the theoretical value, indicating that the forward speed recorded by the telemetry system was correct (Sarri et al., 2017).

5.6.3 Fertilizer Information System

Lavanya et al. (2020) developed an IoT-based system by designing a new nitrogen-phosphorus-potassium (NPK) sensor with light-dependent resistor (LDR) and light-emitting diode (LED) (Figure 5.8). In the system, it is stated that the colorimetric principle is used to monitor and analyze the nutrients in the soil, the data detected by the NPK sensor is sent to the Google cloud database for rapid evaluation, and the fuzzy logic method is applied to detect the nutrient deficiency from the detected data. In the study, the exact value of each detected data was divided into five fuzzy values during blurring: Very low, low, medium, high, and very high. The Mamdani extraction procedure was used to draw a conclusion about the deficiency of N, P, and K present in selected soil tests. According to the results, a warning message was sent to the farmer about the amount of fertilizer to be used at regular intervals. The proposed hardware prototype and software embedded in the microcontroller are reported to be developed on Raspberry Pi 3 using Python, tested in three different soils, and cause linear variation according to the concentration of the soil solution.

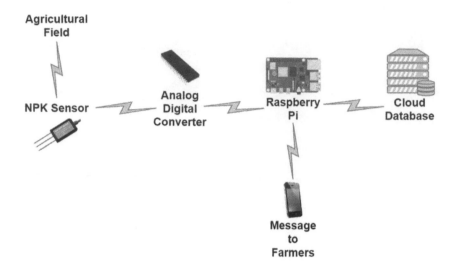

FIGURE 5.8   Block diagram of IoT-based fertilizer information system.

5.6.4 WATER QUALITY MEASUREMENTS

Water is indispensable for living things. Water plays an important role in all metabolic activity. It can be used as drinking water after being filtered in the treatment plant. Water supply to homes and industry is mostly provided by underground pipeline network systems. If there is any damage/leakage in the pipeline system, the water mixes with the pollution and becomes polluted. High dissolved oxygen (DO) value accelerates corrosion in water pipes. Due to this corrosion, drinking water quality is affected, and DO levels in the supplied water need to be measured. Kothari et al. (2021) developed a temperature sensor in a drinking water quality measurement system. In this system, there is a total dissolved solids (TDS) sensor that gives the value of the solid body dissolved in the water, a pH sensor to monitor the content of acidic substances in the water, and a DO sensor used to monitor the dissolved oxygen. The developed system has been tested sequentially in tube well, tap, rain, and reverse osmosis treated water. It has been reported that Arduino Mega2560 is used as a microcontroller and the data from the sensors are transmitted to the database server in Thingspeak using a wireless connection with the help of the GSM SIM 900 module, and this data can be viewed in real time using an Android IoT application connected to the server (Figure 5.9).

5.6.5 SMART MILK-MONITORING PLATFORM

Globalization has facilitated global trade with agricultural outputs. Farmers and breeders can stand out by informing their consumers about high-quality products, the origin of the product, and the processes it goes through in the value chain until it reaches the retail point. For this purpose, technologies such as the IoT, blockchain, and distributed ledger technologies (DLT) that offer tracking and traceability features, can be applied in the agri-food industry. Alonso et al. (2020) stated that by using IoT, edge computing (EC), and DLT in the dairy industry, it is possible to optimize the processes of the producers and ensure product safety and quality by providing source monitoring and traceability in the value chain. Thus, it is possible to make the process transparent and safe, provide detailed information to the consumer about the final process, and guarantee its quality to the consumers. In the study, a platform has been designed for real-time monitoring of the status of dairy cattle and feed grains for the application of IoT, EC, artificial intelligence, and blockchain techniques in smart farming environments through the new global edge computing architecture (Figure 5.10). The edge nodes of the global edge computing architecture filter and preprocess data

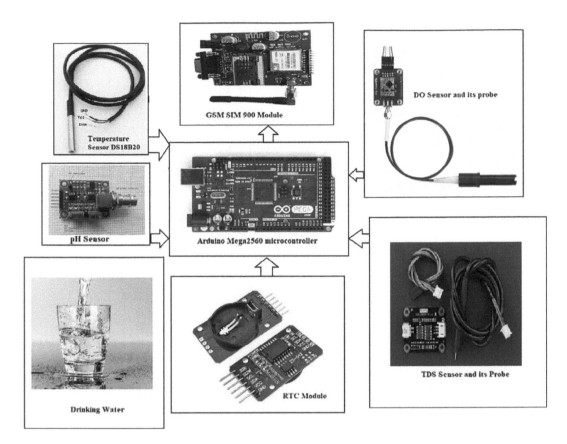

FIGURE 5.9 Block diagram for an IoT sensor-based drinking water quality measurement system (Kothari et al., 2021).

from devices in the IoT layer. They also ensure that duplicate values are discarded due to the retransmission of frames from physical sublayers (ZigBee, wi-fi) to the IoT layer. They can also perform averages and regression data analysis happening on the same edge. In both cases, the amount of data transmitted to the cloud and the cost of transmission are reduced, reducing data traffic costs, as well as the need for computation and storage in the cloud. In addition, it is stated that thanks to the edge elements, users can access the data accumulated from the IoT sensors of the last 30 days, as well as the averages, trends, and alert levels detected in the data models, even if the cloud connection is disconnected.

5.7 AGRICULTURE 5.0

With the continuous development of technology, new hardware and software are constantly emerging. Technology offers new solutions to meet the needs of people and societies. The performance of these new solutions in a wide variety of fields, creating significant differences in concepts compared to previous solutions, opened new eras such as Industry 5.0 and Agriculture 5.0. With Industry 5.0 and Agriculture 5.0, it is possible to develop personal solutions instead of mass production, which is not available in Industry 4.0 and Agriculture 4.0. This new era has just begun and studies have begun today to enable these applications to be made in the future.

One of the most important differences that emerged in the Industry 5.0 and Agriculture 5.0 period is the automation of machines, robots, or systems that will increase human–machine cooperation. The feature of being smart includes the machine and robot being aware of their surroundings during

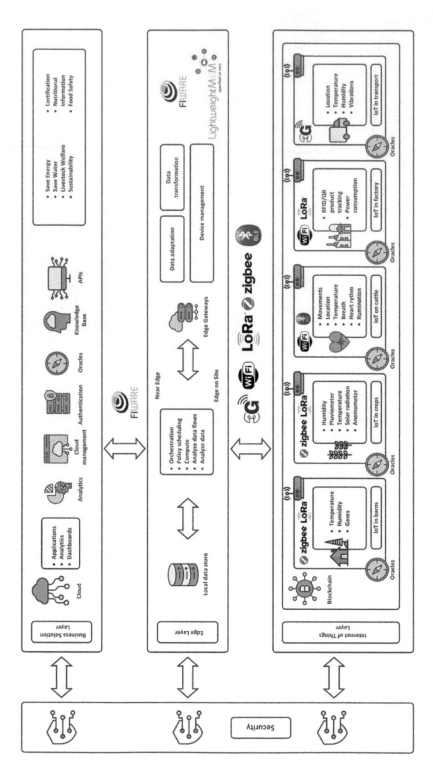

FIGURE 5.10 Diagram of smart milk-monitoring platform (Alonso et al., 2020).

operation and making their own decisions about changes such as changing the movement style or speed according to predetermined situations and applying the decision itself. The autonomous feature is to provide machines and robots with capabilities such as sensing, evaluation, decision-making, control, and fault detection for safe operation and working without human intervention.

With the use of agricultural robots and smart agricultural machinery in agricultural activities, less manpower is needed, as well as increasing productivity and product quality in agricultural production and reducing production costs. However, environmental conditions such as humidity, temperature, and abrasive factors in agricultural areas and technical difficulties such as communication problems at long distances are some of the obstacles to the application of agricultural robots and smart machinery. Agricultural robots can be designed specifically for the application area or conventional agricultural machinery can be turned into autonomous mobile robots by adding systems such as automated steering. In addition, robotic systems that can be placed on mobile agricultural vehicles that can perform operations such as harvesting, spraying, and pest control precisely appear as agricultural robots (Özgüven and Közkurt, 2021).

Industrial revolutions have always produced a breakthrough in the field of agriculture. The paradigm shift toward the implementation of precision farming, digital farming, smart farming, and finally "smart-intelligent-precision farming" has largely succeeded in achieving the goals. These have revolutionized meeting this growing demand for food and preparing for the future. Agriculture has been transformed by technologies at every stage from the past to the present. With Agriculture 5.0, agriculture is completely technology driven. Agriculture 5.0 features data-driven Agriculture 4.0 applications with a boost from robotics and techniques such as artificial intelligence, machine learning, and deep learning. Agriculture 5.0 is still in development, but with the advancement of AI, it will certainly accelerate and become more modern for sustainable agriculture in the future (Ahmad and Nabi, 2021).

Digital solutions for farmers are combined with robotics and artificial intelligence to launch the idea of Agriculture 5.0 in the near future. Thus, agricultural management systems can process farm data in such a way that the results are organized toward customized solutions for each farm. However, to get the most out of Agriculture 5.0 users, ideally, young farmers willing to learn and apply modern technologies to agriculture need to be comprehensively trained and refreshed for future generations. Transitioning to Agriculture 5.0 is on the agenda of most major farm equipment manufacturers for the next decade. Therefore, if agricultural robots are accepted as the next smarter generation of agricultural machines, field equipment manufacturers will play a key role in this move (Saiz-Rubio and Rovira-Más, 2020).

REFERENCES

Ahmad, L. and Nabi, F., 2021. *Agriculture 5.0: Artificial Intelligence, IoT and Machine Learning.* Taylor & Francis Group, LLC, Boca Raton. ISBN: 978-0-367-64608-0.

Alonso, R.S., Sittón-Candanedo, I., García, Ó., Prieto, J. and Rodríguez-González, S., 2020. An Intelligent Edge-Iot Platform for Monitoring Livestock and Crops in a Dairy Farming Scenario. *Ad Hoc Networks*, 98, 102047. https://doi.org/10.1016/j.adhoc.2019.102047.

Anonymous, 2016. Institute for Communication Systems 5G Innovation Centre – University of Surrey. The Flat Distributed Cloud (FDC) 5G Architecture Revolution. Available from https://www.surrey.ac.uk/sites/default/files/5G-Network-Architecture-Whitepaper-(Jan-2016).pdf [Accessed 23.09.2021].

Bagheri, B., Yang, S., Kao, H. and Lee, J., 2015. Cyber-Physical Systems Architecture for Self-Aware Machines in Industry 4.0 Environment, *IFAC-Pap Online*, 48–3, 1622–1627.

Bangerter, B., Talwar, S., Arefi, R. and Stewart, K., 2014. Networks and Devices for the 5G Era. *IEEE Communications Magazine*, 52(2), 90–96.

Bungartz, H.J., Zimmer, S., Buchholz, M. and Pflüger, D., 2014. *Modeling and Simulation: An Application-Oriented Introduction.* Springer Verlag, Berlin Heidelberg.

CISCO, 2014. The Internet of Things Reference Model. http://cdn.iotwf.com/resources/71/IoT_Reference _Model_White_Paper_June_4_2014.pdf. [Accessed 22.09.2021].

DİTAP, 2021. https://ditap.gov.tr/ditap-nedir.html. [Accessed 10.05.2021]. (Turkish).

Gaub, H., 2015. Customization of Mass-Produced Parts by Combining Injection Molding and Additive Manufacturing with Industry 4.0 Technologies, Reinforced Plastics. https://doi.org/10.1016/j.repl.2015 .09.004.

Gilchrist, A., 2016. *Industry 4.0: The Industrial Internet of Things.* Apress, Thailand. ISBN: 978-1-4842-2047-4.

Guillemin, P. and Friess, P., 2009. Internet of Things Strategic Research Roadmap, The Cluster of European Research Projects, Tech. Rep., September 2009. http://www.internet-of-things-research.eu/pdf/IoT Cluster Strategic Research Agenda 2009.pdf. [Accessed 01.04.2016].

IBM, 2021. https://www.ibm.com/tr-tr/analytics/hadoop/big-data-analytics. [Accessed 23.09.2021].

IERC, 2011. Internet of Things: Strategic Research Roadmap. http://www.internet-of-things-research.eu/ about_iot.htm. [Accessed 26.04.2017].

Jazdi, N., 2014. Cyber Physical Systems in the Context of Industry 4.0. 2014 IEEE international conference on automation, quality and testing, robotics. https://doi.org/10.1109/AQTR.2014.6857843.

Kagermann, H., Wahlster, W. and Helbig, J., 2013. Recommendations for Implementing the Strategic Initiative Industrie 4.0. Final Report of the Industrie 4.0 Working Group, Acatech-National Academy of Science and Engineering, Frankfurt.

Khan, J.Y., 2019. Introduction. *Internet of Things (IoT): Systems and Applications.* (Editors: Khan, J.Y. and Yuce, M.R.) Jenny Stanford Publishing Pte. Ltd, Singapore.1–24. ISBN: 978-0-429-39908-4.

Khan, J.Y. and Yuce, M.R., 2019. Preface. *Internet of Things (IoT): Systems and Applications.* (Editors: Khan, J.Y. and Yuce, M.R.) Jenny Stanford Publishing Pte. Ltd, Singapore.xiii–xvi. ISBN: 978-0-429-39908-4.

Kothari, N., Shreemali, J., Chakrabarti, P. and Poddar, S., 2021. Design and Implementation of IoT Sensor Based Drinking Water Quality Measurement System. *Materials Today: Proceedings.* https://doi.org/10 .1016/j.matpr.2020.12.1142.

Kumar, J., 2004. http://www.dr-joyanta-kumar-roy.com/study_meterial/Telemetry%20systems/Telemetry %20basics.pdf. [Accessed 10.04.2021].

Lavanya, G., Rani, C. and Ganeshkumar, P., 2020. An Automated Low Cost IoT Based Fertilizer Intimation System for Smart Agriculture. *Sustainable Computing: Informatics and Systems*, 28, 100300. https:// doi.org/10.1016/j.suscom.2019.01.002.

Mell, P. and Grance, T., 2011. The NIST Definition of Cloud Computing. Special Publication 800–145. USA.

Nayak, P., Kavitha, K. and Rao, C.M., 2020. IoT-Enabled Agricultural System Applications, Challenges and Security Issues. *IoT and Analytics for Agriculture.* (Editors: Prasant Kumar Pattnaik, P.K., Kumar, R., Pal, S. and Panda, S.N.) Springer Nature Singapore Pte Ltd, Singapore. 139–164. ISBN: 978-981-13-9177-4.

Obitko, M. and Jirkovský, V., 2015. Big Data Semantics in Industry 4.0. (Editors: Mařík, V. et al.) HoloMAS 2015, LNAI 9266, pp. 217–229.

Oksanen, T., Linkolehto, R. and Seilonen, I., 2016. Adapting an Industrial Automation Protocol to Remote Monitoring of Mobile Agricultural Machinery: A Combine Harvester with IoT. *IFAC-PapersOnLine*, 49(16), 127–131. https://doi.org/10.1016/j.ifacol.2016.10.024.

Özgüven, M.M. and Közkurt, C., 2021. Agricultural Robots and Smart Agricultural Machinery. International Symposium of Scientific Research and Innovative Studies. 22–25 February 2021. Bandırma-Turkiye, pp. 81–85. ISBN: 978-625-44365-8-1.

Paelke, V., 2014. Augmented Reality in the Smart Factory Supporting Workers in an Industry 4.0. Environment. IEEE Emerging Technology and Factory Automation (ETFA).

Pereira, J.M.D. and Viegas, V., 2019. Industrial Internet of the Things. *Sensors in the Age of the Internet of Things Technologies and Applications.* (Editors: Postolache, O.A., Sazonov, E. and Mukhopadhyay, S.C.) The Institution of Engineering and Technology, London, 239–274.

Saiz-Rubio, V. and Rovira-Más, F., 2020. From Smart Farming towards Agriculture 5.0: A Review on Crop Data Management. *Agronomy*, 10(2), 207. https://doi.org/10.3390/agronomy10020207.

Salkin, C., Oner, M., Ustundag, A. and Cevikcan, E., 2018. A Conceptual Framework for Industry 4.0. *Industry 4.0: Managing the Digital Transformation.* (Editors: Ustundag, A. and Cevikcan, E.) Springer International Publishing, Cham, 3–24. ISBN: 978-3-319-57869-9.

Sarri, D., Martelloni, L. and Vieri, M., 2017. Development of a Prototype of Telemetry System for Monitoring the Spraying Operation in Vineyards. *Computers and Electronics in Agriculture*, 142(Part A), 248–259. https://doi.org/10.1016/j.compag.2017.09.018.

Schwab, K., 2017. *The Fourth Industrial Revolution.* World Economic Forum Crown Publishing Group, New York. ISBN: 9781524758868.

Seidler, H., 2019. Chapter 2 IoT Architecture. *Internet of Things, for Things, and by Things.* (Editor: Chaudhuri, A.) CRC Press Taylor & Francis Group LLC, Boca Raton. 20–51. ISBN: 978-1-138-71044-3.

Sendler, U., 2018. Part II The Basics. *The Internet of Things, Industrie 4.0 Unleashed*. (Editor: Sendler, U.) Springer-Verlag GmbH, Berlin. 15–36. ISBN: 978-3-662-54904-9.

Solanki, V.K., Makkar, S., Kumar, R. and Chatterjee, J.M., 2019. Theoretical Analysis of Big Data for Smart Scenarios. *Internet of Things and Big Data Analytics for Smart Generation*. (Editors: Balas, V.E., Solanki, V.K., Kumar, R. and Khari, M.) Springer Nature, Cham. 1–12. ISBN: 978-3-030-04202-8.

Tang, Y., Dananjayan, S., Hou, C., Guo, Q., Luo, S. and He, Y., 2021. A Survey on the 5G Network and its Impact on Agriculture: Challenges and Opportunities. *Computers and Electronics in Agriculture*, 180(2021), 105895. https://doi.org/10.1016/j.compag.2020.105895.

Tehrani, M., Uysal, M. and Yanikomeroglu, H., 2014. Device-to-Device Communication in 5G Cellular Networks: Challenges, Solutions and Future Directions. *IEEE Communications Magazine*, 52(5), 86–92.

The German Plattform Industrie 4.0, 2015. Result Report of the Platform Industrie 4.0, Editorial Board of BITKOM e. V., VDMA e. V., ZVEI e. V., p. 8.

Triantafyllou, A., Sarigiannidis, P. and Bibi, S., 2019. Precision Agriculture: A Remote Sensing Monitoring System Architecture. *Information*, 10, 348. https://doi.org/10.3390/info10110348.

Uckelmann, D., 2008. A Definition Approach to Smart Logistics. *Next Generation Teletraffic and Wired/Wireless Advanced Networking*. NEW2AN 2008. Lecture Notes in Computer Science, vol. 5174. (Editors: Balandin, S., Moltchanov, D. and Koucheryavy, Y.) Springer, Berlin. 273–284.

USDA, 2019. A Case for Rural Broadband. No. April 2019.

Wittenberg, C., 2015. Cause the Trend Industry 4.0 in the Automated Industry to New Requirements on User Interface. *Human-Computer Interaction, Part III, HCII 2015, LNCS*, vol. 9171. (Editor: Kurosu, M.), Springer, Cham, pp. 238–245.

Zhang, Y., Ren, S., Liu, Y. and Si, S., 2016. Big Data Analytics Architecture for Cleaner Manufacturing and Maintenance Processes of Complex Products. *Journal of Cleaner Production*, 142(Part 2), 1–16.

6 Image Processing and Machine Vision in Agriculture

6.1 WHAT IS IMAGE PROCESSING?

An image processing technique is a method used to turn the image in a photo or video frame obtained with a camera, scanner, or sensors into digital format after recording and to extract some useful information from this digital data with the aid of a set of algorithms. In this technique, images are rearranged with various processes and meaningful results are obtained finally through these processes. During these processes, it is attempted to obtain the descriptive parameters that represent the important data in the image. In this way, defining and separating the features to be measured, correcting image defects, enhancing the visibility of certain features, and thresholding them in the background are performed (Ozguven and Yanar, 2022).

Image processing involves using a computer to change the nature of an image. The first of two modifications is to improve photographic information for human interpretation, and the second is to make it more suitable for autonomous machine perception. These two items represent two separate but equally important aspects of image processing. A procedure that "shows better" an image in the first modification may be the worst procedure to satisfy the second. While humans like their images to be sharp, clear, and detailed, machines prefer their images to be simple and uncluttered (McAndrew, 2016).

An important feature underlying the design of image processing systems is that the tests and trials that are necessary before reaching an acceptable solution can be easily performed. This feature ensures that the ability to formulate approaches and rapidly prototype candidate solutions plays an important role in reducing the cost and time required to arrive at a viable system implementation (Gonzalez et al., 2009).

A general-purpose image acquisition and processing system typically consists of four key components (Jähne, 2005):

1. An image acquisition system. In the simplest case, this could be a CCD camera, flatbed scanner, or video recorder.
2. A device known as a frame grabber for converting the electrical signal of the image acquisition system into a digital image that can be stored.
3. A personal computer or workstation that provides the processing power.
4. Image processing software that provides tools to process and analyze images.

A digital image is a representation of a two-dimensional image that uses a finite number of points, often called picture elements or pixels. Each pixel is represented by one or more numerical values. For monochrome (grayscale) images, a single value representing the intensity of the pixel (usually in the range (0, 255)) is sufficient. Color images usually require three values (for example, representing the amount of red (R), green (G), and blue (B)) (Marques, 2011). Each pixel represents not just a point in the image, but a rectangular region, which is the base cell of the grid. The value associated with the pixel should appropriately represent the average luminance in the corresponding cell. Figure 6.1 shows the same image represented by a different number of pixels. With large pixel sizes, not only is the spatial resolution poor, but gray value discontinuities at pixel edges appear as annoying artifacts that distract us from the content of the image. If the image contains enough pixels, it appears to be continuous. As the pixels get smaller, the effect becomes less pronounced to the point

FIGURE 6.1 Representation of the same image with a different pixels (Yang et al., 2021).

78	76	85	115	118	115	78	61	67
77	73	85	112	122	96	64	59	58
62	71	75	101	90	67	52	55	55
48	64	79	78	76	68	56	56	56
52	70	84	80	76	70	72	78	76
59	66	86	99	85	87	84	89	101
68	66	80	95	104	92	90	91	86
71	71	77	84	102	97	91	82	76
74	71	74	81	81	93	96	89	87
75	75	71	72	77	79	101	112	96
68	78	77	73	78	86	100	110	109
77	78	75	78	83	90	93	109	116

FIGURE 6.2 A digital image and its numerical representation.

where we get the impression of a spatially continuous image. This happens when pixels become smaller than the spatial resolution of our visual system (Jähne, 2005).

An image is represented by a rectangular array of integers. Image dimensions and the number of gray levels are usually integer powers of two. The number in each pixel represents the brightness or darkness (often called intensity) of the image at that point. For example, Figure 6.2 shows an 8 × 8 digital image with one byte per pixel (i.e., 8 bits = 256 gray levels). The quality of an image is highly dependent on the number of samples and gray levels; the more of these two, the better the quality of an image. However, this will also result in a large amount of storage, since an image's storage space is the product of an image's dimensions and the number of bits needed to store the gray levels. A lower resolution image may cause a checkerboard effect or graininess. In an image with high graininess, the image appears to be composed of dots. A 1,024 × 1,024 image may not show much degradation when scaled down to 512 × 512 but may show noticeable graininess when reduced to 256 × 256 and then rescaled back to 1,024 × 1,024 (Shih, 2010).

6.2   IMAGE PROCESSING OPERATIONS

Image processing includes a number of techniques and algorithms. The most representative image processing operations are (Marques, 2011; Tyagi, 2018):

- *Binarization*: For many image processing tasks, it is often necessary to convert a color image or a grayscale image to a two-level (black and white) image to simplify and speed up processing. This process is done to divide the image into two as the foreground and background.
- *Smoothing*: A technique used to blur or soften the details of objects in an image.

- *Sharpening*: Sharpening techniques are used to enable people to view the edges and fine details of objects in an image.
- *Noise removal*: Before image processing, the amount of noise in the images is reduced by using noise removal filters. Depending on the type of noise, different noise removal techniques are used.
- *Deblurring*: An image can appear blurry for many reasons, from misfocusing the lens to insufficient shutter speed for a fast-moving subject. Before image processing, blur is removed by various algorithms.
- *Blurring*: Sometimes it is necessary to blur an image to minimize the importance of texture and fine details in an image, for example when objects can be better recognized by their shape.
- *Edge extraction*: Removing edges from an image is a basic pre-processing step used to separate objects from each other before determining their content.
- *Segmentation and labeling*: The task of segmenting and labeling objects within an image is a prerequisite for most object recognition and classification processes. After objects of interest have been segmented and labeled, their respective properties can be extracted and used to classify, compare, cluster, or recognize said objects.

6.3 STAGES OF IMAGE PROCESSING

Although the areas where the image processing technique is used are different, there are some basic steps performed in many applications. As seen in Figure 6.3, firstly, the data is transferred to the program with image acquisition. It is prepared in a way that can process the photo more easily and quickly with pre-processing. Segmentation separates objects and backgrounds. With feature extraction, the object is represented according to the features (important features), thus reducing the complexity of the representation. By finding connected components and extracting features, analysis calculations are made easier and faster. With the classification, the type of the object is determined, or it is decided whether the object is good or not (MathWorks, 2008).

- *Image acquisition*: Image processing starts with image acquisition. The hardware and software used for acquisition can look quite different depending on the type of image source and interface. This can range from video cameras with suitable frame grabbers, to digital cameras with their own serial or SCSI interfaces, to slide or flatbed scanners with a SCSI or USB interface. Real-world 3D objects are converted into digital images. CCD, CID, CMOS, infrared cameras, ultrasounds, X-rays and radiography, magnetic resonance imaging, satellite imagers, and scanners are used for image acquisition (Qidwai and Chen, 2009).
- *Pre-processing*: The first steps of image processing can include several different processes and are known as image pre-processing. After capturing a digital image, the first

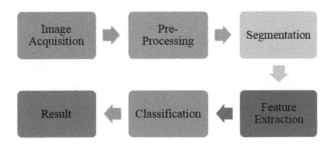

FIGURE 6.3   Basic steps in image processing.

pre-processing steps include two classes of operations – point operations and geometric operations. If there are fundamental distortions of the image formation process, such as nonlinear and non-homogeneous radiometric sensitivity of imaging sensors or geometric distortions of the imaging system, they are corrected in the pre-processing stage. Likewise, dot operations are applied to correct and optimize the image illumination that occurs during image formation, detect insufficiency and overflow, improve and widen contrast, correct non-homogeneous illumination to average images, or perform radiometric calibration (Jähne, 2005).

- *Segmentation*: Image segmentation is the most important and critical step of the image processing technique that divides an image into different regions. The accuracy of the final result of the image analysis depends on the accuracy of the image segmentation. With segmentation, each region becomes homogeneous according to some well-defined characteristics or characteristics. There is no single standard technique for image segmentation. Many image segmentation techniques have been proposed by researchers. Segmentation techniques are based on two properties of the intensity values of the regions: Discontinuity and similarity. In discontinuity-based techniques, an image is segmented according to sudden changes in intensity, while in similarity-based techniques, the image is segmented into various regions based on similarity according to a predefined set of criteria (Tyagi, 2018).

- *Feature extraction and representation*: Feature extraction and representation techniques, methods that increase the visibility of relevant parts or features, are used to extract the necessary information from the image according to the purpose of image processing. The resulting representation can then be used as input to a range of pattern recognition and classification techniques. Therefore, the selection of relevant features in the image is the first step of image processing. While making this selection, several brightness values in the original image can be defined. Thus, pixels in the selected range are brought to the fore and all other pixels are put in the background. In addition, the image can be displayed by distinguishing it as a two-level image using black and white (Ozguven and Yanar, 2022).

- *Classification or pattern recognition*: The purpose of pattern classification techniques is to assign a class to an unknown pattern based on prior knowledge about images or objects and the classes they belong to. Pattern classification techniques are generally divided into two main groups: Statistical and structural (or syntactic). In classification with statistical pattern recognition techniques, it is assumed that each object or class can be represented as a feature vector, and based on distance calculations or probability models, it is decided which class to assign to a particular pattern (Marques, 2011).

6.4 IMAGE PROCESSING APPLICATION EXAMPLES IN AGRICULTURE

Image processing techniques have been widely used in many agricultural activities. Studies have been carried out on many subjects such as disease, pest, and weed detection, determination of plant stresses, yield estimation, monitoring of product development, modeling of irrigation methods, determination of soil properties, monitoring of animal growth, determination of lameness, and determination of pain points and body temperatures of animals. In addition, the experience gained during the implementation of these studies, together with machine learning, deep learning, artificial intelligence, modeling, and simulation applications, has led to the development of real-time and automatic expert systems, autonomous tractors or agricultural machines, and agricultural robotic applications (Altaş et al., 2019). In Table 6.1, many examples of studies conducted by researchers in different agricultural areas are seen.

6.4.1 Detection of Plant Diseases and Pests

In the event of a disease, plants exhibit visual signs in the shape of colorful spots with different shapes and sizes according to the type of disease and in the shape of lines seen on stems and different

TABLE 6.1

Agricultural Applications of Image Processing

Disease detection	Zhang et al., 2017, 2018; Opstad Kruse et al., 2014; Zhang and Meng, 2011; Ashan Salgadoe et al., 2018; Pujari et al., 2015; Mahlein et al., 2013; Camargo and Smith, 2009
Pest and weed detection	Liu et al., 2016; Burgos-Artizzu et al., 2010
Yield estimation	Zhou et al., 2012; Aggelopoulou et al., 2011
Plant identification and detection	Reis et al., 2012
Assessment of vegetation indexes	Candiago et al., 2015
Green area index	Verger et al., 2014
Determination of plant growth variability	Bendig et al., 2013
Follow-up of product development	Yang et al., 2013
Follow-up of root development	Marié et al., 2014
Determination of soil moisture	Robinson et al., 2012
Modeling of irrigation management applications	García-Mateos et al., 2015
Gait analysis and measuring body characteristics	Salau et al., 2015
Determination of body condition score	Coffey and Bewley, 2014
Beef cattle weight determination	Pradana et al., 2016
Lameness detection	Poursaberi et al., 2010
Determination of pain sites	Düzgün and Or, 2009
Monitoring of body temperature	Hoffmann and Schmidt, 2015
GNSS tracking of livestock	Trotter et al., 2010

Source: Altas et al. (2018).

sections or organs of the plants. These symptoms alter color, shape, and size while the disease progresses. With image processing methods, colored objects might be distinguished, and the severity of plant diseases might be determined. Besides image processing methods, expert systems might be improved to allow instant disease diagnosis with machine learning methods (Ozguven, 2020).

Recently, the potential use of image processing and machine learning methods for disease detection in whole plants and/or different plant parts (leaves, stem, fruit, and such) has been comprehensively studied by many researchers (Ozguven and Adem, 2019). Monitoring pests and diseases is an extremely important activity to increase productivity in agriculture. Remote sensing combined with machine learning techniques opens new possibilities for tracking and identifying characteristics such as identifying diseases, pests, water, and nutritional stress (Calou et al., 2020).

Altas et al. (2018) used drone image processing techniques to determine the level of sugar beet leaf spot disease (*Cercospora beticola* Sacc.). Images were taken with a DJI Phantom 3 Advanced brand drone from an area of approximately 200 m^2 where the disease is seen intensely in a local farmer's field where sugar beet is grown. In this study, 12 images showing different levels of development of the disease, taken at different times and under different natural lighting conditions from the field, have been determined by image processing techniques using the Image Processing Toolbox module of the MATLAB program. Using pixel labels, the pixels in the image were separated into colors, resulting in three images (i.e., K = 3) as shown in Figure 6.4, and the disease image was selected from among the three clusters. The results obtained at the end of the study were compared with the observations made by plant protection experts (Table 6.2).

When Table 6.2 is analyzed, it is seen that the results were very close to each other. It was reported that the assessment results acquired by visual evaluation were approximate integer values; the image processing methods gave results of the exact value of the diseased area with a sensitivity that cannot

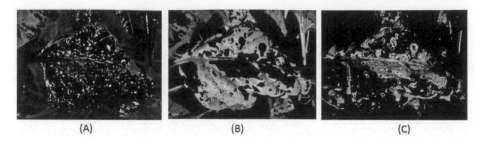

(A)   (B)   (C)

FIGURE 6.4 K-means cluster segmentation of diseased leaf image: (a) Black segment, (b) green segment, (c) brown segment (disease image) (Altas et al., 2018).

TABLE 6.2
Comparison of Image Processing Methods and Plant
Protection Expert Visual Evaluation Results

Image	Image processing (%)	Visual evaluation (%)	Difference (%)
a	100	100	0
b	48	50	–2
c	42	45	–3
d	21	20	+1
e	80	80	0
f	28	30	–2
g	74	75	–1
h	47	50	–3
i	29	30	–1
j	46	50	–4
k	20	20	0
m	51	50	–1

Source: Altas et al. (2018).

be acquired by observation, and the research was achieved successfully. For example, as a result of image processing, the disease severity was calculated as 51% in the area of the image. Since it is almost impossible to read the value of 51% with expert observation, they reported that it was understood that the closest numerical value containing these values, 50%, was chosen by the expert.

Uygun et al. (2020) used image processing techniques to determine the damage of the two-spotted red spider mite (*Tetranychus urticae* Koch) to cucumber plants in a greenhouse and the changes in mite numbers. In the study, firstly, a new farming platform (Figure 6.5) was developed to ensure camera stability to capture quality images. Images of 50 leaves infected with *T. urticae* were captured with the platform for five weeks and 250 images were acquired. Fifty of these obtained images were randomly selected and processed with an image processing algorithm developed using an Image Processing Toolbox module of MATLAB (Figure 6.6), and the results obtained from the image processing algorithm were compared with expert observations. At the end of the study, mite damage in the image processing method was estimated with a 3.91 root mean square error (RMSE). It has been reported that there is a highly significant positive correlation between image processing and expert observation, and the results show that this new image processing method can be successfully used instead of expert observation to determine *T. urticae* damage in greenhouses.

Chowdhury et al. (2021) used computer vision and artificial intelligence to perform image segmentation, which is an important feature for disease detection, for the automatic detection and

FIGURE 6.5 Agriculture platform developed to take images of cucumber leaves working in the greenhouse (Uygun et al., 2020).

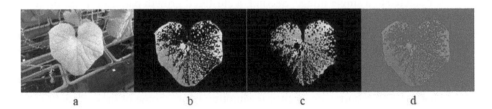

a   b   c   d

FIGURE 6.6 Example of segmentation of damaged leaf image with K-means cluster algorithm: (a) An original leaf image, (b) damaged segment output after K-means cluster, (c) healthy segment output after K-means cluster, (d) damaged segment output in L*a*b* color space output (Uygun et al., 2020).

classification of plant leaf diseases. Researchers have proposed the use of a deep learning architecture based on a new convolutional neural network called EfficientNet for the classification of tomato diseases. The main advantage of this study, which is an example of the studies carried out as a result of the use of image processing with different techniques, is that plant diseases can be identified at an early or incipient stage. In the study, U-net and Modified U-net were used for the segmentation of leaves. Comparative performance of the models for binary classification (healthy and unhealthy leaves), six-class classification (healthy and various diseased leaf groups), and ten-class classification (healthy and various unhealthy leaf types) are also reported. It has been reported that with EfficientNet-B4, an accuracy of 99.89% for ten-class classification was achieved using segmented images, and all the architectures performed better in classifying the diseases when trained with deeper networks on segmented images. Figure 6.7 shows sample images of healthy and different unhealthy tomato leaves from the Plant Village database.

6.4.2 Visual Object Detection

Visual object detection is an essential component in various applications of automated farming. There are several detection problems in the agricultural environment. These difficulties can be

FIGURE 6.7 Sample images of healthy and different unhealthy tomato leaves from the Plant Village database (Chowdhury et al., 2021).

examined in four groups. First, field images can contain dozens of objects with high scale variance. As plants are usually grown in rows, both near and far objects are captured, and many octaves of objects must be detected in a single image. Second, a clustered growth pattern typical of, for example, apple blossoms or tomatoes leads to severe occlusion for many objects. Third, target objects often have a simple shape with little resemblance to background structures, with no distinguishing details. For example, round-shaped objects such as tomatoes and avocados can be mistaken for round leaves. Sticky objects such as cucumbers show a high similarity to some branches and stems. Tomatoes, while plants, are convex and non-skeletal. Finally, the lighting in outdoor conditions creates heavy shadows and requires invariance in shooting hours. Some of these difficulties can be observed in the sample images in Figure 6.8 (Wosner et al., 2021).

Wosner et al. (2021) conducted a study on how to properly implement modern deep networks for detection tasks in agricultural contexts, comparing the performances of the methods and comparing the detection results with human accuracy. In this study, seven different datasets were collected for the testing and three new networks were tested. Experiments revealed that handling small objects and large-scale variance are important points of error, and so a multi-resolution approach was developed for network usage, which significantly improved detection accuracy on most datasets. Examples of identifying the best models for all datasets are presented in Figure 6.9 (banana, flower, avocado) and Figure 6.10 (whole tomato plant, cucumber, buds, and tomato).

6.4.3 DETERMINING THE FERTILIZER DISTRIBUTION PATTERN

There are agronomic, environmental, and economic reasons for the correct distribution of fertilizers to the field in the right amount. Today, preventing fertilizer/seed misapplication is a very important issue for farmers, machine developers, and the agricultural industry. While image processing techniques have great potential for near real-time calibration of the distribution pattern in the field, it is an alternative method for complex and expensive modeling trajectory-based approaches, potentially providing information similar to that obtained from tray measurements. In this regard, having a reliable image processing-based fertilizer spreader calibration system presents several challenges. These include varying lighting conditions, image background used, distance to the target during image acquisition, flexibility to work with different types of agricultural materials (fertilizer and seeds), range of particle sizes tested, detection of particles that often occur at high application rates,

FIGURE 6.8 Detection difficulties in the agricultural setting: (a) Dozens of apple blossoms in cropped image, (b) indistinguishable objects with simple shape and high similarity to leaves (avocadoes), (c) extreme variation in object scale, (d–e) severe obstructions due to clustered growth in tomatoes and cucumbers, (f) tomato plants are non-convex objects, making it difficult to separate neighboring objects (Wosner et al., 2021).

FIGURE 6.9 Examples of detection results: Red bounding boxes are detection results and blue bounding boxes represent ground truth statements (Wosner et al., 2021).

FIGURE 6.10 The left column presents full image examples, showing the cropped portion enlarged from the right image to the left (Wosner et al., 2021).

FIGURE 6.11 Identifying and counting objects in grid cells #18 and #27 (Marcal and Cunha, 2019).

and the combination of multiple images for generating uniformity indices (Marcal and Cunha, 2019).

Marcal and Cunha (2019) developed the Automated Fertilizer Calibration (ACFert) system for use with centrifugal, pendulum, or other types of spreaders that distribute dry granular agricultural materials onto the soil. The ACFert is based on image processing techniques and includes a specially designed mat, which should be placed in the ground for spreaders calibration (Figure 6.11). Each image is processed independently, and two numeric values are provided for each grid element in the image: The amount of fertilizer/number of seeds counted as output and the numeric label. In this study, ACFert's performance was evaluated for automatic granule detection using a series of manual count measurements of nitrate fertilizer and wheat seeds. A total of 185 images obtained with two mobile devices were used with a total of 498 rectangular elements observed and analyzed. The researchers reported that this calibration tool is a very low-cost system that can be used in the field, providing results to support accurate agricultural fertilizer distribution patterns in near real-time for different types of fertilizers or seeds.

6.4.4 DETERMINATION OF AGRICULTURAL PRODUCT CHARACTERISTICS

The size of an agricultural product is an important parameter in determining fruit growth and quality. It can be used as a maturity index to determine the optimum harvest time. Timely harvesting of agricultural products is important for packaging, grading, and sorting. This process should be done easily and economically. In this respect, determining the harvest time by image analysis method is seen as an important tool. The image analysis method has a high accuracy rate but is economical and practical.

Beyaz et al. (2009) evaluated the front, top, and left sides of each pepper by image analysis method on unformed Kahramanmaraş red pepper (*Capsicum annuum* L.). In the study, projection areas were determined from these images (Figure 6.12). Afterward, the effects and image combinations of each image were determined as volume. For volume estimation, the regression coefficients between the projection areas and the volume values were determined. The most appropriate estimation formula was found from the top and left projections. A regression coefficient of 89.7% was given to this formula for volume estimation.

6.4.5 Dynamic Obstacle Detection

Determining and identifying obstacles around autonomous agricultural machinery is of great importance in increasing the safety and working efficiency of agricultural machinery. Xu et al. (2021) used a panoramic camera to quickly detect dynamic obstacles around moving agricultural machinery. A Lucas-Kanade optical flow algorithm is used to detect moving obstacles in these panoramic images, and an external rectangular frame (Figure 6.13) is used to select the foreground moving object. In this study, the main optical flow direction in the segmentation clusters and the distance between clusters determined whether the same foreground moving object was selected. Next, the corresponding combination processing was used, which allows the box to select the exact foreground motion target. By processing 100 frames of images, the result showed that the average time consumption of the proposed method was 0.801 s. The dynamic agricultural machinery detection accuracy rate was 88.06%, the pedestrian detection accuracy was 81.61%, and overall accuracy was 82.93%. Thus, it is reported that the proposed method can meet the requirements of real farmland operations and have good instantaneity and detection results.

6.5 WHAT IS MACHINE VISION?

Machine (computer) vision has the ability to process massive amounts of data quickly without human intervention. Thus, the numerical data of the images taken of an object are processed with algorithms developed for the purpose by various processes such as contrast enhancement and

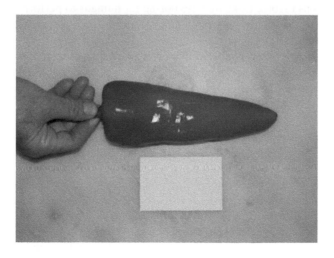

FIGURE 6.12 An example of digital images of Kahramanmaraş red pepper (Beyaz et al., 2009).

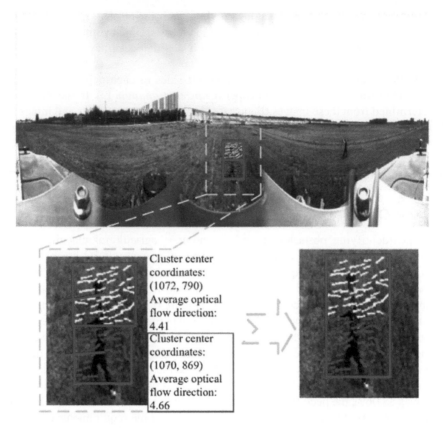

FIGURE 6.13 Accurate extraction of moving targets of the 3610 frame (Xu et al., 2021).

sharpening, patterns are recognized, and robotic or other devices connected to the system are automatically controlled. It is no longer a problem to provide computers with the necessary capability for storing and processing the obtained data, which was a problem in machine vision systems before (Özgüven et al., 2020). Image analysis methods are the main components of a machine vision system. Machine vision systems can be best understood considering different types of applications. Most of these applications involve tasks that are boring for humans to perform, require working in a dangerous environment, are highly computational, or require access to and use of a large database of information (Umbaugh, 2018).

The success of machine vision system development depends on understanding all parts of the imaging chain. The fascination and complexity of machine vision lie in the specific engineering fields within it, such as mechanical engineering, electrical and electronics engineering, optical engineering, and software engineering. Each branch of engineering struggles for a primary role. Interdisciplinary thinking is the foundation for the successful development of a machine vision application, and it defines the challenges and possibilities of machine vision auditing (Hornberg, 2017). All elements of the information processing chain in machine vision are presented in Figure 6.14.

As can be seen from Figure 6.14, although each component is important for the success of the machine vision system, appropriate lighting is of particular importance for the correct inspection of the inspected object. Because a vision system is usually a division within a production system, the processing time must complete its task as quickly as possible without disrupting the process. For this reason, lenses and cameras should be chosen according to the need to capture the image correctly and on time. The processing power of the computer and the selection of interfaces are

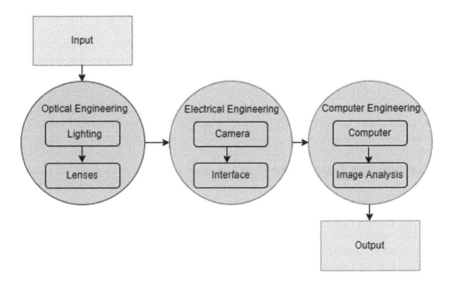

FIGURE 6.14 Information processing chain in machine vision.

important for the success of the system. In addition to the success of the system in the selection of algorithms used in the vision system, the preference for fast processing is decisive in the selection of the algorithm.

6.6 MACHINE VISION APPLICATION EXAMPLES IN AGRICULTURE

There are some difficulties encountered in the use of machine vision systems in determining the quality status of agricultural products. These difficulties are that agricultural products are very diverse and have different characteristics; they have irregular shapes, the lighting cannot be controlled in case of working in natural conditions, and the products come into contact with each other due to their small structure, overlap, or have various similar objects in the background (Özgüven et al., 2020). Due to this, the capability of machine vision systems is best with fruit and nuts (apple, orange, strawberry, hazelnut, tomato, peach, and pear), vegetable control (mushroom, potato, etc.), grain classification and quality assessment (wheat, corn, rice, barley, oats, rye), and food products (it is used in many applications in the evaluation of pizza, bakery products, cheese, meat, and meat products) (Brosnan and Sun, 2004). A classic machine vision system should have good lighting, image acquisition hardware (lens, TV camera, image capture card, and connecting cables), image acquisition software for feature extraction and comprehension, and image processing (Kondo and Kawamura, 2013).

Line scan cameras are preferred to effectively acquire each vertical pixel row of a long horizontal image in sequence with a continuously moving conveyor belt system found in many food processing plants and factories (Davies and Holloway, 2013). They are capable of capturing a 768 × 576-pixel image in color with three bytes per pixel (one for each color channel) in a typical system. Until recently, capturing and display depended on standard TV systems with an image frequency of 25 (if operating in European TV standards) or 30 (US standards) times per second. This equates to a data rate of over 3.3 million bytes per second. Because machine vision data is digital, it is suitable for computer analysis for later use as information control. The data capture device (camera) is inherently contactless. For this reason, it requires simple engineering for its installation, has no interaction with the detected object (for example, suitable for non-destructive testing), and can be placed in the operating environment without any negative effects (for example, for observing animals without disturbing their natural behavior) (Marchant, 2006).

6.6.1 Sorting Agricultural Products by Quality

Market prices of agricultural products are directly related to the quality of the product. In particular, the most common method for classifying fruits is manual sorting. Various factors negatively affect manual classification, such as high labor costs, worker fatigue, inconsistency and changes in ambient light intensity, differences in personal quality perception, and scarcity of trained workers. With the non-destructive use of machine vision in fruit classification, classification is done correctly and with high quality, and labor costs are reduced. Machine vision systems are being developed to automatically classify fruits according to quality criteria such as mature or not, defective or not, shape, size, etc.

Truong Minh Long and Truong Thinh (2020) proposed a novel evaluation of internal quality focusing on the external features as well as the weight of the mango (Figure 6.15). The grading of fruits is handled by four machine learning models. These models are random forest, linear discriminant analysis, support vector machine, and K-nearest neighbors. Models have inputs such as length, width, defect, and weight, and outputs are mango classifications such as grades G1, G2, and G3. The weight signal is received from the load cell and processed to give weight. The signal from the load cell is eliminated with the Kalman filter. Study results show that assessment is more effective than using only one of the external features or weight by combining an expensive non-destructive measurement. As a result, it has been reported that machine learning methods have a high accuracy of over 87.9%, especially the RF model, which has an accuracy of 98.1%.

Arjenaki et al. (2013) developed an efficient machine vision–based sorting system for tomatoes. Sorting parameters included shape (oblong and circular), size (small and large), maturity (color), and defects (Figure 6.16). The developed software evaluated the tomato shape by its eccentricity, the tomato size by the 2D image area, the tomato maturity by the average color, and the tomato defect by the fullness parameter. In this study, a sorting system equipped with machine vision was established to test the system consisting of a CCD camera, a microcontroller, sensors, and a computer according to conveyor belt speed, tomato spacing, and light intensity conditions. After the optimum working conditions were determined, tomato samples were separated according to shape, color, size, and defects using a sorting machine. Defect detection, shape and size algorithm, and overall system accuracy were reported to be 84.4%, 90.9%, 94.5%, and 90.61%, respectively.

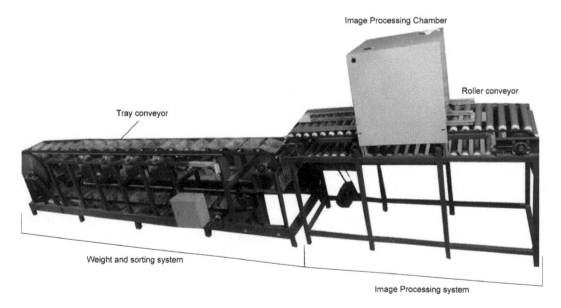

FIGURE 6.15   A mango classification and sorting system with machine vision system (Truong Minh Long and Truong Thinh, 2020).

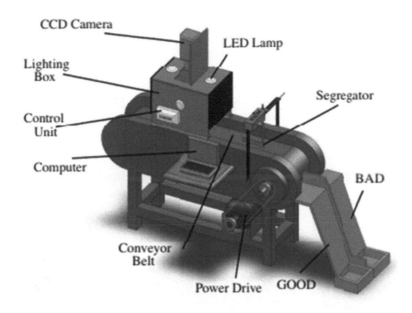

FIGURE 6.16 A machine vision system that sorts tomatoes by quality (Arjenaki et al., 2013).

6.6.2 DETECTION OF PLANT DISEASES AND PESTS

Plant diseases cause economically important income losses in agricultural production all over the world (Savary and Willocquet, 2014; Avelino et al., 2015). To reduce crop loss, plant disease severity must be determined accurately and rapidly. Therefore, determining the outbreak, severity, and progression of diseases in a timely and accurate manner is of great importance for effective integrated disease management (Bock et al., 2010). There is an immediate need to develop faster and more practical methods, which could reduce human errors in the identification of plant diseases, their severity, and their progress, especially in large production areas (Altas et al., 2018).

Liu et al. (2016) developed a machine vision system based on the monitoring of aphid populations and identification of aphid species that cause extensive damage to wheat fields. It has been reported that different aphid densities and colors on the plant are taken into account in the application of this method and that the proposed method is easy and effective to use. It was also stated that the method provides accurate aphid population data and therefore can be used for aphid infestation studies in wheat fields. An example of identifying aphids is shown in Figure 6.17.

6.6.3 DETECTION OF WEEDS

Amziane et al. (2021) proposed an original image formation model that takes into account the variation of illumination during radiance image acquisition with a line scan camera. A new reflection estimation method was derived from this model that takes into account frame-level illumination. It shows how the variation of illumination during the multispectral image acquisition by this instrument impacts the measured radiance that is provided by a Lambertian surface element. In the study, a white diffuser tile mounted on the acquisition system was used to observe a portion of the sensor vertically (Figure 6.18a). The pixel subset WD contains (about 10%) right border pixels that represent the white diffuser (Figure 6.18b). One hundred and nine brightness images with $2,048 \times 2,048$ pixels $\times$ 192 channels at 10-bit depth were obtained under a skylight in a greenhouse. It is an original multispectral image formation model that handles how illumination is associated with both the considered band and pixel since the luminance that is associated with a given spectral band at a pixel is measured in a frame that is acquired at a specific time. The researchers reported experimentally that

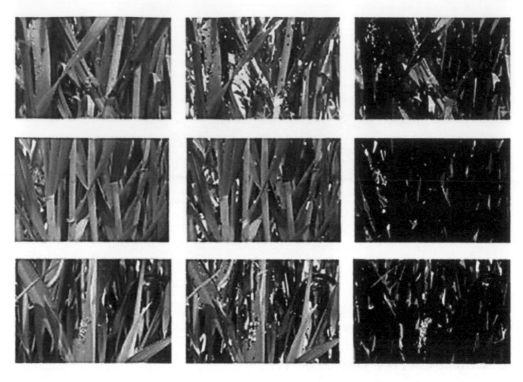

FIGURE 6.17   Detection of aphids in wheat fields with machine vision technique (Liu et al., 2016).

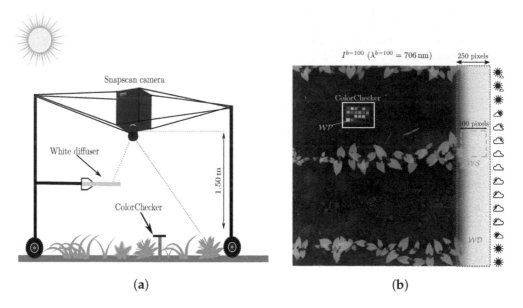

FIGURE 6.18   (a) Camera-mounted acquisition setup. (b) Pixel subsets WD, WS, and WP are shown in red, green, and cyan, respectively (Amziane et al., 2021).

the method was more robust for lighting variation than state-of-the-art methods, and the reflectance properties based on the method were more discriminating for outdoor weed detection and identification. LightGBM (LGBM), a parametric tree-based classifier, and non-parametric quadratic discrimination analysis (QDA) classifiers based on Bayes's theorem were applied for supervised weed detection and identification problems. Figure 6.19 shows the color-coded vegetation pixel classification of test images using the LGBM classifier in weed detection and identification tasks.

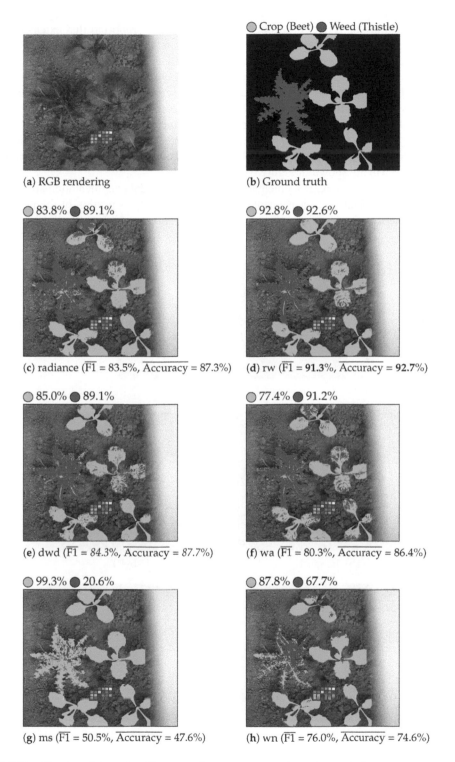

FIGURE 6.19 Crop/weed detection. Beets are shown as green and weeds as red. Accuracy score per class is shown next to each colored circle (class label) for each considered feature (Amziane et al., 2021).

Su et al. (2021) proposed a new data augmentation system based on random image cropping and patching (RICAP), designed to augment data for deep learning–based semantic segmentation and plant/weed classification in agricultural robotics. The researchers reported that extensive experimental evaluations and ablation studies were performed on two datasets from different farms to test their proposed methodology, demonstrating that it can effectively improve segmentation accuracy and that improvements to the original RICAP actually contribute to performance gains. On average, the proposed method increases the average accuracy and average intersection over the union of the deep neural net with the traditional data augmentation (random flipping, rotation, and color jitter) from 91.01 to 94.02 and from 63.59 to 70.77 respectively for the Narrabri dataset, and from 97.99 to 98.51 and from 74.26 to 77.09 respectively for the Bonn dataset. Mechanical and laser tools in the robot perform weeding operations according to the masks produced. The robot and typical images taken in the wheat farm are shown in Figure 6.20.

As can be seen from Figure 6.21, segmentation masks of basic data augmentation are often bubbly, and fine details of plant/weed leaves are missing. Additionally, a portion of the plant leaf in Figure 6.21c was incorrectly classified as grass. A similar situation can be found in the Bonn dataset.

(a) The Digital Farmhand.

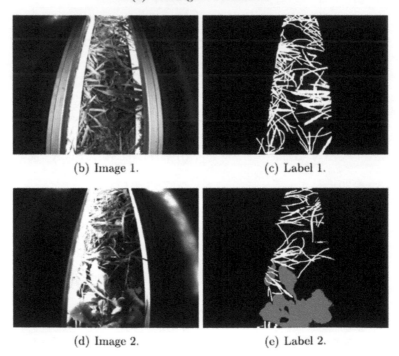

(b) Image 1.   (c) Label 1.

(d) Image 2.   (e) Label 2.

FIGURE 6.20 (a) Digital farmhand robot and typical images taken at the wheat farm. (c) and (e) are manually labeled; (b) and (d) have accuracy labels. (b) taken on a cloudy day. (d) was taken on a sunny day, resulting in overexposure on the left side of the image (Su et al., 2021).

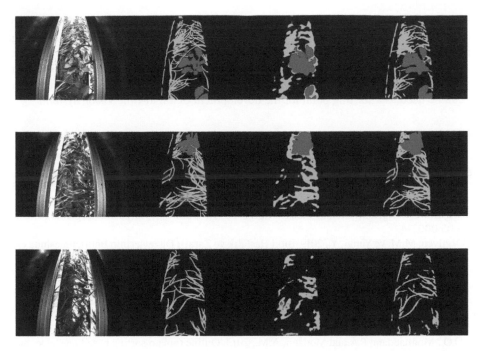

FIGURE 6.21 Basic data augmentation in Narrabri dataset and qualitative comparison of segmentation results with the proposed basic method. In each sub-figure, images from left to right are the raw RGB image, base real partition mask, and partition masks estimated from deep neural network with baseline data augmentation and suggested baseline data augmentation method (Su et al., 2021).

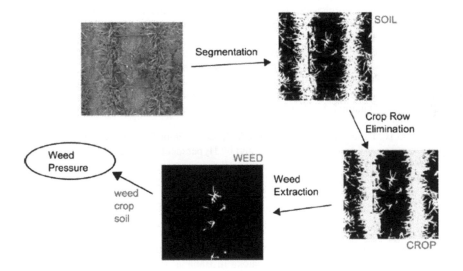

FIGURE 6.22 Stages in the image processing technique of the proposed system (Burgos-Artizzu et al., 2010).

6.6.4 DETECTION OF PLANT AND SOIL

Burgos-Artizzu et al. (2010) conducted a study on several advanced methods based on machine vision for estimating the percentages of weeds, plants, and soils in an image. The image processing was divided into three different stages, in which each different agricultural element is extracted (Figure 6.22). These stages are (1) segmentation of vegetation from non-vegetation (soil), (2) crop row removal (crop), and (3) weed removal (weeds). Researchers have proposed different and inter-changeable methods at each stage that use a set of input parameters whose value can be changed to further refine each processing. A genetic algorithm was used to find the best parameter value and method combination for use in different image sets. It has been reported that the proposed methods give very good results both in the accurate detection of weeds and in low computational complexity, and thus can be used as a starting point for the development of a real-time vision system.

REFERENCES

Aggelopoulou, A.D., Bochtis, D., Fountas, S., Swain, K.C., Gemtos, T.A. and Nanos, G.D., 2011. Yield Prediction in Apple Orchards Based on Image Processing. *Precision Agriculture*, 12, 448–456.

Altas, Z., Ozguven, M.M. and Yanar, Y., 2018. Determination of Sugar Beet Leaf Spot Disease Level (*Cercospora Beticola* Sacc.) with Image Processing Technique by using Drone. *Current Investigations in Agriculture and Current Research*, 5(3), 621–631. https://doi.org/10.32474/CIACR .2018.05.000214.

Altaş, Z., Özgüven, M.M. ve Yanar, Y., 2019. Bitki Hastalık ve Zararlı Düzeylerinin Belirlenmesinde Görüntü İşleme Tekniklerinin Kullanımı: Şeker Pancarı Yaprak Leke Hastalığı Örneği. International Erciyes Agriculture, Animal&Food Sciences Conference 24–27 April 2019- Erciyes University – Kayseri, Turkiye. (Turkish).

Amziane, A., Losson, O., Mathon, B., Dumenil, A. and Macaire, L., 2021. Reflectance Estimation from Multispectral Linescan Acquisitions under Varying Illumination – Application to Outdoor Weed Identification. *Sensors*, 21, 3601. https://doi.org/10.3390/s21113601.

Arjenaki, O.O., Moghaddam, P.A and Motlagh, A.M., 2013. Online tomato sorting based on shape, maturity, size, and surface defects using machine vision. *Turkish Journal of Agriculture and Forestry*, 37, 62–68. https://doi.org/10.3906/tar-1201-10.

Ashan Salgadoe, A.S., Robson, A.J., Lamb, D.W., Dann, E.K. and Searle, C., 2018. Quantifying the Severity of Phytophthora Root Rot Disease in Avocado Trees Using Image Analysis. *Remote Sensing*, 10, 226. https://doi.org/10.3390/rs10020226.

Avelino, J., Cristancho, M., Georgiou, S., Imbach, P., Aguilar, L. and Bornemann, G., 2015. The Coffee Rust Crises in Colombia and Central America (2008–2013): Impacts, Plausible Causes and Proposed Solutions. *Food Security*, 7(2), 303–321.

Bendig, J., Bolten, A. and Bareth, G., 2013. UAV-Based Imaging for Multi-Temporal, Very High Resolution Crop Surface Models to Monitor Crop Growth Variability. Photogrammetrie Fernerkundung Geoinformation, PFG. 2013/6, 0551–0562 Stuttgart.

Beyaz, A., Ozguven, M.M., Öztürk, R. and Acar, A., 2009. Volume Determination of Kahramanmaras Red Pepper (Capsicum Annuum L.) by Using Image Analysis Technique. *Journal of Agricultural Machinery Science*, 5(1), 103–108.

Bock, C.H., Poole, G.H., Parker, P.E. and Gottwald, T.R., 2010. Plant Disease Severity Estimated Visually, by Digital Photography and Image Analysis, and by Hyperspectral Imaging. *Critical Reviews in Plant Sciences*, 29(2), 59–107.

Brosnan, T. and Sun, D.W., 2004. Improving Quality Inspection of Food Products by Computer Vision – A Review. *Journal of Food Engineering*, 61, 3–16.

Burgos-Artizzu, X.P., Ribeiro, A., Guijarro, M. and Pajares, G., 2010. Real-Time Image Processing for Crop/Weed Discrimination in Maize Fields. *Computers and Electronics in Agriculture*, 75, 337–346. https://doi.org/10.1016/j.compag.2010.12.011.

Calou, V.B.C., Teixeira, A.D.S., Moreira, L.C.J., Lima, C.S., De Oliveira, J.B. ve De Oliveira, M.R.R., 2020. The use of UAVs in Monitoring Yellow Sigatoka in Banana. *Biosystems Engineering*, 193, 115–125. https://doi.org/10.1016/j.biosystemseng.2020.02.016.

Candiago, S., Remondino, F., De Giglio, M., Dubbini, M. and Gattelli, M., 2015. Evaluating Multispectral Images and Vegetation Indices for Precision Farming Applications from UAV Images. *Remote Sensing*, 7, 4026–4047. https://doi.org/10.3390/rs70404026.

Camargo, A. and Smith, J.S., 2009. An Image-Processing Based Algorithm to Automatically Identify Plant Disease Visual Symptoms. *Biosystems Engineering*, 102, 9–21.

Chowdhury, M.E.H., Rahman, T., Khandakar, A., Ayari, M.A., Khan, A.U., Khan, M.S., Al-Emadi, N., Reaz, M.B.I., Islam, M.T. and Ali, S.H.M., 2021. Automatic and Reliable Leaf Disease Detection Using Deep Learning Techniques. *AgriEngineering*, 3, 294–312. https://doi.org/10.3390/agriengineering3020020.

Coffey, M. and Bewley, J., 2014. Precision Dairy Farming. https://ruralfuturesconf.agresearch.co.nz/medi-awiki/images/e/e9/Precision_Dairying_NZ_MC_V2.ppt. [Accessed 05.05.2017].

Davies, E.R. and Holloway, R., 2013. Machine Vision in the Food Industry. *Robotics and Automation in the Food Industry Current and Future Technologies*. (Editor: Caldwell, D.G.) Woodhead Publishing Limited, Cambridge, 75–110.

Düzgün, D. ve Or, M.E., 2009. Termal Kameraların Tıpta Veteriner Hekimlikte Kullanımı. TÜBAV Bilim. Cilt:2, Sayı:4, S:468-475. (Turkish).

García-Mateos, G., Hernández-Hernández, J.L., Escarabajal-Henarejos, D., Jaén-Terrones, S. and Molina-Martínez, J.M., 2015. Study and Comparison of Color Models for Automatic Image Analysis in Irrigation Management Applications. *Agricultural Water Management*, 151, 158–166.

Gonzalez, R.C., Woods, R.E. and Eddins, S.L., 2009. *Digital Image Processing using MATLAB*. Gatesmark Publishing, USA. ISBN: 978-0-9820854-0-0.

Hoffmann, G. and Schmidt, M., 2015. Monitoring the Body Temperature of Cows and Calves with a Video-Based Infrared Thermography Camera. *Precision Livestock Farming Applications*. (Editor: Halachmi, I.) Wageningen Academic Publishers, Wageningen. 231–238.

Hornberg, A., 2017. *Preface. Handbook of Machine and Computer Vision: The Guide for Developers and Users*. (Editor: Hornberg, A.) Wiley-VCH Verlag GmbH & Co. ISBN: 978-3-527-41343-0.

Jähne, B., 2005. *Digital Image Processing*. Springer-Verlag, Berlin Heidelberg. ISBN: 978-3-540-24035-8.

Kondo, N. and Kawamura, S., 2013. Postharvest Automation. *Agricultural Automation Fundamentals and Practices*. (Editors: Zhang, Q. and Pierce, F.J.) CRC Press Taylor & Francis Group LLC, Boca Raton. 367–383.

Liu, T., Chen, W., Wu, W., Sun, C., Guo, W. and Zhu, X., 2016. Detection of Aphids in Wheat Fields Using a Computer Vision Technique. *Biosytems Engineering*, 141, 82–93. https://doi.org/10.1016/j.biosystemseng.2015.11.005.

Mahlein, A.K., Rumpf, T., Welke, P., Dehne, H.W., Plümer, L., Steiner, U. and Oerke, E.C., 2013. Development of Spectral Indices for Detecting and Identifying Plant Diseases. *Remote Sensing of Environment*, 128, 21–30.

Marcal, A.R.S. and Cunha, M., 2019. Development of an Image-Based System to Assess Agricultural Fertilizer Spreader Pattern. *Computers and Electronics in Agriculture*, 162, 380–388. https://doi.org/10.1016/j.compag.2019.04.031.

Marchant, J.A., 2006. Section 5.4 Machine Vision in the Agricultural Context, pp. 259–272 of Chapter 5 Precision Agriculture, in CIGR Handbook of Agricultural Engineering Volume VI Information Technology. Edited by CIGR-The International Commission of Agricultural Engineering; Volume Editor, Axel Munack. St. Joseph, Michigan, USA: ASABE. Copyright American Society of Agricultural Engineers. (Çevirmen: Böğrekci İ., Böğrekci İ.; Çeviri Editörleri: Tarhan, S., Ozguven, M.M.).

Marié, C.L., Kirchgessner, N., Marschall, D., Walter, A. and Hund, A., 2014. Rhizoslides: Paper-Based Growth System for Non-Destructive, High Throughput Phenotyping of Root Development by Means of Image Analysis. *Plant Methods*, 10, 13.

Marques, O., 2011. *Practical Image and Video Processing Using MATLAB*. John Wiley & Sons, Inc, Hoboken. ISBN: 978-0-470-04815-3.

MathWorks, 2008. MATLAB for Image Processing. The MathWorks Training Services. By The MathWorks, Inc.

McAndrew, A., 2016. *A Computational Introduction to Digital Image Processing*. CRC Press Taylor & Francis Group LLC, Boca Raton. ISBN: 978-1-4822-4735-0.

Opstad Kruse, O.M., Prats-Montalbán, J.M., Indahl, U.G., Kvaal, K., Ferrer, A. and Futsaether, C.M., 2014. Pixel Classification Methods for Identifying and Quantifying Leaf Surface Injury from Digital Images. *Computers and Electronics in Agriculture*, 108, 155–165.

Ozguven, M.M. and Adem, K., 2019. Automatic Detection and Classification of Leaf Spot Disease in Sugar Beet Using Deep Learning Algorithms. *Physica A-Statistical Mechanics and Its Applications*, 535(122537), 1–8. https://doi.org/10.1016/j.physa.2019.122537.

Ozguven, M.M., 2020. Deep Learning Algorithms for Automatic Detection and Classification of Mildew Disease in Cucumber. *Fresenius Environmental Bulletin*, 29(08/2020), 7081–7087.

Ozguven, M.M. and Yanar, Y., 2022. The Technology Uses in the Determination of Sugar Beet Diseases. *Sugar Beet Cultivation, Management and Processing*. (Editors: Misra, V., Santeshwari and Mall, A.K.) Springer, Singapore.621–642. https://doi.org/10.1007/978-981-19-2730-0_30

Özgüven, M.M., Beyaz, A., Ormanoğlu, N., Aktaş, T., Emekci, M., Ferizli, A.G., Çilingir, İ. ve Çolak, A., 2020. Hasat Sonrası Ürünlerin Korunmasına Yönelik Mekanizasyon Otomasyon Ve Mücadele Teknikleri. Türkiye Ziraat Mühendisliği IX. Teknik Kongresi. Ocak 2020, Ankara. Bildiriler Kitabı-1, s.301–324. (Turkish).

Poursaberi, A., Bahr, C., Pluk, A., Van Nuffel, A. and Berckmans, D., 2010. Real-time Automatic Lameness Detection Based on Back Posture Extraction in Dairy Cattle: Shape Analysis of Cow with Image Processing Techniques. *Computers and Electronics in Agriculture*, 74, 110–119. https://doi.org/10.1016/j.compag.2010.07.004.

Pradana, Z.H., Hidayat, B. and Darana, S., 2016. Beef Cattle Weight Determine by Using Digital Image Processing. The 2016 International Conference on Control, Electronics, Renewable Energy and Communications (ICCEREC).

Pujari, J.D., Yakkundimath, R. and Byadgi, A.S., 2015. Image Processing Based Detection of Fungal Diseases in Plants. *Procedia Computer Science*, 46, 1802–1808.

Qidwai, U. and Chen, C.H., 2009. *Digital Image Processing: An Algorithmic Approach with MATLAB*. CRC Press Taylor & Francis Group LLC, Boca Raton. ISBN: 978-1-4200-7951-7.

Reis, M.J.C.S., Morais, R., Peres, E., Pereira, C., Contente, O., Soares, S., Valente, A., Baptista, J., Ferreira, P.J.S.G. and Bulas Cruz, J., 2012. Automatic Detection of Bunches of Grapes in Natural Environment from Color Images. *Journal of Applied Logic*, 10, 285–290.

Robinson, D.A., Abdu, H., Lebron, I. and Jones, S.B., 2012. Imaging of Hill-Slope Soil Moisture Wetting Patterns in a Semi-Arid Oak Savanna Catchment Using Time-Lapse Electromagnetic Induction. *Journal of Hydrology*, 416–417, 39–49.

Salau, J., Haas, J.H., Junge, W., Leisen, M. and Thaller, G., 2015. Development of a Multi-Kinect-System for Gait Analysis and Measuring Body Characteristics in Dairy Cows. *Precision Livestock Farming Applications*. (Editor: Halachmi, I.) Wageningen Academic Publishers, Wageningen. 55–64.

Savary, S. and Willocquet, L., 2014. Simulation Modeling in Botanical Epidemiology and Crop Loss Analysis. The Plant Health Instructor. 173 p.

Shih, F.Y., 2010. *Image Processing and Pattern Recognition: Fundamentals and Techniques*. John Wiley & Sons, Inc., Hoboken. ISBN: 978-0-470-40461-4.

Su, D., Kong, H., Qiao, Y. and Sukkarieh, S., 2021. Data Augmentation for Deep Learning Based Semantic Segmentation and Crop-Weed Classification in Agricultural Robotics. *Computers and Electronics in Agriculture*, 190, 106418. https://doi.org/10.1016/j.compag.2021.106418.

Trotter, M.G., Lamb, D.W., Hinch, G.N. and Guppy, C.N., 2010. GNSS Tracking of Livestock: Towards Variable Fertilizer for the Grazing Industry. 10th International Conference on Precision Agriculture. 2010; 18–21 July, Denver, USA.

Truong Minh Long, N. and Truong Thinh, N., 2020. Using Machine Learning to Grade the Mango's Quality Based on External Features Captured by Vision System. *Applied Sciences*, 10, 5775. https://doi.org/10.3390/app10175775.

Tyagi, V., 2018. *Understanding Digital Image Processing*. CRC Press Taylor & Francis Group LLC, Boca Raton. ISBN: 978-1-138-56684-2.

Umbaugh, S.E., 2018. *Digital Image Processing and Analysis Applications with MATLAB and CVIPtools*. CRC Press Taylor & Francis Group LLC, Boca Raton. ISBN: 978-1-4987-6602-9.

Uygun, T., Ozguven, M.M. and Yanar, D., 2020. A New Approach to Monitor and Assess the Damage Caused by Two-Spotted Spider Mite. *Experimental and Applied Acarology*, 82(3), 335–346. https://doi.org/10.1007/s10493-020-00561-8.

Verger, A., Vigneau, N., Chéron, C., Gilliot, J.M., Comar, A. and Baret, F., 2014. Green Area Index from an Unmanned Aerial System Over Wheat and Rapeseed Crops. *Remote Sensing of Environment*, 152, 654–664.

Wosner, O., Farjon, G. and Bar-Hillel, A., 2021. Object Detection in Agricultural Contexts: A Multiple Resolution Benchmark and Comparison to Human. *Computers and Electronics in Agriculture*, 189, 106404. https://doi.org/10.1016/j.compag.2021.106404.

Xu, H., Li, S., Ji, Y., Cao, R. and Zhang, M., 2021. Dynamic Obstacle Detection Based on Panoramic Vision in the Moving State of Agricultural Machineries. *Computers and Electronics in Agriculture*, 184, 106104. https://doi.org/10.1016/j.compag.2021.106104.

Yang, C., Everitt, J.H., Du, Q., Luo, B. and Chanussot, J., 2013. Using High-Resolution Airborne and Satellite Imagery to Assess Crop Growth and Yield Variability for Precision Agriculture. *Proceedings of the IEEE*, 101(3): 582–592 March 2013.

Yang, C., Wei, X. and Wang, C., 2021. S-Box Design Based on 2D Multiple Collapse Chaotic Map and their Application in Image Encryption. *Entropy*, 23, 1312. https://doi.org/10.3390/e23101312.

Zhang, M. and Meng, Q., 2011. Automatic Citrus Canker Detection from Leaf Images Captured in Field. *Pattern Recognition Letters*, 32, 2036–2046.

Zhang, S., Wu, X., You, Z. and Zhang, L., 2017. Leaf Image Based Cucumber Disease Recognition Using Sparse Representation Classification. *Computers and Electronics in Agriculture*, 134, 135–141.

Zhang, S., Wang, H., Huang, W. and You, Z., 2018. Plant Diseased Leaf Segmentation and Recognition by Fusion of superpixel, K-Means and PHOG. *Optik*, 157, 866–872.

Zhou, R., Damerow, L., Sun, Y. and Blanke, M.M., 2012. Using Colour Features of Cv.'Gala' Apple Fruits in an Orchard in Image Processing to Predict Yield. *Precision Agriculture*, 13, 568–580.

7 Data Mining in Agriculture

7.1 WHAT IS DATA MINING?

Nowadays, vast and continuous data is collected from a variety of sources in many fields. As the data grows, the complexity increases. Data mining algorithms are used to extract useful data from such enormous data. Data mining is an iterative process in which large volumes of data, from gigabytes to terabytes or sometimes zettabytes, known as data warehouses, are automatically discovered for previously unknown, valuable, and useful patterns and models. Extracting these useful data is for a wide variety of purposes, but generally for decision-making or problem-solving on any subject. Data discovery and extraction consist of estimation and identification studies. As the data grew, the accuracy of the prediction increased.

The goal of predictive data mining is to produce a model, expressed as executable code, that can be used to perform classification, prediction, estimation, or other similar tasks. The purpose of descriptive data mining is to understand the analyzed system by revealing patterns and relationships in large data sets. The relative importance of estimation and identification for particular data mining applications can vary considerably. Estimation and identification objectives can be achieved using the following data mining techniques (Kantardzic, 2020):

- *Classification*: Discovery of a predictive learning function that classifies a data item into one of several predefined classes;
- *Regression*: Discovery of a predictive learning function that maps a data item to a real-valued predictive variable;
- *Clustering*: A common descriptive task attempting to define a limited set of categories or clusters to describe data;
- *Summarization*: An additional descriptive task that includes methods for finding a compact description for a dataset (or subset);
- *Dependency modeling*: Finding a local model that describes significant dependencies between variables in a dataset or part of a dataset or between values of a property;
- *Change and deviation detection*: Discovering the most significant changes in the dataset.

While data mining and machine learning use automated methods to analyze patterns in all kinds of data, data mining focuses on extracting unknown features from historical data. Machine learning focuses on predictions made from learned data based on known features. In fact, the two fields overlap in many ways. Data mining uses many machine learning methods while machine learning uses data mining methods such as pre-processing steps to improve unsupervised learning accuracy (Özgüven, 2019).

7.2 DATA MINING PROCESS

Data mining is not a random application of statistical, machine learning, and other methods and tools. It is a carefully planned and considered process for deciding what will be most useful, promising, and illustrative in the field of analytical techniques. While analyzing the data in data mining, some analytical techniques are used, or it may be decided to look at the data in another way; the data may be changed or another data analysis tool may be applied by going back to the beginning, and either better or different results can be achieved. This relapse can happen many times and different aspects of the data can be investigated by asking a different question of the data with each technique applied. Therefore, it is important to understand that the problem of discovering or pre-dicting dependencies from data or discovering entirely new data is only part of the experimental procedure that uses standard steps to draw conclusions from the data. The general experimental procedure adapted to data mining problems includes the following (Figure 7.1) steps (Kantardzic, 2020):

- *State the problem and formulate the hypothesis*: Most data-based modeling work is per-formed in a specific application domain. Therefore, it requires domain-specific knowl-edge and experience and combined expertise of a data mining model to come up with a meaningful and clear problem statement. In practice, it usually means a close interaction between the data mining specialist and the application specialist. In successful data min-ing applications, this cooperation does not end at the initial stage, it continues throughout the entire data mining process. Therefore, in this step, a modeler usually specifies a set of variables for the unknown dependence and, if possible, a general form of that dependence as an initial hypothesis. At this stage, there may be several hypotheses formulated for a single problem.
- *Collect the data*: This step is about how data is produced and collected. In general, there are two different possibilities. First, the data creation process is under the control of an expert (modeler). This approach is known as a designed experiment. The second possi-bility is when the expert cannot influence the data generation process, and this is known as the observational approach. Most data mining applications assume an observational setting, that is, random data generation. Typically, the sampling distribution is not com-pletely known after data collection or is partially and implicitly given in the data collection procedure. However, it is crucial to understand how data collection affects its theoretical distribution, as such prior knowledge can be very useful for modeling and then for the final interpretation of the results. It is also important to make sure that the data used to predict a model and the data used to test and apply a model later on come from the same unknown sampling distribution. If this is not the case, the predicted model cannot be used success-fully in the final application of the results.
- *Pre-processing the data*: In the observational environment, data is usually taken from existing databases and data warehouses. Data pre-processing usually includes the detection

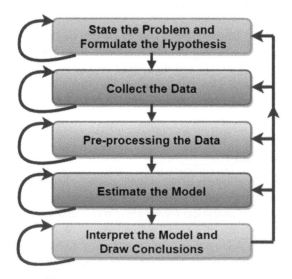

FIGURE 7.1   Data mining process.

(and removal) of outliers and several tasks such as variable scaling and different encoding types. Outliers are unusual data values that are not consistent with most observations. Usually, outliers result from measurement errors, coding errors, and registration errors, and are sometimes natural, abnormal values. Such non-representative samples can seriously affect the model produced later on. Features in the range (0, 1) and (–100, 1,000) will not have the same weight in the applied technique and the features will affect the final data mining results differently. Therefore, it is recommended to scale them up for further analysis and weight both features the same. In addition, application-specific data preparation and data size reduction practices are also carried out at the data pre-processing stage.

- *Estimate the model*: The selection and application of the appropriate data mining technique is the main task at this stage. This process is not simple. Generally in field applications, the application is based on several models, and choosing the best model is the most important task.
- *Interpret the model and draw conclusions*: Data mining models should assist in decision-making. Therefore, these models must be interpretable to be useful. A user does not want hundreds of pages of numerical results because such a result is not understandable. The user cannot summarize, interpret, and use these results for successful decision-making. Modern data mining methods are expected to yield highly accurate results using high-dimensional models. At the same time, the interpretation of these models and the verification of the results with special techniques are important tasks at this stage.

7.3   DATA PRE-PROCESSING

Data pre-processing in data mining is a broad field and consists of several different strategies and techniques that are interrelated in complex ways. These strategies and techniques are divided into two categories: Selecting data objects and attributes for analysis or creating/modifying attributes. In both cases, the goal is to improve data mining analysis in terms of time, cost, and quality. Important data pre-processing applications are given below (Tan et al., 2014):

- *Aggregation*: There are several motivations for aggregation. First, smaller datasets resulting from data reduction require less memory and processing time and thus aggregation may allow the use of data mining algorithms. Second, aggregation can act as a scope or scale change by providing a high-level view of the data rather than a low-level view. Finally, the behavior of groups of objects or attributes is often more stable than individual objects or attributes.
- *Sampling*: Sampling is an approach in which sample size and sampling techniques are used to select a subset of data objects to be analyzed. Sampling is a statistical process. A sample is representative if it has approximately the same property as the original data set, or if the mean of the data objects is the feature of interest, it has a mean close to the original data. If the representativeness is high, it will work almost as well as using the entire dataset.
- *Dimensionality reduction*: Datasets can have thousands or tens of thousands of features. Many data mining algorithms work better if the dimensionality (number of features in the data) is lower. Reducing dimensionality has several benefits. These benefits are that it can eliminate irrelevant features, reduce noise, allow easier visualization of data, and reduce the amount of time and memory required by the data mining algorithm.
- *Feature subset selection*: Another way to reduce dimensionality is to use only a subset of the features. While such an approach would seem to lose information, information is not lost because redundant features duplicate most or all of the information contained in one or more other attributes. A dataset with redundant and irrelevant features can reduce the classification accuracy and quality of the clusters found. The ideal approach to feature selection is to try all possible subsets of features as input to the respective data mining algorithm

and then take the subset that gives the best results. There are three standard approaches to feature selection: Embedded, filter, and wrapper.

- *Feature creation*: It is possible by creating a new feature set that captures important information in a dataset much more effectively. Also, the fact that the number of new features is less than the number of original features allows us to enjoy the benefits of dimensionality reduction. There are three methodologies for creating new attributes: feature extraction, mapping data to a new space, and feature construction.
- *Discretization and binarization*: Some data mining algorithms, especially certain classification algorithms, require data to be in the form of categorical attributes. Algorithms that find association patterns require data to be in the form of binary attributes. Therefore, it is often necessary to convert a continuous attribute to a categorical attribute (discretization), and it may be necessary to convert both continuous and discrete attributes into one or more binary attributes (binarization). Also, if a categorical attribute has many values (categories) or some values appear infrequently, reducing the number of categories by combining some values can be beneficial for certain data mining tasks. As with feature selection, the best discretization and binarization approach is the one that "produces the best results for the data mining algorithm that will be used to analyze the data". It is often impractical to directly apply such a criterion. As a result, discretization or binarization is performed to meet a criterion that is considered to be associated with good performance for the data mining task under consideration.
- *Variable transformation*: Variable conversion refers to a transformation applied to all values of a variable (attribute). For example, if only the size of a variable is important, the values of the variable can be converted by taking the absolute value. There are two important forms of variable transformation: Simple functional transformation and normalization.

7.4 DATA MINING TECHNIQUES

Data mining techniques can be classified into four groups as given here (Bhatia, 2019):

- *Predictive modeling*: Predictive modeling is based on predicting the outcome of an event. It was developed using a supervised learning approach with classification or regression on a model similar to the human learning experience in using observations to build a model of key features of some tasks. This model includes some labeled data, and these data are used to predict the outcome of unknown samples.
- *Database segmentation*: Database segmentation is based on the concept of data clustering and falls under unsupervised learning where data is not labeled. These data are divided into groups or clusters according to their characteristics or attributes. Segmentation is creating a group of similar records that share a set of characteristics.
- *Link analysis*: Link analysis aims to establish links, called relationships, between individual records or sets of records in a database. Link analysis has three specialties: Association discovery, sequential pattern discovery, and similar time sequence discovery. Association discovery finds elements that imply the presence of other elements in the same event. Sequential pattern discovery finds patterns among events. Time sequence discovery is used to determine whether there is a connection between two time-dependent datasets.
- *Deviation detection*: Deviation detection is a relatively new technique in terms of existing data mining tools. It is based on the detection of outliers in the database that show deviations from some previously known expectations and norms. This process can be performed using statistical and visualization techniques.

Many data mining techniques have been developed over the years. Some of these are conceptually very simple, while others are more complex. In data mining, the goal is to divide the data into different categories, each representing some of the properties that the data may have. For example, K-nearest neighbor, neural network, and support vector machines can be used for a classification application, and K-means binary clustering methods can be used for a clustering application. Statistical methods such as principal component analysis and regression techniques are widely used to find patterns in datasets (Mucherino et al., 2009). The schematic representation of the classification of data mining techniques is given in Figure 7.2.

7.4.1 CLASSIFICATION TECHNIQUES

Since classification is one of the most basic operations in the prediction part of data mining, it constitutes an important part of the problems encountered (Ozguven et al., 2019). It consists of classification, model creation, and model use stages. During model creation, it is assumed that each instance belongs to a predefined class specified by the class variable. The set of examples used for model building is called training data. With the models created using classification algorithms, classification rules, decision trees, or mathematical formulas are introduced for the estimation of the class variable. At the stage of model use, the accuracy of the model is estimated by comparing the known class value of the samples in the test data with the class value obtained as a result of the model. The set of examples included during model use is called the test set. The accuracy rate is the percentage of test set samples correctly classified by the model. The test set is independent of the training set, otherwise, there will be overfitting. If the accuracy rate obtained with the use of the model is at an acceptable level, this model is used to classify unknown or future new samples. After the model use step, the performance of the classifier is evaluated by considering the success of the correct classification, speed, stability, scalability, comprehensibility, and structure of the rules (Özcan, 2015).

7.4.1.1 K-Nearest Neighbor

The K-nearest neighbor (K-NN) algorithm is most frequently used for classification, although it can also be used for estimation and prediction. The K-NN is an example of instance-based learning where the training dataset is stored. Thus, a classification for a new unclassified record can be found simply by comparing it to the most similar records in the training set (Larose and Larose, 2015). In general, however, this classification rule is sometimes based on a known sample, so its success can sometimes be poor. The result can be correct when the unknown sample is surrounded by several known samples with the same classification. Instead, the accuracy of the classification may decrease if the surrounding samples have different classifications, for example, when the unknown sample

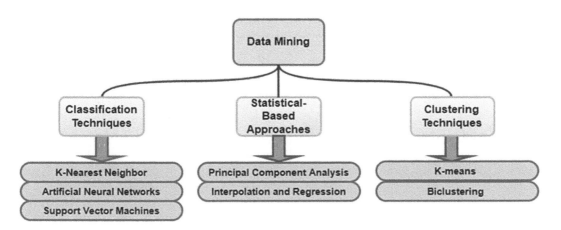

FIGURE 7.2 Schematic representation of the classification of data mining techniques.

is located between samples belonging to two different classes. To increase the level of accuracy, all surrounding samples should be considered, and the unknown sample should be classified accordingly. In general, the classification rule based on this idea assigns any unclassified sample to the class containing most of its K-NN (Cover and Hart, 1967).

Figure 7.3 shows the K-NN decision rule for k = 1 and k = 3 for a set of examples divided into two classes. In Figure 7.3 (left panel), an unknown sample is classified using only one known sample. Since more than one known sample is used in Figure 7.3 (right panel), the parameter k is set to 3 and the closest four samples are taken into account to classify the unknown. Three of them belong to the same class, while only one belongs to the other class. In both cases, the unknown sample is classified as belonging to the class on the left.

The distance function plays a very important role in the success of classification, as in many data mining techniques. The distance function is a function that means the smaller the distance between samples, the more likely the samples belong to the same class. Another important factor is the choice of value for parameter k. This is the main parameter of the method as it represents the number of nearest neighbors considered for the classification of an unknown sample. Usually, it is pre-fixed. If k is very large, classes with a large number of classified instances may overtake small ones, and the results will be biased. On the other hand, if k is very small, the advantage of using many samples in the training set is not taken advantage of. Usually, the value of k is optimized by trials on training and validation sets. Also, in some special cases, it would not be correct to assign a classification based on the k "votes" majority of the nearest neighbors. For example, if the distances of the nearest neighbors vary greatly, an unknown sample can be classified by considering samples farther away from it. Therefore, a more complex approach would be to weight each sample's vote by distance so that the closest samples are of greater importance during classification. Figure 7.4 shows a training set in which the samples are classified into four different classes. Instances belonging to different classes are marked with different symbols. Circled samples are classified using the K-NN rule with k = 3. According to the previous algorithm to find a condensed training set, correctly classified samples can be discarded. For example, sample A is correctly classified: Its three neighbors have the same classification. Example B was incorrectly classified by K-NN. Two neighbors have □ classification and one neighbor has ● classification. Therefore, B was classified as □ while its original classification was ●. Example C is also misclassified. In this case, C is closest to an instance of its own class, but the other two neighbors have a + classification (Mucherino et al., 2009).

7.4.1.2 Artificial Neural Networks

Thanks to their generalization feature, Artificial Neural Networks (ANNs) are able to learn from past events or examples the same as people and can make decisions on new examples that they have never encountered in the future, thanks to the experience they have gained (Öztemel, 2006). The

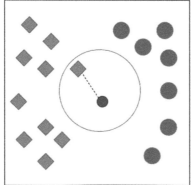
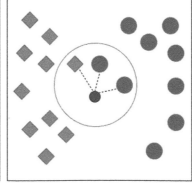

FIGURE 7.3   K-NN decision rule for k = 1 and k = 3.

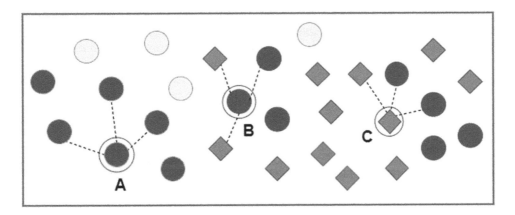

FIGURE 7.4 Examples of correct and incorrect classification with K-NN.

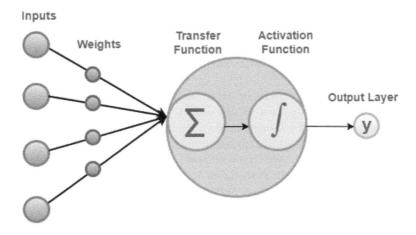

FIGURE 7.5 Real neuron and artificial neuron model.

powerful reasoning and inference capabilities of ANNs in control are used in adaptive control and learning, pattern recognition, classification, and prediction (Huang and Nof, 2001).

The inspiration for ANNs is the understanding that the complex learning systems in animal brains are composed of tightly interconnected clusters of neurons. While the structure of a given neuron is relatively simple, dense networks of interconnected neurons can perform complex learning tasks such as classification and pattern recognition. For example, the human brain contains about 10^{11} neurons, each connected to an average of 10,000 other neurons, creating a total of $1,000,000,000,000,000 = 10^{15}$ synaptic connections. ANNs represent a very basic attempt to emulate the type of nonlinear learning that occurs in naturally occurring neuronal networks (Larose and Larose, 2015). As shown in Figure 7.5, an ANN has five basic elements: Inputs, weights, transfer function, activation function, and outputs. The structure collects the information it receives from the outside in an aggregation function and passes it through the activation function, creates the output, and sends it to other cells over the connections of the network (Öztemel, 2006).

As seen in Figure 7.5, the inputs and the weights show the effect of the information coming to the artificial neuron. The transmission function ($\sum$) calculates the net information coming to a neuron. The activation function determines the values of the outputs to be produced by calculating the net input information coming to the cell (Terzi et al., 2019).

One of the advantages of using ANN is that they are pretty robust to noisy data. Because the network contains many nodes (artificial neurons) with weights assigned to each link, the network can learn to work around these non-informative (or even erroneous) examples in the dataset. However, unlike decision trees, which generate heuristic rules understandable to non-experts, neural networks are relatively opaque to human interpretation, as we will see. Also, neural networks often require longer training times than decision trees and can often take up to several hours (Larose and Larose, 2015). The use of ANNs offers several useful features and capabilities (Kantardzic, 2020):

- *Nonlinearity*: Although an artificial neuron as the basic unit can be a linear or nonlinear processing element, not all ANNs are highly linear. This feature is particularly important for ANN models, which are intrinsically nonlinear real-world mechanisms responsible for generating data for learning.
- *Learning from examples*: An ANN changes the interconnection weights by applying a set of training or learning examples. The ultimate effects of a learning process are the tuned parameters of a network. The parameters are distributed through the main components of the established model and represent implicitly stored information for the problem at hand.
- *Adaptivity*: An ANN has a built-in ability to adapt its interconnect weights to changes in the surrounding environment. In particular, an ANN trained to operate in a particular environment can be easily retrained to cope with changes in environmental conditions. Also, when operating in a non-stationary environment, an ANN can be designed to accept its parameters in real time.
- *Evidential response*: In the context of data classification, an ANN can be designed to provide information not only about which class to choose for a given sample, but also about confidence in the decision being made. This later information can be used to reject ambiguous data, should they arise, and thus improve classification performance or performances of the other tasks modeled by the network.
- *Fault tolerance*: An ANN has the potential to be inherently fault-tolerant or robustly computational. Their performance does not degrade significantly under adverse operating conditions such as neurons being disconnected and noisy or missing data.
- *Uniformity of analysis and design*: Basically, ANNs have universality as information processors. All areas involving ANN applications use the same principles, notations, and steps in methodology.

An ANN architecture is defined by the characteristics of a node and its connectivity in the network. Typically, network architecture is defined by the number of inputs to the network, the number of outputs, the total number of basic nodes, which are usually equal processing elements for the entire network, and their organization and interconnections. An ANN is generally divided into two categories according to the type of interconnections: Feed-forward and recurrent. Examples of ANNs belonging to both classes are given in Figure 7.6 and Figure 7.7. If the process propagates unanimously from the input side to the output side without any looping or feedback, the network should feed forward. If a network has a feedback link that creates a circular path (usually with a delay element as a synchronization component), then the network is recurrent. Although many neural network models have been proposed in both classes, the multilayer feed-forward network with a backpropagation-learning mechanism is the most widely used model in terms of practical applications (Kantardzic, 2020).

The multilayer feed-forward network seen in Figure 7.6 is a data mining model with superior performance rates used for classification prediction. The multilayer feed-forward network model consists of input layer, hidden layer, and output layer components. The hidden layer contains neurons with nonlinear activation functions (Ozguven et al., 2019).

While some applications only have one or only two hidden layers, in some applications it is better to have more than two layers. Input data is provided to the network through the input layer, which

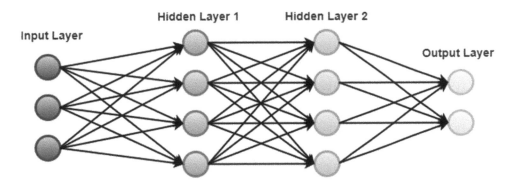

FIGURE 7.6 Feed-forward neural network.

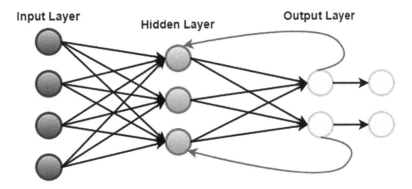

FIGURE 7.7 Recurrent neural network.

sends this information to the hidden layers. Data is processed by hidden layers and the output layer. Each neuron receives output signals from neurons in the previous layer and sends its output to neurons in the next layer. The last layer, the output layer, receives inputs from neurons in the last hidden layer and the neurons provide the output values. The neurons of the input layer do not perform any calculations as they are only allowed to receive the data; they send it to the first hidden layer. Layer by layer, then, the neurons communicate among themselves and process the data they receive. The network can provide the output values after the inputs have propagated from the input layer to the output layer throughout the entire network (Mucherino et al., 2009).

7.4.1.3 Support Vector Machines

It is known that one of the various algorithms used for the correct classification of datasets in data mining is support vector machines (SVM). SVM, which was first proposed by Vapnik (1995), is a supervised learning model in data mining methods used for the classification of linearly separable or nonlinearly separable binary or multiple datasets (Ozguven et al., 2019). SVM is a classifier method that basically performs classification tasks by constructing a linear dividing line or hyperplanes between distributions of points in multidimensional space. In the 2-D view, points on one side of this hyperplane are considered positive and the remaining points are considered negative (Dua and Chowriappa, 2013; Zhang, 2017). This hyperplane is shown in Figure 7.8.

The decision boundary (or hyperplane) is represented using hyphens and the function f(x). The maximum margin limit is calculated with a linear SVM. The region between the two lines defines the margin area. Data points highlighted with black center lines are support vectors. Therefore, the first goal of the algorithm is to maximize this margin of separation. Because the placement of the hyperplane is closely linked to the distribution of the data points, it becomes difficult to adhere to

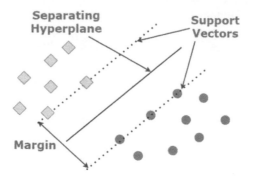

FIGURE 7.8   2-D representation of a linear classifier separating two classes.

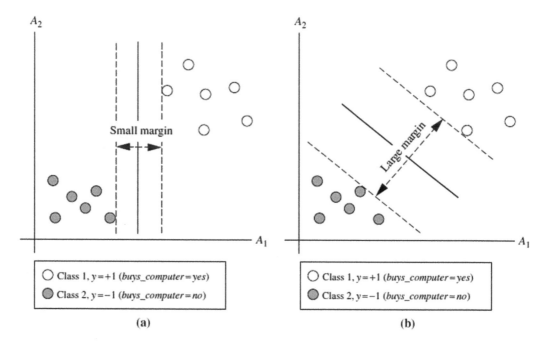

FIGURE 7.9   Two possible separating hyperplanes and their associated margins are seen. Which one is better? The one with the larger margin (b) should have greater generalization accuracy (Han et al., 2012).

a maximum separation margin when the data points overlap in their distribution. When data point distributions overlap, the data is believed to be nonlinear in nature. It is possible to extend linear SVM to suit the nonlinearity of data using kernel functions. As the name suggests, kernel functions are transformation functions that transform a linear classifier into a nonlinear classifier. Such functions consist of mapping the nonlinear data into an abstract feature space where the maximum margin of separation exists (Dua and Chowriappa, 2013).

There are an infinite number of dividing lines that can be drawn. Considering that our data is 3-D (i.e., with three features), we try to find the best separation plane and the best line with the minimum classification error on previously unseen tuples. SVM approaches this problem by seeking the maximum marginal hyperplane (MMH) to express the decision boundary we are looking for regardless of the number of input attributes. Figure 7.9 is given to illustrate two possible separating hyperplanes and their associated margins (Han et al., 2012).

Both hyperplanes in Figure 7.9 can correctly classify all given data groups. However, intuitively, the larger margin hyperplane is expected to be more accurate in classifying future datasets than the smaller margin hyperplane. Therefore (in the learning or training phase), SVM looks for the hyperplane with the largest margin, i.e., MMH. The associated margin gives the largest separation between classes. Margin can be said to be the shortest distance from a hyperplane to one edge of its margin which equals the shortest distance from the hyperplane to the other side of the margin. Here, the margin "edges" are parallel to the hyperplane. When dealing with MMH, this distance is actually the shortest distance from MMH to the nearest training group of either class (Han et al., 2012).

7.4.2 Statistically Based Approaches

Statistics is a discipline concerned with collecting, analyzing, defining, visualizing, and drawing inferences from data. The focus is on defining the properties of a dataset and the relationships that exist between data points. Descriptive statistics describe or visualize key features of the data under study. An example would be to find the best-selling retail product in a store in a given time period. Inferential statistics, on the other hand, are used to draw conclusions that apply to more than just the data studied. This is necessary when the analysis must be performed on a smaller, representative dataset when the actual population is very large or difficult to study. Because the analysis is done on a subset of the total data, the results that can be reached with inferential statistics are never 100% accurate and are instead just probabilistic estimates. Election polls can be given as an example in this regard. These surveys are based on surveying a small percentage of citizens to gauge the emotions of the entire population. Data mining, on the other hand, is the automation of exploratory statistical analysis on large-scale databases, but the term is often used to describe any algorithmic data analysis and information processing, which may include machine learning and deep learning techniques. The purpose of data mining is to extract patterns and information from large-scale datasets so that they can be translated into a more understandable structure for later analysis (Yao et al., 2018). Ranging from one-dimensional data analysis to multivariate data analysis, statistics offer a variety of methods for data mining, including different types of regression and discriminant analysis (Kantardzic, 2020).

7.4.2.1 Principal Component Analysis

Principal component analysis (PCA) is a method used to reduce the size of a given dataset while maintaining the variability present in the set. Each dataset contains information represented by vectors of single variables (usually real, integer, or binary). For example, a geometric point in 3-D space can be represented by a vector with three variables, each associated with one of the three coordinate axes x, y, and z. In general, a sample can be represented by a vector formed by a certain number of variables. The number of such variables defines the length of the vectors in the set and hence the size of the set. In addition, a certain range of variability can be defined for each variable, which determines the range of values that a single variable can take. For example, if the dataset contains 3-D points that have one side and are bounded by a cube with center (0, 0, 0) then three variables representing Cartesian coordinates are constrained to have values in (−1/2, 1/2). The interval defines the range of variability of three variables. The purpose of PCA is to find hidden patterns among the data and transform the original data in a way that highlights their similarities and differences. Once patterns are found, data can be represented as components ordered by relevance. It is possible to discard components of low relevance without loss of significant information later on (Mucherino et al., 2009).

7.4.2.2 Interpolation and Regression

In interpolation and regression techniques for data mining, the goal is to model a given data set with an appropriate mathematical function. Datasets from real applications often contain a discrete number of examples that describe a particular process or event. By applying interpolation or regression techniques, it is attempted to find a function that can describe this phenomenon or process in general. Suppose that the amount of water y in a given soil is monitored at time x. Experimental analysis can be used to obtain y at different times x. Thus, a series of points (x', y') can be defined. As always in real-life applications, the number of experiments is discrete and limited, whereas we are looking for a general function that can always relate x to a water level y. Finding this function using available data (in this case the x' and y' pairs) means finding a model that can provide the correct water level y for x at any time. In this simple example, the points (x', y') belong to a two-dimensional space, so any function defined in R and having values in R can be a good model for the process under consideration (Mucherino et al., 2009).

In mathematics in general, given an independent variable x, a function f provides a value for the corresponding dependent variable y = f(x). These functions must comply with the following specifications. Given a known x' they should be able to provide the corresponding y or a good approximation of y. They also need to be able to generalize: Given an unknown x (no pairs (x, y) are found in the dataset), the y-value provided by the function should be an estimate of the modeled process behavior. The ultimate goal is to find the general rule about x and y. For example, suppose water levels y in a given soil are measured x every hour for ten consecutive hours. The ten pairs (x', y') containing the details of these measurements represent the current dataset. An interpolation or regression function modeling this dataset is required to provide a good estimate of y water levels, even for x-multiplications not included in the data. If this goal is achieved, no further measurements are required, but the process can be monitored using the model obtained (Mucherino et al., 2009).

7.4.3 Clustering Techniques

Organizing data into logical groups is one of the most fundamental approaches to understanding and learning. Cluster analysis is the study of methods and algorithms for naturally grouping or clustering objects based on their measured or perceived intrinsic properties or similarity. For clustering, samples are represented as a measurement vector or more formally as a point in a multidimensional space. Instances within a valid cluster are more similar to each other than an example belonging to a different cluster. Clustering methodology is particularly suitable for investigating interrelationships among samples to make a preliminary assessment of sample structure. Human performances can compete with automated clustering procedures in one, two, or three dimensions, but most real problems involve clustering in higher dimensions. It is very difficult for humans to intuitively interpret data embedded in a high-dimensional space. Clustering is a very difficult problem because data can reveal clusters of different shapes and sizes in an n-dimensional data space. To further complicate the issue, the number of clusters in the data often depends on the resolution (fine and coarse) at which we are viewing the data. Figure 7.10 is given in order to analyze the problem of separating the points into several groups (Kantardzic, 2020). Figure 7.10a shows a series of points distributed in a 2-D plane. The number of N groups is not given beforehand. Figure 7.10b shows the natural clusters of G1 and G2 bordered by curves. Since the number of clusters is not given, there is another

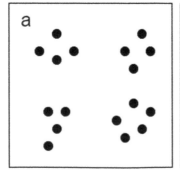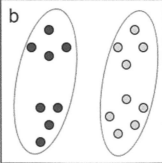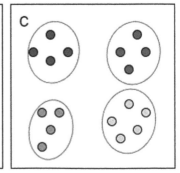

FIGURE 7.10   Cluster analysis of points in 2-D space. (a) Initial data, (b) two datasets, (c) four datasets.

section of four clusters in Figure 7.10c that is as natural as the groups in Figure 7.10b. This kind of arbitrariness for the number of clusters is a big problem in clustering.

7.4.3.1 K-Means

The K-means algorithm defines the centroid of a cluster as the mean value of the points in the cluster. First, it randomly selects k objects in D, each of which initially represents a cluster mean or center. For each of the remaining objects, an object is assigned to the cluster with which it is most similar, based on the Euclidean distance between the object and the cluster mean. The K-means algorithm then iteratively improves the within-cluster variation. For each cluster, it calculates the new mean using the objects assigned to the cluster in the previous iteration. All objects are then reassigned as new cluster centers using updated tools. Iterations continue until the assignment is stable, i.e., the clusters created in the current round are the same as those created in the previous round. The K-means procedure is summarized in Figure 7.11 (Han et al., 2012).

Clustering with K-means partitioning as shown in Figure 7.11a, when considering a set of objects in 2-D space, when k = 3, that is, the user wants the objects to be divided into three clusters. Each object is assigned to a cluster based on the cluster center it is closest to. Such a distribution creates silhouettes surrounded by dotted curves, as shown in Figure 7.11a. Next, the cluster centers are updated. That is, the mean value of each cluster is recalculated based on the available objects in the cluster. Using new cluster centers, objects are redistributed into clusters according to which cluster center is closest. Such a redistribution creates new silhouettes surrounded by dashed curves, as shown in Figure 7.11b. This process is repeated and leads to Figure 7.11c. The process of iteratively reassigning objects to clusters to improve partitioning is called iterative relocation. In the end, reassignment of objects in any cluster does not occur, thus ending the process. The resulting clusters are returned by the clustering process (Han et al., 2012).

7.4.3.2 Biclustering

The purpose of biclustering techniques is to find some of the samples (Sr) and their features (Fr) in biclusters. In this way, not only is a part of the samples obtained, but also the features that cause this partition are identified. As with standard clustering, single clusters Sr and Fr can be labeled from 0 to k-1. Independently, the clusters Sr can be sorted by their own labels, and the same can be done for the clusters Fr. A color or a gray scale can be associated with each label and a pixel matrix can be created. In the rows of this matrix, the clusters Fr are ordered by their labels and the clusters Sr are ordered in the columns. Even though this matrix is built considering the clusters Sr and Fr independently, it gives a graphic visualization of the biclusters (Sr, Fr). The matrix shows a checkerboard pattern in which the biclusters can be easily identified (Mucherino et al., 2009). Biclustering techniques were first proposed to address the need for analyzing gene expression data. The gene expression data or DNA microarray data are conceptually a gene sample/condition matrix in which each

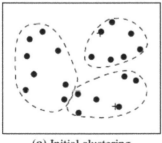 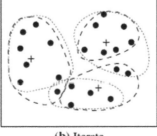 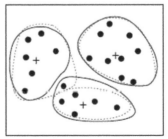

| (a) Initial clustering | (b) Iterate | (c) Final clustering |

FIGURE 7.11   Clustering a set of objects using the K-means method (the mean of each cluster is marked with a +) (Han et al., 2012).

row corresponds to a gene, and each column corresponds to a sample or condition. Each element in the matrix is a real number and records the expression level of a gene under a given condition. Figure 7.12 shows the analysis of a gene expression data matrix in two dimensions (gene dimension and sample/condition dimension) from a clustering perspective (Han et al., 2012).

7.5   DATA MINING PRACTICE EXAMPLES IN AGRICULTURE

By evaluating the hardware, algorithms, and software that emerged with the development of technology together with the existing knowledge and experience in agriculture, it has been possible to facilitate agricultural operations and to bring alternative solutions to the problems that await solution or improvement. The use of technological methods, models, and tools that manage the processes of obtaining, processing, storing, transferring, and using information, as well as portable computers and hardware with high processing and computational power, can be found easily in the field, and their use in field applications has increased. Depending on the experience gained during the increasing use, the development has enabled it to be applied to different areas (Özgüven et al., 2020). To process this information and turn it into useful information, advanced decision mechanisms are used, such as image, audio, and video processing algorithms, artificial neural networks, machine learning, and statistical data analysis, and the autonomous tractor works successfully (Özgüven, 2018).

The sensors were previously used only in automation applications in agricultural applications. In these automation applications, often repetitive tasks were performed in which the data from the sensors were evaluated and the desired output was obtained. The development of data mining and machine learning techniques has led to new very important applications in this field. In these applications, it has been possible to develop a wide variety of expert systems by using various computational techniques that replace human ability for real-time evaluation of data obtained from sensors and decision-making. Especially with real-time monitoring of images and sounds, a wide variety of agricultural applications are made, and necessary business decisions can be made automatically. The advantages of different techniques in data mining and machine learning techniques are used

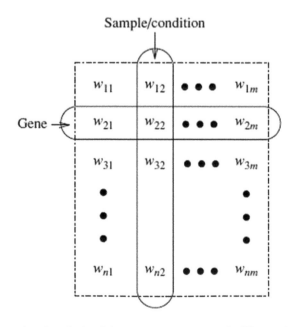

FIGURE 7.12   Two-dimensional analysis of the microarray data matrix (Han et al., 2012).

for problem-solving or increasing system success in various agricultural applications, especially clustering and classification.

The data mining methods are an important research area for agricultural science and a promising tool for problems in the agricultural sector that need solutions and improvement. Through the identification of plant, weed, disease, and soil pattern classes using data mining methods in agriculture, numerous studies have been conducted such as fertility, biomass, chlorophyll, breeding environment and agricultural drought forecasts, various analyses, modeling and simulations, mineral elements, toxic elements, organic and inorganic pollution, and evaluation of biological markers. Summary information about some of the studies carried out is given in Table 7.1 (Ozguven et al., 2019).

7.5.1 Use in Clone Selection

Ozguven et al. (2019) examined the clones that were developed in line with the potato variety breeding program initiated with hybrid combinations, using multilayer perceptron neural network (MLPNN) and support vector machine (SVM) data mining methods, and determined that they were ineligible and to be eliminated by negative selection. A total of 703 potato clones from 12 hybrid combinations were used in the study. To identify the selected clones, two different models were created using three features of each clone (number of tubers and average tuber weight), and two features of each clone (eye depth and eye pit depth) were used to identify clones. A total of four models were used together with two different models created. In the study, experiments were carried out by comparing the sensitivity, specificity, and accuracy rates by using the generated dataset as input to the MLPNN and SVM classifiers for each model. As a result of the experimental studies, the highest success was achieved in two-class models, and it was determined that the MLPNN classifier was more successful in these models. In addition, it has been reported that this study demonstrates that data mining methods can be used for early generation selection in cultivar development studies in plants. In the study, using the models (Model-1, Model-2, Model-3, and Model-4) created based on the data mining approach seen in Table 7.2, the same clones and the datasets of these clones were processed, the results were classified, and the results obtained by the classical methods were classified. The results were very close to the classical results. As a result of the study, it was found that the MLPNN method was more successful in the two-class models with positive and negative selections

TABLE 7.1

Some Studies Using Data Mining Methods in Agriculture

Author	Year	Number of data	Method	Accuracy (%)
Camargo and Smith	2009	45	SVM	93.1
Ribeiro et al.	2011	64	Genetic algorithm (GA)	76–92
Guerrero et al.	2012	106	SVM	93.1
Bhange and Hingoliwala	2015	610	SVM	82
Taghavifar et al.	2015	100	ANN-GA	96.72
VijayaLakshmi and Mohan	2016	150	Kernel-based PSO and FRVM (fuzzy relevance vector machine)	99.87
Nguyen-ky et al.	2017	19,723	Artificial neural network – Bayesian	99
Beucher et al	2017	1,733	ANN-DTC (decision tree classification)	52–54
Liu et al.	2018	262	ANN	99.6
Akbarzadeh et al.	2018	200–7,800	NDVI-based SVM-Gaussian-kernel SVM	70–97
Niell et al.	2018	312	SVM	57
Huang et al.	2018	537	SVM	96

Source: Ozguven et al. (2019).

TABLE 7.2

Success Rates of Experiments on Classification of Potato Clones

Problem	Model	Classifier	Class	Sens.	Spec.	Acc.
Positive selection	Model-1 (selected/other)	MLPNN	2	80	96.96	**94.17**
		SVM		70	97.67	93.12
	Model-2 (other/TY/TC/ATW including selected)	MLPNN	4	32.26	93.02	86.35
				61.76	93.02	
				69.44	93.02	
		SVM		48.39	93.2	**88.51**
				76.47	93.2	
				75	93.2	
Negative selection	Model-3 (eliminated/other)	MLPNN	2	78.05	99.36	**96.87**
		SVM		73.17	99.52	96.44
	Model-4 (other/ED/EPD/TS including eliminated)	MLPNN	4	45.46	96.78	89.62
				25.64	96.78	
				42.86	96.78	
		SVM		68.18	96.14	**91.18**
				41.03	96.14	
				61.91	96.14	

Source: Ozguven et al. (2019).

(Model-1 and Model-3), and the success rates were found as 94.2% and 96.9%, respectively. It was found that the SVM method was more successful in four-class models (Model-2 and Model-4) and the success rates were found 88.5% and 91.2%, respectively.

7.5.2 MONITORING ANIMAL EATING BEHAVIORS

Campos et al. (2018) presented a method for classifying different grass-eating behaviors by the surface electromyography (sEMG) signal of the masseter muscle. The main hypothesis tested in the study is whether rumination and food eaten can be distinguished from sEMG signal features using machine learning techniques. The three scenarios examined were differentiation (IR) between ruminant and grazing, food identification (FC) for four different foods, and both conditions combined (FCR). The entire data collection, instrumentation, and pattern recognition system is summarized in Figure 7.13. In the study, a new segmentation technique was developed and applied to automatically subdivide the chewing motion signal and evaluated by eight features extracted from seven classifier signals (LDA, QDA, SVM, MLPNN, RBF-NN, KNN, and MLPNN) and combined into five sets. Despite the similar characteristics of grasses, it has been reported that the results of food recognition were found to be reasonable and that feeding and rumination were distinguished with relatively high accuracy.

7.5.3 ANIMAL IDENTIFICATION

Livestock animal identification is of great importance to achieve precision animal production, as it is a prerequisite for modern livestock management and automated behavioral analysis. Regarding cow identification, computer vision-based methods have been widely considered due to their non-contact and practical advantages. Hu et al. (2020) propose a new non-contact cow identification method based on the fusion of deep part features. In the study, first, a set of side-view images of the cows were captured, and the YOLO object detection model was applied to locate the cow object

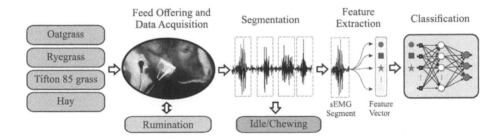

FIGURE 7.13 Animal sEMG data acquisition system (Campos et al., 2018).

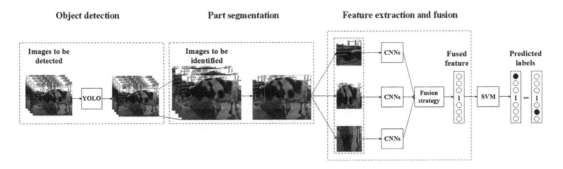

FIGURE 7.14 Flow chart of the proposed cow identification method (Hu et al., 2020).

in each original image, which was then divided into three parts, head, trunk, and legs, with a part segmentation algorithm using frame differentiation and segmentation span analysis. Next, three independent CNNs were trained to extract deep features from these three parts and a feature fusion strategy was designed to fuse the features. Finally, an SVM classifier trained with the fused features was used to identify each cow. The proposed method achieved 98.36% cow identification accuracy in a dataset containing side-view images of 93 cows, outperforming existing studies. Experimental results showed the effectiveness of the proposed cow identification method and good potential for this method in the individual identification of other farm animals. Figure 7.14 shows the flow chart of the proposed cow identification method.

7.5.4 IDENTIFYING ANIMAL SOUNDS

Because animal sounds in breeding contain information about animal welfare and behavior, automatic sound detection has the potential to facilitate a continuous acoustic monitoring system for use in a range of PLF applications. Bishop et al. (2019) presented a multipurpose livestock voice classification algorithm using voice-specific feature extraction techniques and machine learning models. In the study, to test the multipurpose nature of the algorithm, three separate datasets of farm animals consisting of sheep, cattle, and shepherd dogs were created, voice data were extracted from continuous recordings made in situ at three different operational farming operations reflecting the actual distribution conditions, Mel-frequency cepstral coefficients and discrete wavelet transform–based (DWT) features were made, and the classification was determined using an SVM model. At the end of the study, high accuracy was obtained for all datasets (sheep: 99.29%, cattle: 95.78%, dog: 99.67%). The classification performance alone was insufficient to determine the most appropriate feature extraction method for each dataset; computational timing results of DWT-based features were significantly faster (14.81% to 15.38% reduction in execution time), and a highly sensitive livestock vocalization classification algorithm has been developed, which forms the basis of an automated livestock vocalization detection system. Figure 7.15 shows the spectrogram examples of

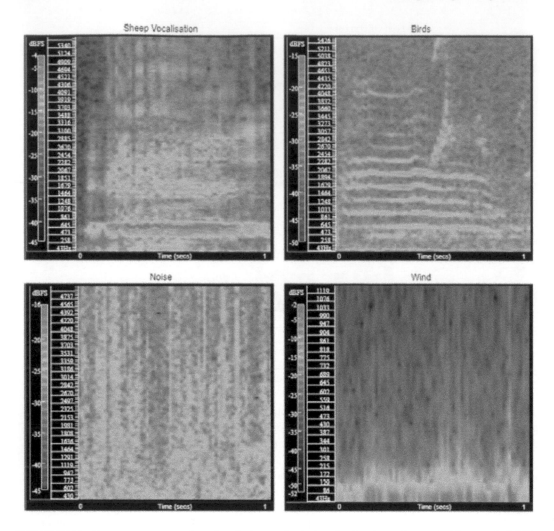

FIGURE 7.15   Spectrogram examples of each class and subclass for the sheep dataset (Bishop et al., 2019).

each class and subclass for the sheep dataset, and Figure 7.16 shows an overview of the proposed multipurpose live animal vocalization detection algorithm.

7.5.5   DETECTION OF PLANT DISEASE AND PEST DAMAGE

Orchi et al. (2022) provide a contemporary overview of research carried out over the past decade in the field of disease detection of different crops using machine learning, deep learning, image processing techniques, the Internet of Things, and hyperspectral image analysis. They also conducted a comparative study of various techniques applied for crop disease detection. The different techniques used for the automatic detection of leaf diseases and the procedure adopted in the identification process of leaf diseases are shown in Figure 7.17.

Calou et al. (2020) used high-spatial-resolution drone images to monitor the extent of Yellow Sigatoka damage in a banana plant, following basic assumptions about the identification, classification, quantification, and estimation of phenotypic factors. It has been reported that the monthly flights were conducted on a commercial banana plantation using a drone equipped with a 16-megapixel RGB camera; five classification algorithms were used to identify and quantify damage, and field assessments were made according to traditional methodology. The researchers found that the

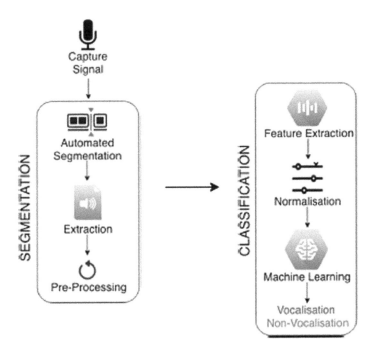

FIGURE 7.16 Overview of the proposed multipurpose livestock vocalization detection algorithm (Bishop et al., 2019).

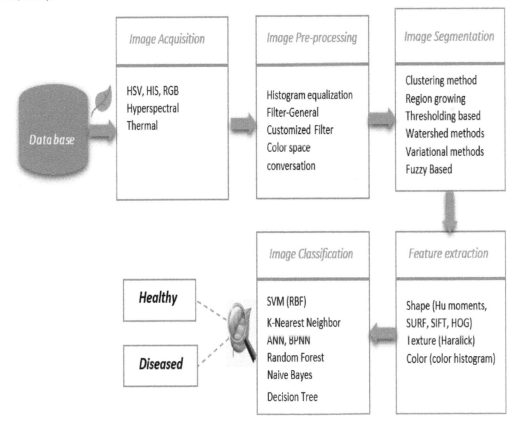

FIGURE 7.17 Different approaches for the identification of leaf diseases (Orchi et al., 2022).

SVM algorithm achieved the best performance (99.28% overall accuracy and 97.13 Kappa Index), followed by the ANN and minimum distance algorithms, in quantifying the disease. The SVM algorithm was more effective than other algorithms compared to the traditional methodology used to estimate the extent of Yellow Sigatoka, demonstrating that the tools used for monitoring leaf spots can be handled with RS, machine learning, and high spatial-resolution RGB images. Figure 7.18 shows the damage levels of Yellow Sigatoka pests on banana leaves.

7.5.6 Detection of Plant Pests

Tetila et al. (2020) developed a computer vision system for the acquisition of soybean pests by UAV images and their identification and classification using deep learning (Figure 7.19). Researchers used a dataset of 5,000 images captured under real growing conditions and evaluated the performance of five deep learning architectures, Inception-v3, Resnet-50, VGG-16, VGG-19, and Xception, for the classification of soybean pest images. The experimental results showed that the deep learning architectures trained with fine-tuning can lead to higher classification rates in comparison to other approaches, reaching accuracies of 93.82%. Deep learning architectures outperformed traditional feature extraction methods, such as SIFT and SURF, with the bag-of-visual-words approach, the semi-supervised learning method OPFSEMImst, and supervised learning methods used to classify images, such as SVM, K-NN, and random forest.

FIGURE 7.18   Damage levels of Yellow Sigatoka on banana leaves (Calou et al., 2020).

(a) Image acquisition (b) SLIC segmentation (c) Image dataset

Input image   CONVOLUTION + RELU   POOLING   CONVOLUTION + RELU   POOLING   FULLY CONNECTED   SOFTMAX

(d) Feature learning (e) Classification

FIGURE 7.19 System developed to detect soybean pests with UAV images using deep learning (Tetila et al., 2020).

7.5.7 DETECTION OF WEEDS

Correct detection and control of weeds in agricultural areas is a necessary procedure to increase plant yield and prevent herbicide pollution. Identifying weeds and precision-spraying only weeds are the ideal solution for weed eradication. For this purpose, superior features of imaging systems are used. With the advent of UAVs, the ability to acquire images of the entire agricultural area at very high spatial resolution and at low cost becomes possible, and the resulting input data meets high standards for weed localization and weed management. The similarity of weeds to plants can sometimes be a problem in the automatic detection of weeds. Color information alone is not sufficient to distinguish between plants and weeds. Multispectral cameras provide luminance images with high spectral resolution, and spectral reflectance is estimated in several narrow spectral bands for detection. Supervised and unsupervised learning methods can be used to solve this problem automatically.

In Sabzi and Abbaspour-Gilandeh (2018), potato plants and three common weed species (*Chenopodium album*, *Secale cereale* L., and *Polygonum aviculare* L.) were located and identified using a new machine vision system. The developed system consists of a video processing subsystem that can detect green plants in any frame, and a machine learning subsystem to classify weeds and potato plants. A hybrid approach consisting of artificial neural networks and a particle swarm optimization algorithm was used for classification. It has been reported that this approach can optimize number layers, neurons in each layer, network functions, weighting, and slopes. In this study, image capture was performed under controlled lighting conditions using white LED lamps. After shooting, the plants were classified and 30 color, texture, and shape features were extracted from each. A decision tree was used to select the six most important traits for the difference between potato plants and weeds. Finally, the ANN-PSO method was applied to classify the entries as potato plants or weeds and a comparison was made using a Bayesian classifier. Experimental results have been reported to reach 99.0% and 71.7% accuracy in the ANN-PSO and Bayes training set, respectively, and 98.1% and 73.3% in the test group.

Louargant et al. (2018) developed an automated image processing technique to distinguish between crop and weed pixels, combining spatial and spectral information extracted from four-band multispectral images. In the study, image data was captured at 3 m above ground with a camera mounted on a manually held pole. For each image, the field of view was approximately 4 m × 3 m and the resolution was 6 mm/pix. For each image, a specific training dataset was used by a supervised classifier (SVM) to classify only those pixels that could not be distinguished correctly using the initial spatial approach (Hough transform). Then, inter-row pixels were classified as weeds and in-row pixels as crops or weeds based on their spectral characteristics. When the proposed method was evaluated on 14 images captured in maize and sugar beet fields, the average value of weed detection rate was determined as 89% for the spatial and spectral combination method, 79% for the spatial method, and 75% for the spectral method (Figure 7.20).

7.5.8 Yield Estimate

Grasslands and pastures provide a wide range of ecosystem services worldwide. To protect these ecosystems and to better plan studies, it is necessary to develop highly sensitive models that consider the spatial nature of the ecosystem's structure, processes, and functions. Pecina et al. (2021) estimated the above-ground biomass at very high spatial resolution at nine study sites (Figure 7.21). The study used a combination of UAV-derived datasets to generate vegetation indices and micro-topographic models, and a random forest algorithm was used to generate above-ground biomass maps and evaluate the contribution of each predictive variable. The researchers reported that the developed model successfully predicted biomass with very high accuracy, the structural heterogeneity

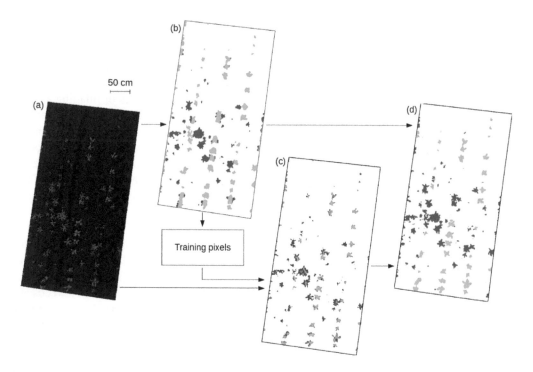

FIGURE 7.20 Example of the spatial and spectral combination results using an SVM classifier. (a) Multispectral orthoimage, (b) crop (green) and weed (red) location deduced from spatial information, (c) weed (green) and crop (red) location deduced from spectral information, (d) weed (green) and crop (red) location deduced from the combination of spatial and spectral information (Louargant et al., 2018).

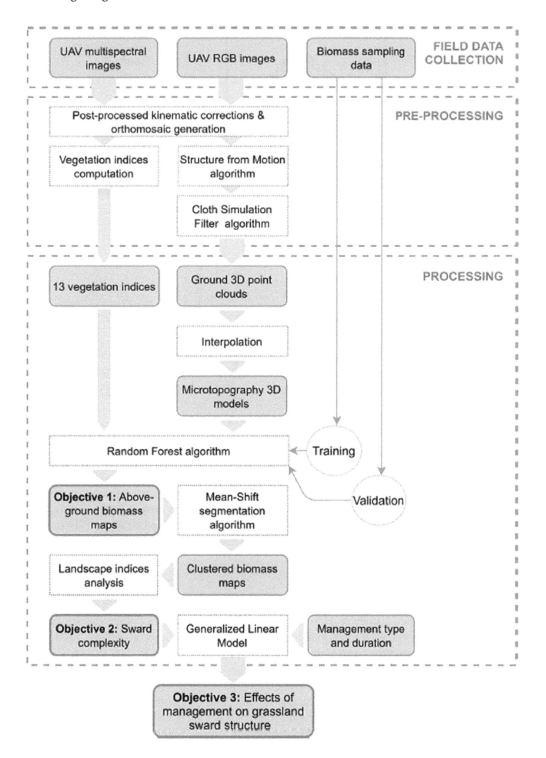

FIGURE 7.21   Flow chart showing the methodological steps performed in the study (Peciña et al., 2021).

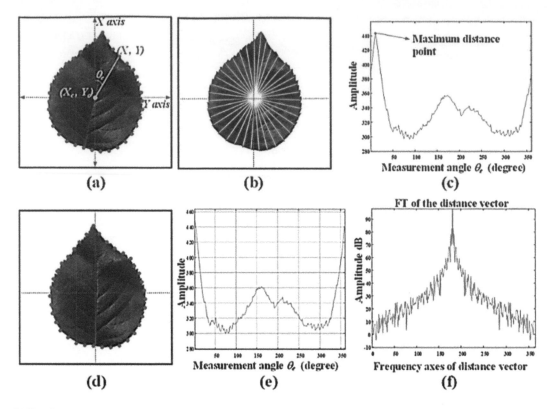

FIGURE 7.22   (a) The center and edges of the non-oriented leaf, (b) distances from origin to edges for non-oriented leaf, (c) 1-D distance vector of non-oriented leaf, (d) the edges of the oriented leaf, (e) 1-D distance vector of oriented leaf, (f) FT of the oriented leaf in logarithmic scale (Yigit et al., 2019).

of the grassland was evaluated using UAV-derived datasets and vegetation indices, and when the results were subsequently related to management history in each study area, it was determined that continuous, monospecific grazing management tends to simplify grassland structure. These results also indicate that UAV-based research can serve as reliable grassland monitoring tools and help develop site-specific management strategies.

7.5.9   IDENTIFICATION OF PLANT LEAVES

Yigit et al. (2019) conducted a study on the visual automatic identification of plant leaves using artificial intelligence techniques such as artificial neural network, naive Bayes algorithm, random forest algorithm, K-nearest neighbor, and support vector machine. In the study, data from 637 healthy leaves from 32 different plant species were used. Twenty-two visual features of each leaf were obtained by image processing technique and these 22 visual features were evaluated in four groups for size, color, texture, and pattern. To investigate the effects of these groups on classification performance, 15 possible different combinations from four groups were created, and then the models were trained with 510 leaves' data and tested for accuracy with 127 leaves' data. According to the test results, the SVM model with 92.91% accuracy was found to be the most successful descriptor (Figure 7.22). In addition, the researchers tested the SVM model to identify diseased and defective leaves. Five hundred and thirty-six randomly selected leaves corresponding to 80% of 637 healthy and 33 diseased-defective leaves were used for training and the remaining 134 leaves were used for testing. It was reported that at the end of the test, it was defined with a precision of 92.53% and the most effective factor in the result was tissue.

REFERENCES

Akbarzadeh, S., Paap, A., Ahderom, S., Apopei, B. and Alameh, K., 2018. Plant Discrimination by Support Vector Machine Classifier Based on Spectral Reflectance. *Computers and Electronics in Agriculture*, 148, 250–258.

Beucher, A., Møller, A.B. and Greve, M.H., 2017. Artificial Neural Networks and Decision Tree Classification for Predicting Soil Drainage Classes in Denmark. *Geoderma*. http://doi.org/10.1016/j.geoderma.2017.11.004.

Bhange, M. and Hingoliwala, H.A., 2015. Smart Farming: Pomegranate Disease Detection Using Image Processing. *Procedia Computer Science*, 58, 280–288.

Bhatia, P., 2019. *Data Mining and Data Warehousing: Principles and Practical Techniques*. Cambridge University Press, New York. ISBN: 978-1-108-72774-7.

Bishop, J.C., Falzon, G., Trotter, M., Kwan, P. and Meek, P.D., 2019. Livestock Vocalisation Classification in Farm Soundscapes. *Computers and Electronics in Agriculture*, 162(2019), 531–542. https://doi.org/10.1016/j.compag.2019.04.020.

Calou, V.B.C., Teixeira, A.D.S., Moreira, L.C.J., Lima, C.S., De Oliveira, J.B. ve De Oliveira, M.R.R., 2020. The use of UAVs in Monitoring Yellow Sigatoka in Banana. *Biosystems Engineering*, 193, 115–125. https://doi.org/10.1016/j.biosystemseng.2020.02.016.

Camargo, A. and Smith, J.S., 2009. An Image-Processing Based Algorithm to Automatically Identify Plant Disease Visual Symptoms. *Biosystems Engineering*, 102, 9–21.

Campos, D.P., Abatti, P.J., Bertotti, F.L., Hill, J.A.G. and da Silveira, A.L.F., 2018. Surface Electromyography Segmentation and Feature Extraction for Ingestive Behavior Recognition in Ruminants. *Computers and Electronics in Agriculture*, 153, 325–333. https://doi.org/10.1016/j.compag.2018.08.033.

Cover, T.M. and Hart, P.E., 1967. Nearest Neighbor Pattern Classification, IEEE Transactions on Information Theory IT-13 (1).

Dua, S. and Chowriappa, P., 2013. *Data Mining for Bioinformatics*. CRC Press Taylor & Francis Group LLC, Boca Raton. ISBN: 978-1-4200-0430-4.

Guerrero, J.M., Pajares, G., Montalvo, M. Romeo, J. and Guijarro, M., 2012. Support Vector Machines for Crop/Weeds Identification in Maize Fields. *Expert Systems with Applications*, 39(2012), 11149–11155.

Han, J., Kamber, M. and Pei, J., 2012. *Data Mining: Concepts and Techniques*. Morgan Kaufmann Publishers is an imprint of Elsevier, Waltham. ISBN: 978-0-12-381479-1.

Hu, H., Dai, B., Shen, W., Wei, X., Sun, J., Li, R. and Zhang, Y., 2020. Cow Identification Based on Fusion of Deep Parts Features. *Biosystems Engineering*, 192(2020), 245–256. https://doi.org/10.1016/j.biosystemseng.2020.02.001.

Huang, C.Y. and Nof, S.Y., 2001. Automation Technology. *Handbook of Industrial Engineering: Technology and Operations Management*. (Editor: Salvendy G.) A Wiley-Interscience Publication, John Wiley & Sons, Inc, New York, 155–176. ISBN: 0471-33057-4.

Huang, T., Yang, R., Huang, W., Huang, Y. and Qiao, X., 2018. Detecting Sugarcane Borer Diseases Using Support Vector Machine. *Information Processing in Agriculture*, 5(2018), 74–82.

Kantardzic, M., 2020. *Data Mining: Concepts, Models, Methods, and Algorithms*. IEEE Press, John Wiley & Sons, Inc, Hoboken. ISBN: 978-1-119-51604-0.

Larose, D.T. and Larose, C.D., 2015. *Data Mining and Predictive Analytics*. John Wiley & Sons, Inc., Hoboken. ISBN: 978-1-118-11619-7.

Liu, Z.W., Liang, F.N. and Liu, Y.Z., 2018. Artificial Neural Network Modeling of Biosorption Process Using Agricultural Wastes in A Rotating Packed Bed. *Applied Thermal Engineering*, 140, 95–101.

Louargant, M., Jones, G., Faroux, R., Paoli, J.-N., Maillot, T., Gée, C. and Villette, S., 2018. Unsupervised Classification Algorithm for Early Weed Detection in Row-Crops by Combining Spatial and Spectral Information. *Remote Sensing*, 10, 761. https://doi.org/10.3390/rs10050761.

Mucherino, A., Papajorgji, P.J. and Pardalos, P.M., 2009. *Data Mining in Agriculture*. Springer Science+Business Media, LLC, Dordrecht. ISBN: 978-0-387-88614-5.

Nguyen-ky, T., Mushtaq, S., Loch, A., Reardon-Smith, K., An-Vo, D.A., Ngo-Cong, D. and Tran-Cong, T., 2017. Predicting Water Allocation Trade Prices Using a Hybrid Artificial Neural Network Bayesian Modelling Approach. *Journal of Hydrology*. https://doi.org/10.1016/j.jhydrol.2017.11.049.

Niel, S., Jesús, F., Díaz, R., Mendoza, Y., Notte, G., Santos, E., Gérez, N., Cesio, V., Cancela, H. and Heinzen, H., 2018. Beehives Biomonitor Pesticides in Agroecosystems: Simple Chemical and Biological Indicators Evaluation Using Support Vector Machines (SVM). *Ecological Indicators*, 91(2018), 149–154.

Orchi, H., Sadik, M. and Khaldoun, M., 2022. On Using Artificial Intelligence and the Internet of Things for Crop Disease Detection: A Contemporary Survey. *Agriculture*, 12, 9. https://doi.org/10.3390/agriculture12010009.

Ozguven, M.M., Yilmaz, G., Adem, K., Kozkurt, C., 2019. Use of Support Vector Machines and Artificial Neural Network Methods in Variety İmprovement Studies: Potato Example. *Current Investigations in Agriculture and Current Research*, 6(1), 706–712. https://doi.org/10.32474/CIACR.2019.06.000229.

Özcan, T., 2015. Veri Madenciliği Ders Notu. İstanbul Üniversitesi Açık ve Uzaktan Eğitim Fakültesi. İstanbul. (Turkish).

Özgüven, M.M., 2018. Hassas Tarım. Akfon Yayınları, Ankara. ISBN: 978-605-68762-4-0. (Turkish).

Özgüven, M.M., 2019. Teknoloji Kavramları ve Farkları. International Erciyes Agriculture, Animal & Food Sciences Conference 24–27 April 2019- Erciyes University – Kayseri, Turkiye. (Turkish).

Özgüven, M.M., Türker, U., Akdemir, B., Çolak, A., Acar, A.İ., Öztürk, R. ve Eminoğlu, M.B., 2020. Tarımda Dijital Çağ. Türkiye Ziraat Mühendisliği IX. Teknik Kongresi. Ocak 2020, Ankara. Bildiriler Kitabı-1, s.55–78. (Turkish).

Öztemel, E., 2006. Yapay Sinir Ağları. Papatya Yayıncılık, 2. Baskı, ISBN: 978-975-6797-39-6. İstanbul. (Turkish).

Peciña, M.V., Bergamo, T.F., Ward, R.D., Joyce, C.B. and Sepp, K., 2021. A Novel UAV-Based Approach for Biomass Prediction and Grassland Structure Assessment in Coastal Meadows. *Ecological Indicators*, 122, 107227. https://doi.org/10.1016/j.ecolind.2020.107227.

Ribeiro, A., Ranz, J., Burgos-Artizzu, X.P., Pajares, G., Sanchez del Arco, M.J. and Navarrete, L., 2011. An Image Segmentation Based on A Genetic Algorithm for Determining Soil Coverage by Crop Residues. *Sensors*, 11, 6480–6492. https://doi.org/10.3390/s110606480.

Sabzi, S. and Abbaspour-Gilandeh, Y., 2018. Using Video Processing to Classify Potato Plant and Three Types of Weed Using Hybrid of Artificial Neural Network and Particncle Swarm Algorithm. *Measurement*, 126, 22–36.

Taghavifar, H., Mardeni, A. and Hosseinloo, A.H., 2015. Appraisal of Artificial Neural Network-Genetic Algorithm Based Model for Prediction of The Power Provided by the Agricultural Tractors. *Energy*, 93, 1704–1710.

Tan, P.N., Steinbach, M. and Kumar, V., 2014. *Introduction to Data Mining*. Pearson Education Limited, Harlow. ISBN: 978-1-292-02615-2.

Terzi, İ., Özgüven, M.M., Altaş, Z., Uygun, T., 2019. Tarımda Yapay Zeka Kullanımı. International Erciyes Agriculture, Animal & Food Sciences Conference 24–27 April 2019 - Erciyes University – Kayseri, Turkiye. (Turkish).

Tetila, E.C., Machado, B.B., Astolfi, G., De Souza Belete, N.A., Amorim, W.P., Roel, A.R. and Pistori, H., 2020. Detection and Classification of Soybean Pests Using Deep Learning with UAV Images. *Computers and Electronics in Agriculture*, 179, 105836. https://doi.org/10.1016/j.compag.2020.105836.

Vapnik, V., 1995. *The Nature of Statistical Learning Theory*. Springer, New York.

VijayaLakshmi, B. and Mohan, V., 2016. Kernel-Based PSO and FRVM: An Automatic Plant Leaf Type Detection Using Texture, Shape, and Color Features. *Computers and Electronics in Agriculture*, 125, 99–112.

Yao, M., Jia, M. and Zhou, A., 2018. *Applied Artificial Intelligence: A Handbook for Business Leaders*. (Editor: Zhang, N.) Topbots Inc, USA. ISBN: 978-0-9982890-2-1.

Yigit, E., Sabanci, K., Toktas, A. and Kayabasi, A., 2019. A Study on Visual Features of Leaves in Plant Identification using Artificial Intelligence Techniques. *Computers and Electronics in Agriculture*, 156, 369–377. https://doi.org/10.1016/j.compag.2018.11.036.

Zhang, A., 2017. Data Analytics: Practical Guide to Leveraging the Power of Algorithms, Data Science, Data Mining, Statistics, Big Data, and Predictive Analysis to Improve Business, Work, and Life.

8 Artificial Intelligence, Machine Learning, and Deep Learning in Agriculture

8.1 WHAT IS ARTIFICIAL INTELLIGENCE?

Artificial intelligence (AI) is the acquisition of the perception, learning, past experience, and thinking ability of human intelligence by computers, machines, or systems, and it is the ability to make decisions and take the necessary action in the face of predictable or unpredictable new situations. During this decision-making process, these variables are taught to artificial intelligence, the parameters of which are evaluated by human intelligence, and formulas are created with the help of computer models that can make comments similar to human mental functions in order to ensure decision-making. Thus, a thinking and decision-making model is created with computer software, similar to the human mindset (Özgüven, 2019).

Decision-making, which is the basic function of human intelligence, is based on the ability to evaluate a situation. No decision can be made without measuring the pros and cons and factoring in the parameters. Mankind takes great pride in its ability to evaluate. However, in most cases, a machine can do better. Chess represents the pride of humanity in the strategy of thinking. No single chess player today can beat the best chess engines. One of the outstanding core capabilities of a chess engine is the evaluation function, and it takes into account many parameters more precisely than humans. Therefore, AI often requires solving a difficult problem that even an expert in a particular field cannot express mathematically. Machine learning sometimes means finding a solution to a problem that people don't know how to explain. Deep learning involving complex networks solves even more difficult problems. Adequate mathematical knowledge is required to understand AI concepts. The best way is to look at everyday life and make a mathematical model of it. Mathematics is a prerequisite, not an option in AI. Next is to write solid source code or implement a cloud platform machine learning solution. What the source code is about, its potential, and its limitations should be known while the work is being done (Rothman, 2018).

AI methods used in AI applications can be grouped as follows (Alpaydın, 2004):

- *Classification*: It is the process of finding in which class the new data will take place if the class of historical data is specified.
- *Clustering*: It is the process of separating the data into clusters according to their similarities when the class of historical data is not specified or known.
- *Regression (curve fitting)*: In cases where historical data consists of continuous numerical values, it is the process of producing a curve model from these values.
- *Feature determination*: If the historical data is too much, the features that determine the class of this data are determined. During this determination process, a subset of existing features can be created, or new features can be created from their combination.
- *Relationship inference*: It is the determination of the most common data by analyzing the coexistence of data with other data.

There is conceptual confusion about AI. The reason for this is that the field of AI covers subjects that require expertise in a wide variety of technical disciplines and a wide variety of techniques are

DOI: 10.1201/b23229-8

used together in developed systems, machines, robots, etc. The developed systems are generally based on automatic operation and the application of various automation systems. However, this does not mean that all systems have AI. It is also difficult to distinguish between systems with or without AI, due to the availability of similar techniques. After these general AI explanations, it is necessary to talk about the different approaches of today's AI experts about AI. According to these approaches, AI is divided into three categories:

- *Artificial narrow or weak intelligence (ANI)*: The ANI approach is associated with the Massachusetts Institute of Technology, and any system that exhibits intelligent behavior is seen as an example of AI. In the ANI approach, regardless of whether the work performs its task like humans, it focuses on whether a program or system works correctly, and the success evaluation is made only by looking at the performance of the created system. Researchers argue that the reason for AI research is to solve difficult problems, and there is no need to look at how problems are actually solved (Campesato, 2020). The ANI approach can only perform narrowly defined tasks that do not have cognitive and thinking abilities, and it consists only of predefined functions that are fed into the system. Examples of applications developed with the ANI approach are Siri, Alexa, self-driving cars, Alpha Go, and humanoid Sophia (Singh et al., 2022).
- *Artificial general or strong intelligence (AGI)*: This approach is associated with Carnegie-Mellon University and is thought of as biological plausibility AI. According to this approach, when a system exhibits intelligent behavior, its performance should be based on the same methodologies used by humans. For example, it is aimed to achieve success by simulating the human hearing system for a hearing AI system. Researchers are interested in the structure of the systems they build, and they argue that computers can have a sense of consciousness and intelligence by having knowledge of heuristics, algorithms, and AI programs (Campesato, 2020). AGI is the evolutionary stage of artificial intelligence, which has the ability to think and make decisions like humans. The full development of such an AI system could bring about the end of humanity. Man is not in a position to compete with such a system, which is the most powerful in terms of thinking and intelligence (Singh et al., 2022).
- *Artificial super intelligence (ASI)*: ASI is currently a hypothetical situation that could only happen in the near future if humanity continues to advance in the development of AI. The ASI could be the AI stage in such a situation where the AI system capacity would exceed all limitations and suppress humans (Singh et al., 2022).

If we classify AI according to the functions that it performs, it can be classified as follows (Singh et al., 2022):

- *Reactive machine AI*: This type of AI system takes the input as up-to-date data and performs the task according to the current situation. Reactive machine AI cannot analyze and predict future tasks or actions in any particular situation. These systems always run on narrowly spaced predefined tasks.
- *Limited memory AI*: These AI systems mostly rely on the data they store and often perform the task by examining the task that was performed before. They also perform the estimation of future tasks by the machine.
- *Theory of mind AI*: These systems are the more advanced type of AI used to make predictions, and they play an important role in psychology. These AI systems always perform tasks related to emotional intelligence and thoughts that the individual can grasp in any situation.
- *Self-aware AI*: These systems, which are the most advanced and have entered the last stage of AI, are the systems where the AI system consists of consciousness, and everything is self-aware. These are expected to be realized in the near future.

8.2 SUB-BRANCHES OF ARTIFICIAL INTELLIGENCE

In the field of AI, there are sub-branches where a wide variety of engineering applications can be made in the real world. These sub-branches can be classified as follows:

- *Expert systems*: Expert systems are the first realization of research in the field of AI in the form of software technology. Especially for application software developers in the medical and engineering disciplines, the decision-making process has begun to be addressed with the use of symbols instead of numbers. Tasks belonging to the category of classification and diagnosis were the first to benefit from the emergence of expert system technology. Expert systems are computer programs designed to act as experts to solve a problem in a particular domain. The program uses the domain knowledge encoded in it and a specific control strategy to arrive at solutions. An expert system is called a system, not a program. This is because the expert system encompasses several different components such as knowledge base, inference mechanisms, explanation ability, etc. All these different components interact together to simulate the problem-solving process by a recognized expert in a domain (Krishnamoorthy and Rajeev, 1996).
- *Artificial neural networks*: Thanks to their generalization feature, artificial neural networks (ANN) can learn from past events or examples just like people, and thanks to these experiences, they can make decisions on new examples that they have never encountered in the past. ANN has five basic elements: Inputs, weights, transfer function, activation function, and outputs. The structure collects the information it receives from the outside in an aggregation function, creates the output by passing it through the activation function, and sends it to other cells over the connections of the network (Öztemel, 2006).
- *Genetic algorithms*: Genetic algorithms are based on the survival of new individuals born from parents with good genes, adapting to the conditions, and the inability of individuals with bad genes to survive. The fitness function in genetic algorithms is a technique that aims to organize self-learning and decision-making systems using operators such as crossover and replacement to generate new solutions (Elmas, 2010).
- *Fuzzy logic*: Fuzzy logic is a technique that works with verbal variables very close to the human thought system without the need for a mathematical model and can be easily applied to complex problems expressed in language (Özgüven, 2019). As in human logic, instead of long/short, hot/cold, fast/slow, black/white, it works according to intermediate values such as long/medium-long/medium/medium-short, hot/warm/less cold/cold/very cold, etc. The fuzzy set redefines the concept of uncertainty by eliminating definite transitions and determines membership degrees for all individuals in the universe. Thus, individuals can belong to larger and smaller values indicated by their membership degrees in the fuzzy set. These membership degrees are expressed with real values between 0 and 1 (Nabiyev, 2016; Elmas, 2018).
- *Ant colony algorithm*: The ant colony algorithm is based on the swarm cooperation of ants in search of food. Every ant looking for food leaves a secretion called a pheromone, which disappears over time and guides the ants coming from behind, on the way to its target and returning to the nest. For this reason, ants taking the shortest path leave more pheromones. The ants coming from behind choose this short route where the smell is high, and thus the ant swarm reaches the food in the shortest way (Akçetin and Çelik, 2015).
- *Machine learning*: Machine learning, first of all, by looking at the historical data with various algorithms and methods, determines the mathematical model that will determine the complex pattern between the data, then makes a prediction about what is desired to be estimated over the data (Özgüven, 2019).
- *Deep learning*: Deep learning uses many layers of nonlinear processing units for feature extraction and transformation from large amounts of labeled training data. Each successive

layer uses the output from the previous layer as input (Deng and Yu, 2014). It is basically based on learning from the representation of data. When we refer to representation for an image, we can think of a vector of intensity values per pixel or features such as edge clusters and special shapes. Some of these features better represent the data (Song and Lee, 2013).

AI is used to enable decision-making in the face of new predictable or unpredictable situations. There are differences between machine learning and deep learning. While processing is done in a single layer in machine learning, deep learning is processed in many layers at the same time. The difference between deep learning algorithms and machine learning algorithms is that there is a huge amount of labeled data and, due to the complex structure of the data, they need GPU-based computers and hardware with very high computational power to process this data. In machine learning, the relevant features are manually extracted from the images, and these features are used to create a model that categorizes the objects in the image. In deep learning, the relevant features are automatically extracted from the images, and it is learned how to do a task such as classification automatically (Özgüven, 2019).

There are many other approaches to AI that can be used alone or in combination with machine learning and deep learning to improve performance. For example, the ensemble learning method combines different machine learning models or blends deep learning models with rules-based models. The most successful applications of machine learning to enterprise problems use ensemble approaches to produce results superior to any single model. There are four broad categories of ensemble learning: Bagging, boosting, stacking, and bucketing. Bagging requires training the same algorithm on different subsets of data and includes popular algorithms such as random forest. Boosting involves training a set of models, where each model prioritizes learning from examples where the previous model failed. In stacking, the output of many models is pooled. In bucketing, more than one model is trained for a particular problem and the best one is dynamically selected for each particular input (Yao et al., 2018).

8.3   WHAT IS MACHINE LEARNING?

Machine learning, a field of research at the intersection of statistics, artificial intelligence, and computer science, is all about extracting information from data. The application of machine learning methods has become ubiquitous in daily life in recent years. Machine learning algorithms are at the center of many modern websites and devices, from automatic suggestions on which movies to watch, what food to order, or what products to buy, to personalized online radio and recognizing your friends in your photos (Müller and Guido, 2017).

Machine learning includes two important mathematical entities as models and features. Models are trying to understand the world through data. It's like trying to piece together reality using a noisy, incomplete puzzle with lots of extra pieces. A mathematical data model is a process of formulating relationships between data, model, and task. The mathematical formulas used for this purpose relate numerical quantities to each other. However, raw data is usually non-numeric. Properties are used as a part that connects numeric and non-numeric data. Attributes are a numerical representation of raw data related to the task at hand. If the model does not have enough informative features in the training process, the model cannot perform the final task. If there are too many features or most of them are irrelevant, the model will be more expensive and harder to train. Good features make the next modeling step easier and the resulting model becomes more capable of completing the desired task. Bad features may require a much more complex model to achieve the same level of performance (Zheng and Casari, 2018). Figure 8.1 shows the machine learning workflow.

Some standard machine learning tasks are as follows (Mohri et al., 2018):

- *Classification*: Classification is the process of assigning a category to each item. The number of categories in such tasks is usually under a few hundred, but in some difficult tasks, it can be much more and even unlimited, as in OCR, text classification, or speech recognition.

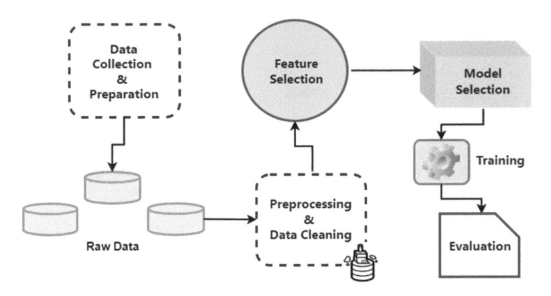

FIGURE 8.1 Machine learning workflow.

- *Regression*: Regression is the process of estimating a true value for each item. In regression, the penalty for an incorrect guess depends on the size of the difference between the actual and predicted values.
- *Ranking*: Ranking is the process of learning to rank items according to some criteria. A web search, for example, returning web pages related to a search query, is an example of canonical ranking. Many other similar sequencing problems arise in the context of information extraction or the design of natural language processing systems.
- *Clustering*: Clustering is the process of dividing a set of items into homogeneous subsets. Clustering is often used to analyze very large datasets. For example, in the context of social network analysis, clustering algorithms attempt to identify natural communities within large groups of people.
- *Dimensionality reduction or manifold learning*: The dimensionality reduction or manifold learning process consists of transforming the initial representation of the elements into a lower dimensional representation while preserving some properties of the initial representation. A common example is the pre-processing of digital images in computer vision tasks.

8.4 MACHINE LEARNING PROCESSES

The goal of machine learning is to make accurate predictions or decisions for an input that has never been seen before and to develop efficient algorithms that automate this prediction and decision-making process. While developing algorithms, criteria for what an expert should consider when making a decision should be taken into account during the creation of rules. In the application of machine learning methods, there may be complexities such as computational complexity, training complexity, and trained algorithm implementation complexity. The performance of an algorithm is evaluated according to the test error. In addition, while an algorithm is running, there may be many test points and quick decisions are required at these points. Therefore, the test process should have a low computational load. In machine learning, data is randomly divided into a training sample, validation sample, and test sample, as in Figure 8.2.

The size of each of these samples depends on a number of different considerations. For example, the amount of data allocated for validation depends on the number of hyperparameters of the

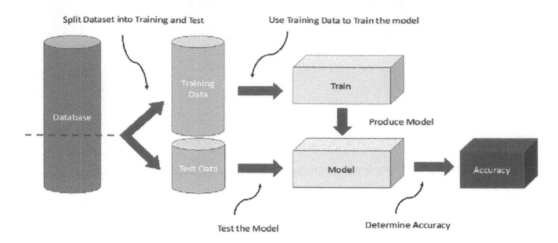

FIGURE 8.2   Stages of the machine learning process (Imbalzano et al., 2022).

algorithm represented here by the vector. Also, when the labeled sample is relatively small, the amount of training data is often chosen to be larger than the test data, since the learning performance is directly dependent on the training sample. Then, the relevant features are associated with the examples. This is a critical step in the design of machine learning solutions. Useful features can effectively guide the learning algorithm, while weak or uninformative features can be misleading. Although critical, the choice of features is largely left to the user. This selection reflects the user's prior knowledge of the learning task, which can have a significant impact on performance results in practice. Selected features are used to train the learning algorithm by adjusting the values of the free parameters. For each value of these parameters, the algorithm chooses a different hypothesis from the hypothesis set. In the validation example, the one with the best performance is selected. Finally, using this hypothesis, the labels of the samples in the test sample are estimated. The performance of the algorithm is evaluated using the task-related function to compare the estimated and actual tags. Therefore, the performance of an algorithm is judged by its test error, not its error in the training sample (Mohri et al., 2018).

The machine learning process is listed here (Marsland, 2015):

- *Data collection and preparation*: When a machine learning study is made about a problem, ready-made data sets or new data sets collected from scratch can be used. By combining the new data generation phase with the next step of feature selection, only the necessary data is collected. This can be done by putting together a fairly small dataset with all the features you believe might be useful and experimenting before selecting the best features and collecting and analyzing the full dataset. The large amount of data on the subject of the study makes it difficult to collect data because it requires taking many measurements or the data are in various places and formats. Finally, the amount of data must be considered. Machine learning algorithms need a certain amount of data, preferably without too much noise, but increasing dataset size introduces increased computational load.
- *Feature selection*: Identifying the most useful features for the problem being studied always requires prior knowledge of the problem and the data. In addition to identifying features that are beneficial to the learner, it is also necessary that features can be collected without significant cost or time and that they are resistant to noise and other data corruption that may occur during the collection process.
- *Algorithm choice*: An appropriate algorithm (or algorithms) is selected considering the data set. Knowledge of the principles underlying each algorithm and examples of their use is what is required for this.

- *Parameter and model selection*: Most algorithms have parameters that need to be set manually or that require experimentation to determine appropriate values.
- *Training*: Given the dataset, algorithm, and parameters, training should be simply the use of computational resources to build a data model to predict the outputs of new data.
- *Evaluation*: Before a system is developed, the accuracy of untrained data needs to be tested and evaluated. This usually includes a comparison with experts in the field and the selection of appropriate metrics for this comparison.

8.5 MACHINE LEARNING METHODS

Learning strategies in machine learning methods are divided into three groups (Figure 8.3):

- *Supervised learning*: Supervised algorithms learn from labeled data. In supervised learning, there are defined inputs and outputs. In a supervised learning problem, given a sample of input-output pairs, called the training sample (or training set), the task is to find a deterministic function that maps any input to an output that can predict future input-output observations, minimizing the errors as much as possible (Camastra and Vinciarelli, 2008; Kulkarni et al., 2020).
- *Unsupervised learning*: Unsupervised algorithms work with unlabeled data. Without target values, only the relationship between input values is attempted to be revealed. With the help of this relationship, values close to each other are clustered. The new entry will belong to whichever of these clusters it is associated with. If it is possible to find a clustering such that the similarities of objects in a cluster are much greater than the similarities between objects in different clusters, it tries to extract structure from the training sample so that the entire cluster can be represented by a representative data point. In addition to clustering algorithms, in unsupervised learning techniques, there are algorithms whose purpose is to represent high-dimensional data in low-dimensional spaces and try to preserve the original information of the data (Camastra and Vinciarelli, 2008; Atalay and Çelik, 2017; Kulkarni et al., 2020).
- *Reinforcement learning*: Instead of a consultant giving the target output, a criterion is used by which the output obtained is evaluated as true or false against the given input. The representative is rewarded for the right decisions; for wrong decisions, the representative

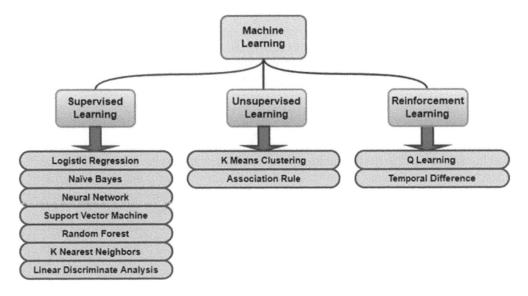

FIGURE 8.3 Classification of machine learning methods.

is punished. Therefore, the goal for an agent would be to maximize rewards and achieve target status. Reinforcement learning (RL) has its roots in control theory. It considers a dynamic environment scenario as data, resulting in condition-action-reward triplets. The difference between reinforcement and supervised learning is that in RL there is no optimal action in a given situation, but the learning algorithm must define an action to maximize the expected reward over time. A brief description of the data is the strategy that maximizes the reward. The problem with RL is learning what to do to maximize a given reward, i.e., how to match situations with actions. Because there is no optimal action in a given situation, one of the biggest challenges of the RL algorithm is finding a balance between exploration and use. To maximize reward (or minimize punishment), a learning algorithm should choose actions that have been tried in the past and have been found to be effective in producing rewards, i.e., it should leverage its current knowledge. On the other hand, to discover these actions, the learning algorithm must select actions that have not been tried in the past and thus explore the state space (Camastra and Vinciarelli, 2008; Atalay and Çelik, 2017; Kulkarni et al., 2020).

Machine learning supervised learning methods are explained here (Kulkarni et al., 2020):

- *Logistic regression*: Logistic regression, which is a supervised learning technique, is a classification technique rather than a regression technique. Logistic regression is mostly applied to discrete classes such as male/female, cat/dog, and pass/fail. Logistic regression models its output as a logistic sigmoid function that matches discrete classes. There are three types of logistic regression. The first is binary logistic regression: This type of logistic regression models only two outcomes (for example, pass and fail). The second is multinomial logistic regression: This type of logistic regression models three or more outcomes without any ordering (e.g., cats, dogs, and sheep). The third is ordinal logistic regression: This type of logistic regression models three or more outcomes with ordering (for example, product ratings from 1 to 5). To apply logistic regression, the dataset should ideally support some features. These features are that independent variables are independent of each other, there is consideration of appropriate and significant variables to model logistic regression, and logistic regression usually requires large sample sizes.
- *Naive Bayes*: Naive Bayes, a probabilistic classifier based on Bayes's theorem, is a supervised learning technique. It is a classification technique based on the independence of estimators (independent variables). The term naive implies that the emergence of a particular feature is independent of other features. The Bayesian term says that the basis of naive Bayes is Bayes's theorem. Bayes's theorem gives the conditional probability of event X given that event Y occurs, as stated in Equation 8.1.

$$P(X \mid Y) = P(Y \mid X).P(X)/P(Y) \tag{8.1}$$

Here $P(X|Y)$ is the conditional probability of X given Y. $P(Y|X)$ is the conditional probability of Y given X. $P(X)$ is the probability of event X and $P(Y)$ is the probability of event Y. The best feature of naive Bayes is that it can process any number of independent variables, whether they are independent or continuous. From a set of variables $X = \{x1, x2, ..., xn\}$ we want to obtain the posterior probability of the event Cj among the result set $C = \{c1, c2, ..., cn\}$, where X is the set of predictors and C is the categorical levels found in the dependent variables. The final posterior probability is as given in Equation 8.2.

$$p(C_j|X) \propto p(C_j) \prod_{k}^{n} p(x_n|C_j) \tag{8.2}$$

The advantages of naive Bayes are: It sets up the model quickly and makes quick decisions. It can handle both continuous and discrete data. It is highly scalable in terms of the number of predictors and data points.

- *Neural network*: A neural network, commonly known as an artificial neural network, is a system for processing information inspired by biological neural networks. It consists of a large number of processing elements known as neurons that interact with each other to solve a problem. Neural networks are divided into three categories: Multi-layer perceptrons (MLPs), convolutional neural networks (CNNs), and recurrent neural networks (RNNs).

  Figure 8.4 shows the MLP classifier, which consists of at least three neuron layers. MLPs are suitable for classification problems where inputs are assigned a class or label. LPs are fully connected; each layer is connected to another layer with some weight. A neural network works in two stages: feedforward and backpropagation. The following steps are executed in the feed-forward phase: (1) The values in the inputs are multiplied by the weights. (2) Hidden layer neurons have an activation function such as sigmoid or ReLU. These are applied to the values received from the input layers. (3) The processed values are then given to the output layers. These are the actual outputs of the algorithm. The outputs produced may not be correct. In this case, the backpropagation phase is initiated. The following steps are performed in the backpropagation stage: (1) The error is calculated by calculating the difference between the output and the input. This difference is known as a loss and the corresponding function is the loss function. An example of a loss function is the mean squared error. (2) Calculate a partial derivative of the error function known as gradient descent. If the slope of the error function is positive, the weights can be decreased or increased. This exercise reduces the overall error, and the corresponding function is called the optimization function.

- *Support vector machine (SVM)*: SVM is a supervised learning technique that helps with data classification and regression analysis. SVM tries to find a hyperplane that divides the data into related classes. At the simplest level, a hyperplane is a straight line that divides data into corresponding classes. Support vectors are data points closest to a hyperplane. The distance from the nearest data point to the hyperplane is known as the margin, and the goal is to have a wider margin for optimal classification. The conversion of non-separable data points to transferable data points is possible with kernel functions. For example, Equations 8.3 and 8.4 need to be calculated to get the maximum margin separation in linear SVM.

$$W^T X_i + b > +1 \qquad (8.3)$$

$$W^T X_i + b < -1 \qquad (8.4)$$

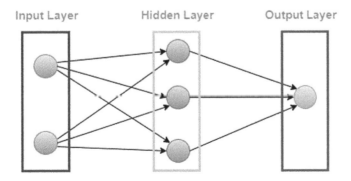

FIGURE 8.4 Three-layer multi-layer perceptron.

Here Equation 8.3 is assumed for all positive cases. Equation 8.4 is taken into account for negative cases. $\|W\|^2$ should be as small as possible. The separator is then defined as a set of points for Equation 8.5.

$$Wx + b = 0 \tag{8.5}$$

The advantages of an SVM are that it works well in high-dimensional spaces and is capable of avoiding problems caused by bias and overfitting. The disadvantages are that it experiences performance issues when working with larger datasets and is affected by the presence of noise in datasets.

- *Random forest*: Random forest falls under the category of supervised learning. Random forest creates multiple trees and combines them for an accurate decision. Therefore, the random forest is used to construct multiple decision trees. For example, a decision tree follows a top-down approach in which root nodes initiate the binary fission process with some criteria for reduction in entropy. Here entropy is the measure of unpredictability in the data. Root node N is a dataset with different classes or data labels. It is divided into right and left subgroups. The purpose of these subsets is to have the purity of the class. If this is performed on the corresponding node with entropy 0, only one type of label remains. Otherwise, division continues to determine the remaining labels. The advantages of random forest are that the problem of excessive learning is reduced, it has the ability to produce accurate predictions for large datasets, and it has the ability to evaluate missing data. The disadvantages are that trees are sensitive to data changes in the training data and too many trees can slow down the algorithm.

- K-nearest neighbors (K-NN): K-neighbors or K-nearest neighbors (K-NN) is a supervised learning algorithm. In this type of example-based learning, training observations are kept as part of the model. It is also known as competitive learning because competition between data samples is used for prediction. It is called lazy learning because the decision to build the model is delayed until the prediction time. It is also called non-parametric K-NN because it makes no assumptions about data distributions. K-nearest neighbors is used for classification. To measure the distance, the Euclidean distance formula is used as in Equation 8.6.

$$d(a,b) = \sum_{i=1}^{n} (b_i - a_i)^2 \tag{8.6}$$

The Euclidean distance between two points provides the length of the path connecting them. K-NN believes that samples with similar inputs belong to the same class as shown in Figure 8.5(a) and Figure 8.5(b). The new sample and + and − represent training data samples. The K value determines the nearest neighbors to be considered. When $K = 1$, the label of the new sample will be that of the 1 nearest neighbor.

When $K = 3$, the majority vote is accepted. Assuming two tags are X and one tag is Y, the new tag will be X according to the majority system. The advantages of K-NN are that it is relatively fast for small training datasets, it makes no data assumptions and works well for nonlinear data, and it can be applied for both classification and regression. The disadvantages of K-NN are that it is computationally expensive as it preserves the training data and is sensitive for larger datasets.

- *Linear discriminate analysis (LDA)*: LDA is a supervised learning technique. It is used as a size-reduction technique. LDA was developed as a classification technique. LDA tries to increase the distance between two clusters and reduce the within-cluster distance. "W" is chosen to maximize the ratio of distribution between classes and within-class distribution. The interclass distribution matrix is defined in Equation 8.7.

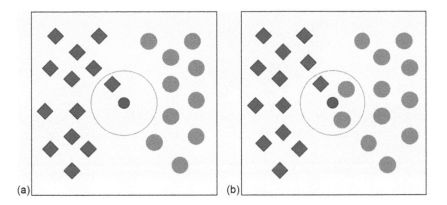

FIGURE 8.5   (a) $K = 1$ in KNN, (b) $K = 3$ in KNN.

$$S_B = \sum_{i=1}^{c} N_i \left(\mu_i - \mu \right) \left(\mu_i - \mu \right)^T \tag{8.7}$$

Here μi is the mean of class Xi, Ni is the number of samples in the class, $Xi\mu$ is the overall mean of the data, and c is the total number of classes. The in-class distribution matrix is given in Equation 8.8.

$$S_W = \sum_{i=1}^{c} \cdot \sum_{x_k \in x_i} \left(x_k - \mu_i \right) \left(x_k - \mu_i \right)^T \tag{8.8}$$

Here xk is a subset of the sample class Xi. The generalized eigenvalue problem of the matrix is given in Equation 8.9.

$$S_w^{-1} S_B \tag{8.9}$$

Equation 8.9 is solved to calculate eigenvectors w that maximize Equation 8.10.

$$w = \arg\max \frac{\left| W^T S_B W \right|}{\left| W^T S_w W \right|} \tag{8.10}$$

The eigenvectors are sorted by decreasing the eigenvalues and the k eigenvectors with the largest eigenvalues are selected to construct a transformation matrix. The transformation matrix should be used to convert the original dataset to a lower-dimension space. The advantages of LDA are that it is simple and mathematically robust and it produces results with accuracy comparable to more complex models. The disadvantages of LDA are that it is susceptible to overfitting and is not well suited for nonlinear problems.

- *K-means clustering*: The main purpose of the K-means clustering method is to divide a set of n observations into k distinct clusters in such a way that each point belongs to the nearest cluster with the shortest distance from the corresponding cluster mean or centroid.

  Suppose there are n $x1, x2, ..., xn$ observation points in a d-dimensional vector space. When it is desired to divide these observations into k clusters ($S1, S2, ..., Sk$) with their centroid mean ($\xi_1, \xi_2, ..., \xi_k$) as shown in Figure 8.6, so that the clusterwise sum of squares, also called the within-cluster sum of squares, can be minimized, i.e. (Yang, 2019):

FIGURE 8.6   K-means algorithm and cluster centers.

$$\sum_{j=1, x_i \in S_i}^{k} \left\| x_i - \xi_j \right\|^2 \tag{8.11}$$

Here $1 < k \leq n$ and typically $k \ll n$.

There are some major problems with this method. Choosing k points as the initial center of gravity is not efficient. In the worst case, k randomly selected points may belong to the same cluster. A possible solution is to choose k points with the greatest distances from each other. This is usually accomplished by starting from a random point, then trying to find the second point as far as possible from the first point, and then trying to find the third point as far as possible from the previous two points. This continues until the first k points are initialized. This method is an improvement over the previous random selection method, but there is no guarantee that the selection of these starting points will lead to the best clustering solutions. So, some sort of random reboot and multiple runs are required. On the other hand, the algorithm complexity of this clustering method is typically $O(n^{kd+1} \log(n))$ where d is the size of the data. Even for $n = 106$, $k = 3$, and $d = 2$, this would be $O(10^{43})$ which is computationally extremely expensive. However, it's worth noting that such complexity is only theoretical, and these methods can sometimes work surprisingly well in practice (at least for small datasets). In the worst cases, such an algorithm complexity can become NP-hard. For a given dataset it is difficult to know what k should be used since k is a hyperparameter. Ideally, the initial selection k should be close enough to the actual number of intrinsic classes in the data, and some adjustments can then be made around this initial estimate. But in practice, this may require some experience or other methods to understand the data. Additionally, clustering distances can also be used to check whether k is a suitable choice in many cases (Yang, 2019).

- *Association rule*: Suppose we observe eight customers each purchasing one or more apples, beer, chips, and nappies (Flach, 2012).

Transaction	Items
1	Nappies
2	Beer, crisps
3	Apples, nappies
4	Beer, crisps, nappies
5	Apples
6	Apples, beer, crisps, nappies
7	Apples, crisps
8	Crisps

Each transaction in this table contains a set of items. Conversely, for each item we can list the operations it is involved in: Operations 1, 3, 4, and 6 for nappies, operations 3, 5, 6, and 7 for apples, and so on. We can also do this for product sets: E.g., in transactions 2, 4, and 6, beer and chips were purchased together; it covers the transaction set {2, 4, 6} of the item set {beer, chips}. There are 16 such pen sets (including the empty set that covers all operations); a lattice is created using the subset relationship between transaction sets as a partial order (Figure 8.7). Let us call the number of transactions covered by an item set I. Its support, denoted Supp(*I*) (sometimes called frequency), deals with common sets of items that exceed a certain *f*0 support threshold. The support is uniform: It can never increase when moving down a path in the item set cage. This means that the set of frequently used item sets is convex and is completely determined by the lower bound of the largest item sets: In this example, this is maximal frequent item sets, *f*0 = 3: {apples}, {beer, chips}, and {nappies}. So at least three transactions involved apples; at least three related nappies; at least three contained both beer and chips; and any other combination of products was purchased less (Flach, 2012).

- *Q-learning*: This provides an algorithm for estimating the optimal state-action value function $Q*$ in the case of an unknown model. The optimum policy or policy value can be derived directly from $Q*$ in the following ways: $\pi*(s) = \text{argmax}_{a \in A} Q* (s, a)$ ve $V* (s) = \text{max}_{a \in A} Q* (s, a)$. Assuming a deterministic reward function to simplify the presentation, the Q-learning algorithm is based on equations that yield the optimal state-action value function $Q*$ (8.12) (Mohri et al., 2018):

$$Q^*(s,a) = \mathbb{E}\left[r(s,a)\right] + \gamma \sum_{s' \in S} \mathbb{P}\left[s'|s,a\right] V^*(s')$$

$$= \mathbb{E}_{s'}\left[r(s,a) + \gamma \max_{a' \in A} Q^*(s',a)\right]. \tag{8.12}$$

As for the policy values, the distribution model is unknown. Therefore, the Q-learning algorithm consists of these main steps: sampling a new state *s'* and updating the policy values according to (Mohri et al., 2018):

$$Q(s,a) \leftarrow (1-\alpha)Q(s,a) + \alpha\left[r(s,a) + \gamma \max_{a'A} Q^*(s',a')\right] \tag{8.13}$$

Here the parameter is a function of the number of visits to state *s*. The algorithm can be viewed as a stochastic formulation of the value iteration algorithm presented in the previous section. At each training step (epoch), an action is selected from the current *s* states using a policy π derived from *Q*. The choice of the π policy is arbitrary, as long as it guarantees an infinite number of visits to each pair (*s*, *a*). The reward received and the observed *s'* state are then used to update the *Q* (8.13) (Mohri et al., 2018).

Here the parameter is a function of the number of visits to state *s*. The algorithm can be viewed as a stochastic formulation of the value iteration algorithm presented in the previous section. At each training step (epoch), an action is selected from the current *s* states using a policy π derived from *Q*. The choice of the π policy is arbitrary, as long as it guarantees an infinite number of visits to each pair (*s*, *a*). The reward received and the observed *s'* state are then used to update the *Q* (8.13) (Mohri et al., 2018).

- *Temporal difference*: For any situation and action, it is desired to learn the expected cumulative reward starting from that situation with that action. This is an expected value because it is an average over all sources of randomness in the rewards and future states. The expected cumulative rewards of two consecutive state-action pairs are related by the Bellman equation and are used to back up rewards to previous actions from later actions,

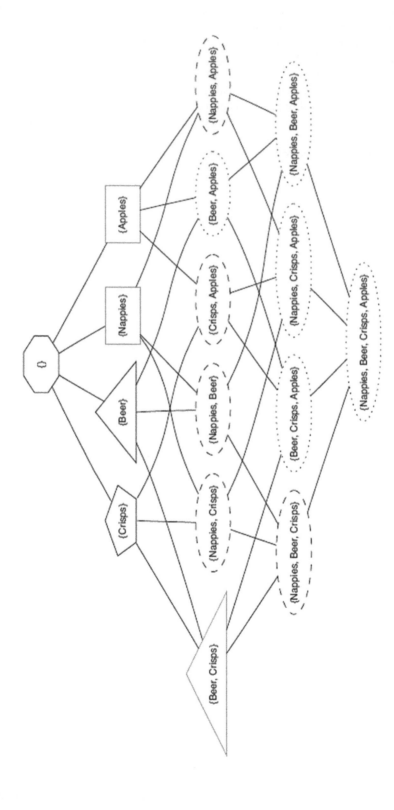

FIGURE 8.7 An item set lattice. Item sets in dotted ovals cover a single transaction; in dashed ovals, two transactions; in triangles, three transactions; and in polygons with n sides, n transactions. The maximal item sets with support three or more are indicated in green (Flach, 2012).

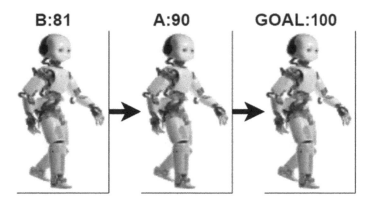

FIGURE 8.8 Temporal difference learning through reward backup.

as follows. Let's take the last move of the robot to the target; for reaching the goal, let's say we get a reward of 100 units (Figure 8.8). While in state A, we get a real prize of 100 if we go right. In case B just before that, if we take the right action (i.e., go to the right), we come to A, and here with one more action we can get the real reward, so it's as if going to B also has a reward. However, it is discounted (here by a factor of 0.9) as it is one step ahead and is a simulated internal reward, not a real one. The real reward for going from B to A is zero; the internal reward of 90 indicates how close we are to receiving the real reward. Now consider the situation and action just before that. In this case, although it didn't give us an immediate reward (because we were still one step away from the target), we did one more action that puts us in the situation where we got the full reward of 100. So let's reduce this reward by a factor of 0.9 because the reward is in the future and the future is never certain – in other words, "A bird in the hand is worth two in the bush". Thus, it can be said that the state-action pair one step ahead of the goal has an internal reward of 90. Note that in the case that happens, real reward there is still zero, because the goal is still not reached, but it is internally rewarded for reaching a state that is only one step away from the goal. Similarly, in a situation that is only one step away from the target, the person before this action receives an intrinsic reward of 81, while it is possible to continue assigning intrinsic values to all previous actions. Of course, this is just for a trial episode. Due to uncertainties, multiple trials are required, each observing different rewards, visiting different subsequent states, and averaging all these intrinsic reward estimates. This is called temporal difference learning. The internal reward estimate for each state-action pair is represented by Q, and the algorithm that updates them is called Q-learning (Alpaydın, 2016).

8.6   MACHINE LEARNING PERFORMANCE METRICS

In machine learning training, the problem is posed, examples are presented, each carefully marked with the label or class that the algorithm must learn, the algorithm is trained for a while, and the resulting output is just an opportunity (or a risk in another way) to propose a solution and get a correct answer. This output is a probability and does not mean that the results will provide the desired solution. Therefore, the results need to be verified.

The performance criteria used to validate the results are listed here (Marsland, 2015):

- *Overfitting*: The training process in machine learning is somewhat complicated. As we learn, we also want to know how well the algorithm generalizes, and it should be ensured that enough training is done for this. Overtraining is just as dangerous as undertraining. The degree of variability in most machine learning algorithms is enormous. For example, there are many weights for a neural network, and each of these can vary. So be careful. If training is done for too long, it means that besides the actual function, information about noise and inaccuracies in the data is learned and the data is memorized. Therefore, the model taught will be very complex and cannot be generalized. Figure 8.9 illustrates this by plotting the predictions (as the curve) of some algorithms at two different points in the learning process. On the left of the figure, the curve fits the overall trend of the data well (generalized to the underlying general function), but the training error would still not be that close to zero, as the training data passed close but not through it. As the network continues to learn, it will eventually produce a much more complex model with a lower training error (close to zero). This means that the training samples memorize the training data, including any noise components. The learning process should be stopped before the algorithm overfits. This means that we need to know how well it generalizes at each step. For this, training data is not used because overfitting cannot be detected. In addition, test data is not used as it will be used in post-tests. Therefore, a third data set, the validation set, is used to validate the learning. This is part of model selection and it is necessary to choose the right parameters for the model in order to generalize as well as possible.

- *Training, testing, and validation sets*: Three types of datasets are needed: The training set to actually train the algorithm, the validation set to track how well it did as it learns, and the test set to produce the final results. This becomes costly in terms of data, especially since for supervised learning, target values have to be added and it is not always easy to get the right labels. Obviously, every algorithm will need a reasonable amount of data to learn from. The exact need varies, but the more data the algorithm sees, the more likely it is to see examples of each possible input type. But more data also increases the computation time to learn. A similar approach applies to validation and test sets. In general, the exact ratio of training from testing to validation data is up to individuals. However, 50:25:25 or 60:20:20 ratios are typical in studies with large amounts of data. How you do the splitting can also be important. Choosing the first few points as the training set, the next test set, etc., the results will be pretty bad as the training doesn't see all the classes. This can be addressed by first randomly reordering the data or randomly assigning each data point to one of the clusters as shown in Figure 8.10. If you don't really have the training data and there is a separate validation set or there is concern that the algorithm will not be trained

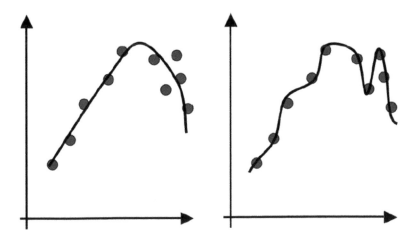

FIGURE 8.9 The effect of overfitting is that instead of finding the generating function as shown on the left, the neural network perfectly matches the inputs (right), including the noise inside them. This reduces the generalization capabilities of the network (Marsland, 2015).

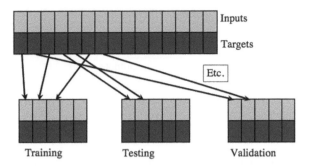

FIGURE 8.10 Splitting the dataset for training, validation, and testing (Marsland, 2015).

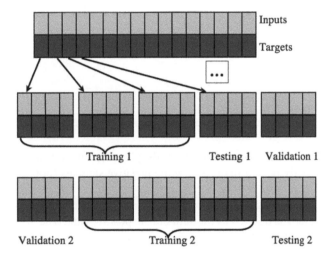

FIGURE 8.11 Solving the data shortage problem by training many models with excluded multiplex cross-validation (Marsland, 2015).

enough then it is also possible to do multi-level cross-validation, which is somewhat left out (Figure 8.11).

- *The confusion matrix*: No matter how much data is used to test the trained algorithm, a method such as a confusion matrix should still be used to decide whether the result is good or not. The confusion matrix is made of a square matrix containing all possible classes in both horizontal and vertical directions. Classes are listed as predicted outputs at the top of a table, then down the left-hand side as the targets. For example, the element of the matrix at (i, j) tells us how many input patterns were put into class i in the targets, but in class j by the algorithm. Anything on the front diagonal (the diagonal from the top left corner of the matrix to the bottom right corner) is the correct answer. Assuming there are three classes C1, C2, and C3, the output is class C1 when the target is C1, then when the target is C2 and so on until the table is filled (Table 8.1).

  Table 8.1 shows that most samples for the three classes were classified correctly, but two samples of the C3 class were misclassified as C1. For a small number of classes, this is a nice way to look at the output. When only one number is desired, it is then possible to divide the sum of the items in the leading diagonal by the sum of all the items in the matrix, which gives the fraction of correct answers. This is known as accuracy.

- *Accuracy metrics*: More can be done to analyze results than just measuring accuracy. Given the possible outputs of the classes, they can be arranged in a simple table like Table 8.2.

TABLE 8.1

Confusion Matrix

	Outputs		
	C_1	C_2	C_3
C_1	5	1	0
C_2	1	4	1
C_3	2	0	4

Source: Marsland (2015).

TABLE 8.2

Possible Outputs of Results

True positives	False positives
False negatives	True negatives

Source: Marsland (2015).

Here a true positive is an observation correctly placed in class 1, a false positive is an observation incorrectly placed in class 1, while negative samples (both true and false) are those in class 2.

The entries in the leading diagonal of this table are correct and those outside the diagonal are false. Accuracy is defined as the sum of true positives and true negatives divided by the total number of samples. Here # stands for "number" and TP stands for true positive, etc.:

$$Accuracy = \#TP + \#FP / \#TP + \#FP + \#TN + \#FN \qquad (8.14)$$

The problem with accuracy is that it doesn't tell us everything about the results, as it converts four numbers to one. There are two complementary pairs of metrics that can help interpret the performance of a classifier. These are sensitivity and specificity, and precision and recall. Sensitivity is the ratio of the number of true positive samples to the number classified as positive while specificity is the same ratio for negative samples. Precision is the ratio of true positive samples to the number of true positive samples, while recall is the ratio of the number of correct positive examples out of those that were classified as positive, which is the same as sensitivity. Looking at Table 8.2, it can be seen that sensitivity and specificity sum the columns for the denominator, while precision and recall sum the first column and the first row.

$$Sensitivity = \#TP / \#TP + \#FN \qquad (8.15)$$

$$Specificity = \#TN / \#TN + \#FP \qquad (8.16)$$

$$Precision = \#TP / \#TP + \#FP \qquad (8.17)$$

$$Recall = \#TP / \#TP + \#FN \qquad (8.18)$$

Together, any of these pairs of measurements provide more information than accuracy. For example, considering precision and recall, one can see that they are inversely proportional to some degree, that is, if the number of false positives increases (meaning the algorithm uses a broader definition of that class), then the number of false negatives usually decreases and vice versa. It can be combined to give a single measure, the F1 measure, which for precision and recall can be written as:

$$F1 = 2 \times Precision \times Recall \: / \: Precision + Recall \tag{8.19}$$

and in terms of numbers of false positives, etc. (it can be seen that it calculates the mean of false samples):

$$F1 = \# TP \: / \: \# TP + \left(\# FN + \# FP \right) / 2 \tag{8.20}$$

- *The receiver operator characteristic (ROC) curve*: These measures can be used to evaluate a particular classifier, or to compare the same classifier or completely different classifiers with different learning parameters. In this case, the ROC curve is useful. The ROC curve is a graph of the percentage of false positives on the x-axis versus true positives on the y-axis, an example of which is shown in Figure 8.12. Running the classifier once produces a single point on the ROC graph, and a perfect classifier would have a point at (0, 1) (100% true positive, 0% false positive), whereas the anti-classifier who got it all wrong would be (1.0), that is, the closer the result of a classifier is to the upper left corner, the better the classifier performed. Any classifier on a diagonal from (0, 0) to (1, 1) behaves exactly at chance level (assuming positive and negative classes are equally common), and therefore probably a lot of learning effort is wasted. To compare selections of classifiers or parameter settings for the same classifier, the point farthest from the "chance" line along the diagonal can be calculated. But it's normal to calculate the area under the curve (AUC) instead. If you have only one point for each classifier, the curve is trapezoidal from point (0, 0) to point (1, 1). If there are more points, they are only included sequentially along the diagonal line.

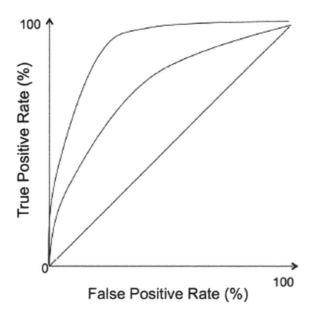

FIGURE 8.12   An example of an ROC curve. The diagonal line represents exactly chance, so anything above the line is better than chance, and the farther from the line the better. Represents a more accurate method that is further away from the diagonal line from the two curves shown (Marsland, 2015).

The key to getting a curve instead of a point on the ROC curve is to use cross-validation. If ten-fold cross-validation is used, there are ten classifiers with ten different test sets and also "ground truth" labels. The true labels can be used to produce a ranked list of different cross-validation-trained results, which can be used to specify a curve through the ten data points on the ROC curve corresponding to the results of this classifier. It is possible to compare results by generating an ROC curve for each classifier.

- *Unbalanced datasets*: For accuracy, it is implicitly assumed that there is the same number of positive and negative samples in the dataset. However, this is often not true. If not, we can calculate the balanced accuracy by dividing the sum of the sensitivity and specificity by 2. However, for this, Matthew's correlation coefficient (MCC) is a more accurate measure and is calculated as:

$$MCC = \#TP \times \#TN - \#FP \times \#FN \,/ \\ \sqrt{\left[\left(\#TP + \#FP \right)\left(\#TP + \#FN \right)\left(\#TN + \#FP \right)\left(\#TN + \#FN \right) \right]} \tag{8.21}$$

If any of the parentheses in the denominator is 0, the entire denominator is set to 1. This provides a balanced calculation of accuracy. As a final note on these evaluation methods, if there are more than two classes and it is useful to distinguish the different error types, then the calculations get a little more complicated because instead of one set of false positives and one set of false negatives, you have some for each class. In this case, the specificity and recall are not the same. However, it is also possible to construct a series of results where one class is used positively and everything else is used negatively, and this is repeated for each of the different classes.

- *Measurement precision*: The word precision is used, although with a different meaning, in evaluating the accuracy of a machine learning algorithm as a measurement system. The inputs are fed, the resulting outputs are looked at, and something about the algorithm can be measured, even before comparing the outputs to the target values. If a set of similar inputs were fed, then similar outputs would be expected for them as well. This measure of the algorithm's variability, also known as precision, indicates how repeatable the algorithm's predictions are. It may be helpful to think of precision as something like the variance of a probability distribution. In this case, it shows how widely the mean is spread. However, just because an algorithm is precise does not mean it is correct. For example, if the algorithm always gives the wrong prediction, it can definitely be wrong. A measure of how well the algorithm's predictions match reality is known as accuracy and can be defined as the average distance between the correct output and the prediction. Accuracy usually doesn't make much sense for classification problems unless there is a concept that certain classes are similar to each other.

8.7   WHAT IS DEEP LEARNING?

Although machine learning algorithms work very well on a wide variety of important problems, they have failed to solve central problems in AI such as speech recognition or object recognition. The evolution of deep learning was due in part to the inability of traditional algorithms to generalize well on such AI tasks. Generalizing to new examples becomes exponentially more difficult when working with high-dimensional data, and the mechanisms used to generalize in traditional machine learning fall short of learning complex functions in high-dimensional spaces. Such spaces often also impose high computational costs. Deep learning was designed to overcome these and other barriers (Akgul et al., 2019).

Deep learning is a new field of machine learning that has gained popularity in the recent past. The word "deep" refers to architectures with multiple hidden layers (deep networks) to learn different properties of input data. The number of layers used to model the data determines the depth of the model. Machine learning usually focuses on learning one or two layers of data, while deep learning learns from training data in tens or hundreds of layers. Machine learning techniques are limited in the way they process natural data in its raw form. It requires significant domain expertise to come up with a feature extractor that converts the raw data to the appropriate internal representation or feature vector. For example, a learning system such as a classifier can detect or classify patterns in the input. In deep learning, raw data is entered into the learning algorithm without first extracting features or defining a feature vector. Instead of creating a set of rules and algorithms to extract features from raw data, it involves learning these features automatically during the training process, without manual work (Figure 8.13). Therefore, deep learning is capable of learning suitable features on its own, requiring little guidance from the user. In deep learning, problems are solved in a hierarchy with each layer built on top of the others. Each layer performs some transformations on the input it receives from previous layers. The lower layers of the model encode some basic representations of the problem, while the higher-level layers build on these lower layers to create more complex concepts (Wani et al., 2020).

Deep learning has emerged from research in artificial intelligence and machine learning. Figure 8.14 shows the relationship between artificial intelligence, machine learning, and deep learning. The field of artificial intelligence was born in a workshop at Dartmouth College in the summer of 1956. The workshop conducted research on many topics such as mathematical theorem proving, natural language processing, planning for games, computer programs that can be learned from examples, and neural networks. The modern field of machine learning draws on computers that can learn from examples and neural network research. Machine learning involves the development and evaluation of algorithms that enable a computer to extract or learn functions from a dataset. When the machine learning algorithm finds a function that is accurate enough (in terms of the outputs it

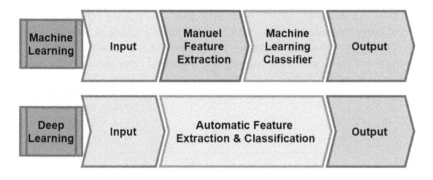

FIGURE 8.13   Comparison of machine learning and deep learning methods.

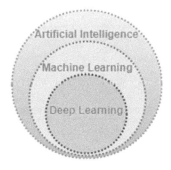

FIGURE 8.14   The relationship between artificial intelligence, machine learning, and deep learning.

produces that match the correct outputs listed in the dataset) for the problem we are trying to solve, the training process is complete and the final model is returned by the algorithm. Deep learning, on the other hand, enables data-driven decisions by identifying and extracting patterns from large datasets that accurately map from complex sets of inputs to good decision outcomes (Kelleher, 2019).

On March 27, 2019, Yoshua Bengio, Geoffrey Hinton, and Yann LeCun jointly received the ACM A.M. Turing award. The award was given to recognize their contribution to making deep learning the key technology driving the modern AI revolution. The ACM A.M. Turing prize, often described as the "Nobel Prize for Computers", is a prize worth $1 million. The announcement of the award highlighted the recent surprising breakthroughs deep learning has made in computer vision, robotics, speech recognition, and natural language processing, as well as the profound impact that these technologies have on society and deep learning–based artificial intelligence, which billions of people now use daily through smartphone apps. The announcement also highlighted that deep learning provides scientists with powerful new tools that result in scientific breakthroughs in various fields such as medicine and astronomy. This award to these researchers reflects the importance of deep learning to modern science and society. The transformative effects of deep learning on technology will increase in the coming decades with the development and adoption of deep learning continuing to be driven by the ingenious cycle of ever-larger datasets, the development of new algorithms, and improved hardware. These trends do not stand still, and how the deep learning community responds to them will drive growth and innovation in this field for years to come (Kelleher, 2019).

8.8   COMMON ARCHITECTURAL PRINCIPLES OF DEEP NETWORKS

The deep learning model discovers what parameters to give weight to according to the structure of the data. Although the deep learning algorithm also performs data-based learning, the learning process does not work with a single mathematical model as in standard machine learning algorithms, but with calculations developed in a structure similar to network diagrams expressed as a neural network. It is a machine learning algorithm because every deep learning algorithm learns from data. However, not every machine learning algorithm is a deep learning algorithm. Therefore, deep learning is a specific type of machine learning. Deep learning is a subset of machine learning in which layered neural networks with high computing power and large datasets can create powerful mathematical models (Karakuş, 2021).

The basic components of deep networks are listed as follows (Patterson and Gibson, 2017) (Figure 8.15):

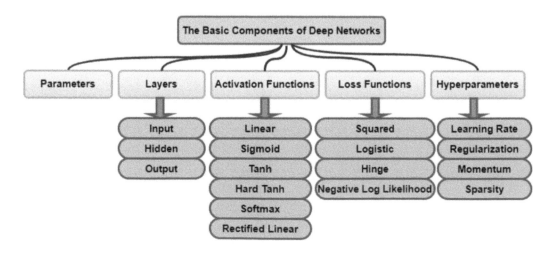

FIGURE 8.15   The basic components of deep networks.

- *Parameters*: The parameters are related to the parameter vector x in the equation $Ax = b$ in basic machine learning. Parameters in neural networks are directly related to the weights of the connections in the network. The point outputs of matrix A and parameter vector x are taken to obtain the current output column vector b. The closer the result vector b is to the true values in the training data, the better the model. Optimization methods such as gradient descent are used to find good values for the parameter vector to minimize the loss in the training dataset.
- *Layers*: The layers are classified as:

  Input layer: It consists of input data (vectors) fed into the network. The number of neurons in an input layer is typically the same as the input feature to the network. Input layers are followed by one or more hidden layers. In classical feedforward neural networks, the input layers are fully connected to the next hidden layer, but in other network architectures, the input layer may not be fully connected. Weight values on connections between layers are how neural networks encode learned information extracted from raw training data.

  Hidden layer: This is key to allowing neural networks to model nonlinear functions.

  Output layer: It is the layer where the answer or prediction of the developed model is taken. Depending on the setup of the neural network, the final output may be a revalued output (regression) or a set of possibilities (classification). This is controlled by the type of activation function we use on neurons in the output layer. The output layer typically uses a softmax or sigmoid activation function for classification.
- *Activation functions*: The functions that govern the behavior of the artificial neuron are called activation functions. Activation functions are used to propagate the output of nodes of one layer to the next layer. The transmission of this input is known as forward propagation. Activation functions transform the combination of inputs, weights, and biases. The outputs of these transformations are the input for the next node layer. Most nonlinear transformations used in neural networks convert the data into a suitable range, such as 0 to 1 or −1 to 1. An artificial neuron is said to become active when it transmits a non-zero value to another artificial neuron. Many enable functions belong to a class of logistic transformations, which when plotted looks like an S. This class of functions is called sigmoidal. The sigmoid function class includes several variations, one of which is known as the sigmoid function. Some useful activation functions in neural networks are:

  Linear: It is an identity function that passes the signal unchanged.

  Sigmoid: Sigmoids reduce extreme values or outliers in data without removing them.

  Tanh: Tanh is a hyperbolic trigonometric function. Unlike the sigmoid function, the normalized tanh range is −1 to 1. The advantage of tanh is that it can more easily deal with negative numbers.

  Hard tanh: Similar to tanh, hard tanh imposes hard limits on the normalized range. Anything greater than 1 is converted to 1, and anything less than −1 is converted to −1.

  Softmax: Softmax is a generalization of logistic regression as it can be applied to continuous data (rather than binary classification) and can include multiple decision boundaries. It manages polynomial tagging systems. Softmax is a function often found in the output layer of a classifier.

  Rectified linear: Rectified linear is a transform that activates a node only if the input is above a certain amount. The output is zero when the input is below zero, but when the input goes above a certain threshold it has a linear relationship with the dependent variable $f(x) = \max(0, x)$. Rectified linear units (ReLU) are cutting-edge technology as they have proven to work in many different situations.
- *Loss functions*: Loss functions quantify the match between the predicted output (or label) and the exact result output. Loss functions are used to determine the penalty for misclassifying an input vector. Thus, searching for the ideal state of the neural network is equivalent

to finding the parameters (weights and bias) that will minimize the "loss" from errors. The loss functions are:

Squared loss: When working on a regression model that requires a real-valued output, the squared loss function is used, as in the case of ordinary least squares in linear regression.

Logistic loss: Logistic loss is used when probabilities are of greater interest than strict classifications.

Hinge loss: It is the most commonly used loss function when the network needs to be optimized for strict classification.

Negative log-likelihood: The product of the probabilities turns into the sum of the log of the probabilities. The logarithm is a monotonically increasing function. Therefore, minimizing the negative log probability is equivalent to maximizing the probability.

- *Loss functions for regression*: These include:

Mean squared error loss (MSE): When working on a regression model that requires a real-valued output, the square loss function is used, as in the ordinary least squares case in linear regression. MSE is highly sensitive to outliers and this is something to consider when choosing a loss function. When wanting to consider outliers, the median is more of a concern.

Mean absolute error loss (MAE): It is the mean absolute error loss.

Mean squared log error loss (MSLE): It is the average square log error.

Mean absolute percentage error loss (MAPE): It is the average absolute percentage error loss.

- *Loss functions for classification*: When building neural networks for classification problems, the focus is often on adding probabilities to these classifications. These different scenarios require different loss functions. These functions include:

Hinge loss: It is the most commonly used loss function when the network needs to be optimized for strict classification and is traditionally called a 0–1 classifier.

Logistic loss: It is used when probabilities are of more interest than strict classifications.

Negative log-likelihood: The product of the probabilities turns into the sum of the log of the probabilities.

- *Loss functions for reconstruction*: A neural network is trained to reconstruct the input as closely as possible. The key here is to fine-tune the scenario so that the network is forced to learn commonalities and features in the dataset. In one approach, the number of parameters on the network is limited such that the network is forced to compress and then regenerate the data. Another frequently used approach is to corrupt the input with meaningless "noise" and train the network to ignore the noise and learn from the data.

  Optimization methods: Training a model in machine learning involves finding the best set of values for the model's parameter vector. Machine learning can be thought of as an optimization problem in which the loss function is minimized according to the parameters of the predictive function (based on our model).

- *Hyperparameters*: In machine learning, the adjustment parameters made to make networks work better and faster are called hyperparameters. Hyperparameters are concerned with controlling optimization functions and model selection during training with the learning algorithm. Hyperparameter selection focuses on learning the structure of the data as quickly as possible while ensuring that the model neither underfits nor overfits the training dataset. The hyperparameters are listed as follows:

Learning rate: The learning rate affects the amount of time that parameters are adjusted during optimization to minimize the error of the neural network's predictions. It is a coefficient that scales the size of the steps a neural network takes to the parameter vector x as it traverses the missing function space. During backpropagation, the error gradient is multiplied by the learning rate, and then the last iteration of a link weight is updated by the product to arrive at a new weight. The learning rate determines how much of the gradient is desired to be used for the next step of the algorithm. A large error and a steep gradient combine with the learning rate to produce a large step. The closer you get to the minimum error and the flatter the gradient, the shorter the step size tends to be. Time is of the essence when neural network training can take weeks on large datasets. If results cannot be expected for another week, a moderate learning rate is chosen (e.g., 0.1) and attempts should be made to achieve the best speed and accuracy at the same time.

Regularization: Regularization helps the effects of out-of-control parameters by using different methods to minimize parameter size over time. The main purpose of regularization is to control overfitting in machine learning. In mathematical notation, regularization is seen, represented by the lambda coefficient, which controls the balance between finding a good fit and keeping the value of certain feature weights low as exponents on the features increase. The regularization coefficients L1 (8.22) and L2 (8.23) help combat overfitting by reducing certain weights. In the L1 regularization, we penalize the absolute value of the weights. The L1 regulation term is highlighted in a red box. In L2 regularization, the regularization is the sum of the squares of all feature weights. L2 regularization forces the weights to be small but does not make them zero and makes a non-sparse solution. Smaller value weights lead to simpler hypotheses, and simpler hypotheses are the most generalizable. Irregular weights with several higher-order polynomials in the feature set tend to overfit the training set. As the input training set size grows, the effect of regularization decreases and the parameters tend to increase in size. The redundancy of features relative to the training set examples leads to overfitting in the first place. Bigger data is the ultimate regulator.

$$L\left(x,y\right) \equiv \sum\nolimits_i^n (y_i - h_\theta\left(x_i\right))^2 + \boxed{\lambda \sum\nolimits_{i=1}^n \left|\theta_i\right|} \qquad (8.22)$$

$$L\left(x,y\right) \equiv \sum\nolimits_{i=1}^n (y_i - h_\theta\left(x_i\right))^2 + \boxed{\lambda \sum\nolimits_{i=1}^n \left|\theta_i^2\right|} \qquad (8.23)$$

Momentum: Momentum helps the learning algorithm get out of the way of getting stuck in case of an error in the search space and helps the updater find the channels that go to the minimum in the error area. Momentum is for the learning rate what the learning rate is for the weights, and with momentum, better quality models are produced.

Sparsity: The sparse hyperparameter assumes that only a few properties are relevant for some inputs. For example, if a network is assumed to be able to classify one million images, any of these images will be displayed with a limited number of features. But to effectively classify millions of images, a network must be able to recognize many more features that are often invisible. Therefore, sparse features can limit the number of nodes that enable and inhibit a network's learning ability. In response to sparseness, bias forces neurons to activate, and activations remain around an average that prevents the network from becoming stuck.

8.9   DEEP LEARNING ARCHITECTURES

Deep learning is a class of representative machine learning methods with multiple representation levels. It consists of several simple but nonlinear modules, each of which transforms the representation from the previous levels (starting with the raw input) to a representation at a higher, somewhat more abstract level. With enough combination of such transformations, very complex properties and implications can be learned. In general, all deep learning methods can be classified into one of three different categories as follows (Hosseini et al., 2020):

- *Convolutional neural networks (CNNs)*: CNNs are inspired by biological processes and are designed to mimic neural connectivity found in the brain's visual cortex. They require much less data pre-processing compared to traditional image classification algorithms that require manually designed pre-processing filters (Hosseini et al., 2017). CNNs have a wide range of applications in image and video recognition, recommendation systems, image classification, medical image analysis, and natural language processing. CNNs differ from traditional neural networks in that they perform convolution rather than standard matrix multiplication in at least one of their layers (Figure 8.16).

  Each layer of a convolutional network usually performs three steps. In the first step, the layer performs multiple convolutions in parallel to generate a series of linear activations. The second step (often referred to as the detector stage) involves running linear activations via nonlinear activation functions. The purpose of this is to infer a nonlinear mapping to the output when needed. The third stage is a pooling function stage that makes further changes to the output of the layer. The purpose of a pooling function is to vary the output of a particular location in the network based on statistical values of nearby outputs (Krizhevsky et al., 2012; Hosseini, 2016; Hosseini, 2018).

- *Pre-trained unsupervised networks (PUNs)*: Data generation and feature extraction are very important applications in deep learning as there are often limited training data available. Synthetic data can be generated based on the original dataset using deep learning architectures such as generative adversarial networks (GANs) and autoencoders to expand the initial dataset to provide a larger dataset to train the network. Both of these architectures are PUN implementations. PUN is a deep learning model that uses unsupervised learning to train each of the hidden layers in a neural network to better fit the dataset. The unsupervised learning algorithm is used to train each layer independently one by one while using the pre-trained layer as input. After pre-training at each layer, a fine-tuning step is performed on the entire network using supervised learning. PUN types include autoencoders, deep belief networks (DBN), and GAN (Patterson and Gibson, 2017).

- *Recurrent and recursive neural networks (RNNs)*: Recurrent neural networks are used in video analysis, captioning, natural language processing, and music analysis because of their ability to process sequential data and capture features.

  Recursive neural networks, on the other hand, are a set of nonlinear adaptive models used to process data of variable length. They are particularly experts at handling data structure inputs. Recursive networks feedback the state of the network into itself, which can be seen as a loop. They are primarily suitable for image and sentence processing. The

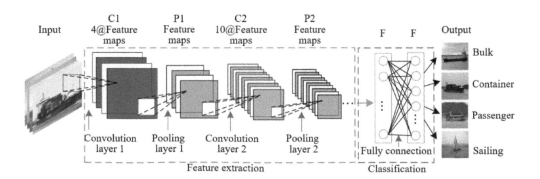

FIGURE 8.16   Representation of convolutional neural networks (CNN) (Ren et al., 2019).

architecture of recursive neural networks allows users not only to identify the components of the input data, but also to quantitatively determine the relationships between them (Eraslan et al., 2019).

8.10 FUNDAMENTALS OF CONVOLUTIONAL NEURAL NETWORKS

Looking into deep learning, it may be surprising to see that there is a lot of legacy technology. But surprisingly, everything is working like never before because researchers have finally managed to get some simple, legacy solutions to work together. For example, new activations like rectified linear units (ReLU), a neural network function that converts negative values to zero without touching positive values, are not actually new. However, ReLU appears to be used in new ways. As a result, deep learning can automatically filter, process, and transform big data. Also, the image recognition capabilities that popularized deep learning a few years ago are not new either. In recent years, deep learning has gained great acceleration thanks to the ability to code specific features into architecture using CNNs. By making it possible to configure CNNs into complex architectures that can improve their learning from a lot of useful data, it is starting to come as a surprise to outperform humans on certain recognition tasks (Mueller and Massaron, 2019).

Fundamentals of convolutional neural networks are classified as follows (Michelucci, 2019):

- *Kernels and filters*: One of the main components of CNNs are filters, which are square matrices with dimensions $nK \times nK$, where nK is an integer, usually a small number like 3 or 5. Sometimes filters are also called kernels. The kernel usage comes from the operations such as sharpening, blurring, embossing, etc. found in classical image processing techniques. All these operations are done with kernels. Following are sample filters for four different operations.

  Kernel to allow detection of horizontal edges:

  $$\mathfrak{J}_H = \begin{pmatrix} 1 & 1 & 1 \\ 0 & 0 & 0 \\ -1 & -1 & -1 \end{pmatrix}$$

  Kernel to allow detection of vertical edges:

  $$\mathfrak{J}_V = \begin{pmatrix} 1 & 0 & -1 \\ 1 & 0 & -1 \\ 1 & 0 & -1 \end{pmatrix}$$

  Kernel to allow detection of edges when brightness changes drastically:

  $$\mathfrak{J}_L = \begin{pmatrix} -1 & -1 & -1 \\ -1 & 8 & -1 \\ -1 & -1 & -1 \end{pmatrix}$$

  Kernel to blur edges in an image:

  $$\mathfrak{J}_B = -\frac{1}{9} \begin{pmatrix} 1 & 1 & 1 \\ 1 & 1 & 1 \\ 1 & 1 & 1 \end{pmatrix}$$

- *Convolution*: The first step to understanding CNNs is to understand convolution. First, in the context of neural networks, convolution is done between tensors. The process takes two tensors as input and produces one tensor as output. Each element from one tensor is multiplied by the corresponding element of the second tensor (element in the same position) and then all values are added together. The operation is usually denoted by the operator *. Two examples of tensors, both 3×3 dimensional, can be given to see how it works. Convolution is done by applying the following formula:

$$\begin{pmatrix} a_1 & a_2 & a_3 \\ a_4 & a_5 & a_6 \\ a_7 & a_8 & a_9 \end{pmatrix} * \begin{pmatrix} k_1 & k_2 & k_3 \\ k_4 & k_5 & k_6 \\ k_7 & k_8 & k_9 \end{pmatrix} = \sum_{i=1}^{9} a_i k_i \tag{8.24}$$

In this case, the result is simply the sum of each element a_i multiplied by the corresponding element k_i. In more typical matrix form, this formula with double sum can be written as:

$$\begin{pmatrix} a_{11} & a_{12} & a_{13} \\ a_{21} & a_{22} & a_{23} \\ a_{31} & a_{32} & a_{33} \end{pmatrix} * \begin{pmatrix} k_{11} & k_{12} & k_{13} \\ k_{21} & k_{22} & k_{23} \\ k_{31} & k_{32} & k_{33} \end{pmatrix} = \sum_{i=1}^{3} \sum_{j=1}^{3} a_{ij} k_{ij} \tag{8.25}$$

Convolution is usually done between a tensor and a kernel, which we can denote here as A. Typically, the cores are small, 3×3 or 5×5, and input tensors A are normally larger. For example, in image recognition, input tensors A are images that can have dimensions as high as $1{,}024 \times 1{,}024 \times 3$. Here $1{,}024 \times 1{,}024$ is the resolution and the final size is the number of three color channels, RGB values. In advanced applications, images may have higher resolution. To understand how to apply convolution when there are matrices of different sizes, an example of a 4×4 matrix A is.

$$A = \begin{pmatrix} a_1 & a_2 & a_3 & a_4 \\ a_5 & a_6 & a_7 & a_8 \\ a_9 & a_{10} & a_{11} & a_{12} \\ a_{13} & a_{14} & a_{15} & a_{16} \end{pmatrix}$$

A kernel K 3×3 that we will take for this example:

$$K = \begin{pmatrix} k_1 & k_2 & k_3 \\ k_4 & k_5 & k_6 \\ k_7 & k_8 & k_9 \end{pmatrix}$$

What to do here is to start from the top left corner of matrix A and choose a 3×3 region. The place to be taken in the sample is marked in bold.

$$A = \begin{pmatrix} \mathbf{a_1} & \mathbf{a_2} & \mathbf{a_3} & a_4 \\ \mathbf{a_5} & \mathbf{a_6} & \mathbf{a_7} & a_8 \\ \mathbf{a_9} & \mathbf{a_{10}} & \mathbf{a_{11}} & a_{12} \\ a_{13} & a_{14} & a_{15} & a_{16} \end{pmatrix}$$

Then, if this smaller matrix is convoluted between $A1$ and K, its result is denoted by $B1$:

$$B_1 = A_1 * K = a_1k_1 + a_2k_2 + a_3k_3 + k_4a_5 + k_5a_5 + k_6a_7 + k_7a_9 + k_8a_{10} + k_9a_{11} \qquad (8.26)$$

Next, the 3×3 region selected in matrix A of a column needs to be shifted to the right and the items marked in bold therein should be selected:

$$A = \begin{pmatrix} a_1 & \mathbf{a_2} & \mathbf{a_3} & \mathbf{a_4} \\ a_5 & \mathbf{a_6} & \mathbf{a_7} & \mathbf{a_8} \\ a_9 & \mathbf{a_{10}} & \mathbf{a_{11}} & \mathbf{a_{12}} \\ a_{13} & a_{14} & a_{15} & a_{16} \end{pmatrix}$$

This will give the second submatrix $A2$:

$$A_2 = \begin{pmatrix} a_2 & a_3 & a_4 \\ a_6 & a_7 & a_8 \\ a_{10} & a_{11} & a_{12} \end{pmatrix}$$

Then the convolution between this smaller matrix $A2$ and K is performed again:

$$B_2 = A_2 * K = a_2k_1 + a_3k_2 + a_4k_3 + a_6k_4 + a_7k_5 + a_8k_6 + a_{10}k_7 + a_{11}k_8 + a_{12}k_9 \qquad (8.27)$$

The 3×3 region can no longer be shifted to the right because the end of the A matrix has been reached. For this reason, it is shifted down one line and started again from the left side. Next selected region:

$$A_3 = \begin{pmatrix} a_5 & a_6 & a_7 \\ a_9 & a_{10} & a_{11} \\ a_{13} & a_{14} & a_{15} \end{pmatrix}$$

Again, the convolution of $A3$ with K is performed.

$$B_3 = A_3 * K = a_5k_1 + a_6k_2 + a_7k_3 + a_9k_4 + a_{10}k_5 + a_{11}k_6 + a_{13}k_7 + a_{14}k_8 + a_{15}k_9 \qquad (8.28)$$

As the last step, the 3×3 selected region is shifted to the right of one column and convolution is made again. Selected region:

$$A_4 = \begin{pmatrix} a_6 & a_7 & a_8 \\ a_{10} & a_{11} & a_{12} \\ a_{14} & a_{15} & a_{16} \end{pmatrix}$$

Convolution will give the result:

$$B_4 = A_4 * K = a_6k_1 + a_7k_2 + a_8k_3 + a_{10}k_4 + a_{11}k_5 + a_{12}k_6 + a_{14}k_7 + a_{15}k_8 + a_{16}k_9 \qquad (8.29)$$

Since there is no longer room to shift right and down in the 3×3 region, the calculated four values $B1$, $B2$, $B3$, and $B4$ will form the resulting tensor of the convolution operation that gives us the tensor B:

$$B = \begin{pmatrix} B_1 & B_2 \\ B_3 & B_4 \end{pmatrix}$$

Our 3×3 region has always been moved one column to the right and one row down. The same process can be applied when the tensor A is larger. In such a case, a larger B tensor will be obtained. The number of rows and columns, 1 in this example, are called steps and are usually denoted by s. When step s = 2, it means that the 3×3 region will be shifted two columns to the right and two rows down at each step. Also, another important issue is the size of the selected region in the input matrix A in the example. The dimensions of the selected region shifted in the operation must be the same as the kernel used. If using 5×5 cores, it is necessary to select a 5×5 region in A. In general, given an $nK \times nK$ core, an $nK \times nK$ region is selected in A.

Step s in the context of a neural network with convolution is a process that takes a tensor A of dimensions $nA \times nA$ and a kernel K of dimensions $nK \times nK$ and outputs a matrix B of dimensions $nB \times nB$.

$$n_B = \left\lfloor \frac{n_A - n_K}{s} + 1 \right\rfloor \tag{8.30}$$

Let's start formally explaining the situation with step $s = 1$. The algorithm generates a new tensor B from an input tensor A and a kernel K according to the formula.

$$B_{ij} = \left(A * K\right)_{ij} = \sum_{f=0}^{n_k - 1} \sum_{h=0}^{n_k - 1} A_{i+f, j+h} K_{i+f, j+h} \tag{8.31}$$

- *Pooling*: Pooling is the second fundamental operation in CNNs. To understand the pooling process, a 4×4 matrix will be given as an example:

$$A = \begin{pmatrix} a_1 & a_2 & a_3 & a_4 \\ a_5 & a_6 & a_7 & a_8 \\ a_9 & a_{10} & a_{11} & a_{12} \\ a_{13} & a_{14} & a_{15} & a_{16} \end{pmatrix}$$

To achieve maximum pooling, a region of size $nK \times nK$ must be defined, similar to that done for convolution. If $nK = 2$ is taken, an $nK \times nK$ region of 2×2 from A is selected, starting from the upper left corner of the matrix A.

$$A = \begin{pmatrix} \mathbf{a_1} & \mathbf{a_2} & a_3 & a_4 \\ \mathbf{a_5} & \mathbf{a_6} & a_7 & a_8 \\ a_9 & a_{10} & a_{11} & a_{12} \\ a_{13} & a_{14} & a_{15} & a_{16} \end{pmatrix}$$

From the selected items $a1$, $a2$, $a5$, and $a6$, the maximum pooling operation maximum value is selected, and the result is indicated by $B1$.

$$B_1 = \max_{i=1,2,5,6} a_i \tag{8.32}$$

Then the 2×2 window is shifted two columns to the right.

$$A = \begin{pmatrix} a_1 & a_2 & \mathbf{a_3} & \mathbf{a_4} \\ a_5 & a_6 & \mathbf{a_7} & \mathbf{a_8} \\ a_9 & a_{10} & a_{11} & a_{12} \\ a_{13} & a_{14} & a_{15} & a_{16} \end{pmatrix}$$

The max-pooling algorithm will give the result denoted by *B2* and choose the maximum of the values.

$$B_2 = \max_{i=3,4,7,8} a_i \tag{8.33}$$

At this point, since the 2×2 region can no longer be shifted to the right, it is shifted down two rows and the process starts again from the left side of *A*. Items marked in bold are selected and the maximum is taken as *B3*.

$$A = \begin{pmatrix} a_1 & a_2 & a_3 & a_4 \\ a_5 & a_6 & a_7 & a_8 \\ \mathbf{a_9} & \mathbf{a_{10}} & a_{11} & a_{12} \\ \mathbf{a_{13}} & \mathbf{a_{14}} & a_{15} & a_{16} \end{pmatrix}$$

Here step *s* is similar to the step definition in convolution and is the number of rows or columns in which the region is moved as items are selected. Finally, elements *a11*, *a12*, *a15*, and *a16* located in the last 2×2 region in the bottom-bottom of *A* are selected. Then the maximum is taken and called *B4*. With the values obtained in this operation, an output tensor is formed with the four values *B1*, *B2*, *B3*, and *B4* in the example:

$$B = \begin{pmatrix} B_1 & B_2 \\ B_3 & B_4 \end{pmatrix}$$

In the example, $s = 2$. This operation takes as input a matrix *A*, a step *s*, and a kernel size or size *nK* of the selected region and returns a new matrix *B*:

$$n_B = \left\lfloor \frac{n_A - n_K}{s} + 1 \right\rfloor \tag{8.34}$$

Finally, another way of pooling, although not as widely used as maximum pooling, is average pooling. It can be found on certain network architectures. Here, instead of returning the maximum of the selected values, the average is returned.

- *Padding*: When dealing with images, it is not optimal to get results from a convolution operation with dimensions different from the original image. In these cases, the filling process is applied. Rows of pixels are added above and below the final images, and columns of pixels are added to the right and left of the final images. Thus, the resulting matrices are the same size as the original. Some strategies fill the inserted pixels with zeros, values of the nearest pixels, etc. Using *p* padding as a reference, the final dimensions of matrix *B* in both convolution and pooling are given as:

$$n_B = \left\lfloor \frac{n_A + 2p - n_K}{s} + 1 \right\rfloor \tag{8.35}$$

- *Building blocks of CNN*: Convolution and pooling operations are used to create the layers used in CNNs. CNNs typically have the following layers:
  a. *Convolution layers*: A convolution layer takes a tensor as input (it can be 3-D due to three color channels). For example, an image applies a certain number of kernels, typically adding a bias of 10, 16, or more. For example, to introduce nonlinearity to the result of the convolution, it applies the ReLU activation functions and produces an output matrix *B*. The number of parameters a convolution layer has is independent of the size of the input image. This fact helps reduce overfitting, especially when dealing with large input images.
  b. *Pooling layers*: The pooling layer is usually indicated by POOL and a number. For example, POOL1. Takes one tensor as input and returns another tensor as output after pooling the input. A pooling layer has no parameters to learn from but offers additional hyperparameters: *nK* and stride. In general, one of the reasons to use pooling is usually to reduce the dimensionality of tensors. Therefore, no padding is used in the pooling layers.
  c. *Fully connected layers*: It is a layer where neurons are connected to all neurons of the previous and next layers. In CNNs, the convolution and pooling layers are often put one after the other. A convolutional layer is always followed by a pooling layer. A pooling layer is sometimes called a layer together because it has no learnable weights and is viewed as a simple operation associated only with the convolution layer.
- *Number of weights in a CNN*: It is important to indicate where the weights in a CNN are in different layers.
  a. *Convolution layer*: The parameters learned in a convolution layer are the filters themselves. For example, if there are 32 filters, each 5×5 in size, $32 \times 5 \times 5 = 832$ learnable parameters are taken, because for each filter, there is also a bias term that must be added. It should be noted that this number does not depend on the size of the input image. In a typical feedforward neural network, the number of weights in the first layer depends on the input size, but not here. The number of weights in a convolution layer is generally given as:

$$n_C . \, n_K . \, n_K + n_C \qquad\qquad (8.36)$$

  b. *Pooling layer*: The pooling layer has no learnable parameters and as stated, this is why it is typically associated with the convolution layer. There are no learnable weights in this layer.
  c. *Dense layer*: The weights in the dense layer are known from conventional feedforward networks. So the number depends on the number of neurons and the number of neurons in the previous and next layers.

8.11   IMAGE CLASSIFICATION AND DETECTION

Rapid advances in artificial intelligence research enable new applications to be created every day in different fields that were not possible a few years ago. Learning these tools will enable the invention of new products and applications on their own. Although not working on developing computer vision (CV) systems that solve complex image classification and object detection problems on their own, useful deep learning algorithms and architectures can be learned. Each network has specific features that distinguish it from the others and certain problems that are tried to be solved, which are the reasons for their development. Each successful CNN architecture solves a certain limitation in the previous one. Simpler CNNs such as AlexNet and VGGNet can be used for simple to medium complexity problems, and deeper networks such as Inception and ResNet for very complex problems. When it comes to CNNs, some common design options should be looked at first. At first, it may seem like there are too many choices to make. But when something new is learned in deep learning, it gives us more hyperparameters to design. That's why it's good to narrow down our choices by looking at some common patterns established by leading researchers in the field. So instead of making them completely random, their motivations can be understood and started where they end (Elgendy, 2020):

- *Pattern 1. Feature extraction and classification*: Convolutional networks typically consist of two parts, a feature extraction section consisting of a series of convolutional layers and a classification section consisting of a series of fully connected layers (Figure 8.17). This is almost always the case with ConvNets, starting with LeNet and AlexNet, to the latest CNNs like Inception and ResNet that have emerged over the past few years.
- *Pattern 2. Image depth increases, and dimensions decrease*: The input data in each layer is an image. With each layer, a new convolution layer is applied over a new image. This pushes us to think of an image in a more general way. First, each image appears to be a 3D object with height, width, and depth. The depth is called the color channel. Depth is 1 for grayscale images and 3 for color images. In later layers, images still have depth. However, these are not colors per se, they are feature maps representing features extracted from previous layers. Therefore, as you go deeper into the network layers, the depth increases. In Figure 8.18, the depth of an image is equal to 96 and represents the number of feature maps in this layer. This is a pattern that can always be seen: Image depth increases, and dimensions decrease.

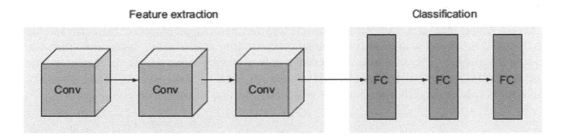

FIGURE 8.17 Convolutional networks often involve feature extraction and classification (Elgendy, 2020).

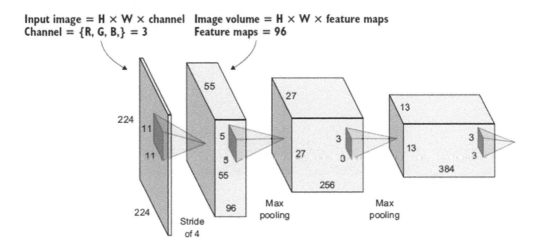

FIGURE 8.18 Image depth increases and dimensions decrease (Elgendy, 2020).

- *Pattern 3. Fully connected layer*: Fully connected layers are usually not as rigid a pattern as the previous two models, but they are very useful to know. Typically, all fully connected layers in a network either have the same number of hidden units or a decrease in each layer. It is rare to find a network where the number of units in fully connected layers increases with each layer. Research has shown that keeping the number of units constant does not damage the neural network. Therefore, this may be a good approach if you want to limit the number of choices that need to be made when designing the network. This way, all one has to do is select a set of units per layer and apply it to all connected layers.

Classical CNN architectures have the same classical architecture stacking convolutional and pool layers with different configurations for their layers. LeNet consists of first and second convolutional layers, followed by a pooling layer, and five weight layers consisting of three convolutional and two fully connected layers. AlexNet is deeper than LeNet. It includes eight weight layers, consisting of five convolutional and three fully connected layers. VGGNet has solved the problem of setting up hyperparameters of the convolution and pooling layers by creating a uniform configuration for their use across the entire network. Inception tried to solve the same problem as VGGNet. Instead of deciding which filter size to use and where to add the pool layer, Inception says "Let's use them all". ResNet followed the same approach as Inception and created residual blocks that, when stacked, make up the network architecture. ResNet has attempted to solve the problem of vanishing gradients that plateau or degrade learning when training very deep neural networks. The ResNet researchers introduced skip connections that allow information to flow from previous layers in the network to later layers, creating an alternative shortcut path for the gradient to flow through. The fundamental breakthrough of ResNet was that it allowed us to train extremely deep neural networks with hundreds of layers.

In image classification, it is assumed that there is only one main target object in the image and the only focus of the model is to identify the target category. However, in many cases, more than one target in the image is of interest. It is desirable not only to classify them, but also to obtain specific positions in the image. In computer vision, reference is made to tasks such as object detection. Figure 8.19 explains the difference between image classification and object detection tasks.

Image classification

Object detection
(classification and localization)

Cat   Cat, Cat, Duck, Dog

FIGURE 8.19   Image classification versus object detection tasks. In classification tasks, the classifier outputs the class probability (cat), while in object detection tasks, the detector outputs the bounding box coordinates that localize the detected objects (four boxes in this example) and their predicted classes (two cats, one duck, and one dog) (Elgendy, 2020).

TABLE 8.3

Differences between Image Classification and Object Detection

Image classification	Object detection
The goal is to predict the type or class of an object in an image.	The goal is to predict the location of objects in an image via bounding boxes and the classes of the located objects.
• Input: An image with a single object	1. Input: An image with one or more objects
• Output: A class label (cat, dog, etc.)	2. Output: or > e or more bounding boxes (defined by coordinates) and a class label for each bounding box
• Example output: Class probability (for example, 84% cat)	3. Example output for an image with two objects: -box1 coordinates (x, y, w, h) and class probability -box2 coordinates and class probability
	Note that the image coordinates (x, y, w, h) are as follows: (x and y) are the coordinates of the bounding-box center point, and (w and h) are the width and height of the box.

Source: Elgendy (2020).

Object detection is a computer vision task that includes both major tasks. The general framework of object detection systems consists of four main components: Region recommendations, feature extraction and predictions, non-maximum suppression, and evaluation metrics. Table 8.3 summarizes the difference between localizing one or more objects in an image and classifying each object in an image. This is done by drawing a bounding box around the object defined by the prescribed class. This means that the system does not only predict the class of the image, as in image classification tasks, but also predicts the coordinates of the bounding box. This is a challenging computer vision task because it requires both successful object localization to place and draw a bounding box around each object in an image, and object classification to predict the correct localized object class. Object detection is widely used in many fields. For example, in autonomous driving technology, routes need to be planned by locating vehicles, pedestrians, roads, and obstacles in a captured video image. Robots often perform such tasks to detect targets of interest. Systems in the security area are also required to detect anomalous targets such as intruders or bombs. Object detection algorithms are evaluated using two main metrics, frames per second (FPS) to measure the speed of the network and the mean average precision (mAP) to measure the precision of the network. The three most popular object detection systems are the R-CNN network family, SSD, and YOLO network family. There are three main variations of the R-CNN network family, namely R-CNN, fast R-CNN, and faster R-CNN. R-CNN and fast R-CNN use a selective search algorithm to recommend RoIs, while faster R-CNN is an end-to-end deep learning system that uses a region bidding network to recommend RoIs. The YOLO network family includes YOLOv1, YOLOv2 (or YOLO9000), and YOLOv3. R-CNN is a multi-stage detector that divides the process of estimating the object class and object class of the bounding box into two different stages. SSD and YOLO are single-stage detectors. In these methods, the image is passed over the network once to estimate the objectivity score and object class. In general, single-stage detectors tend to be less accurate than two-stage detectors, but are significantly faster (Elgendy, 2020).

8.12 ARTIFICIAL INTELLIGENCE, MACHINE LEARNING, AND DEEP LEARNING APPLICATION EXAMPLES IN AGRICULTURE

The technical processes that emerged in line with the technological advances contribute to the economical, sustainable, and productive industry, which are the goals of plant and animal production. Artificial intelligence techniques have become an important tool in facilitating agricultural operations and in bringing alternative solutions to the problems that need to be solved or improved. Thanks to the developed algorithms and software, numerous studies have been carried out by researchers on plant production planning, classification of plants, yield estimation, detection of plant disease, pests, and weeds, determination of route in agriculture robots, determination of appropriate environmental conditions in the greenhouse, enterprise decision-making, irrigation management, determination of product rotation, selection of optimum fertilizer and instrument in agricultural production, determination of animal diseases, preparation of appropriate feed rations, and determination of animal behavior. In Table 8.4, examples of studies carried out with artificial intelligence techniques in the field of agriculture are given (Terzi et al., 2019). Omomule et al. (2020) reported that 127 artificial intelligence studies were conducted on animal husbandry in their review article. In addition, Bao and Xie (2022) reported that 73 articles on AI in dairy or beef cattle, 74 articles focused on intelligent pig breeding, and 33 articles on poultry were made in their review article. Artificial intelligence applications in livestock are shown in Figure 8.20.

Liakos et al. (2018) reported that there are 40 studies conducted by researchers on machine learning in agriculture and published in different journals. Eight of these articles are about livestock management, 4 are about water management, 4 are about soil management, and 24 articles are related to crop management. Figure 8.21 shows the application areas of machine learning in agriculture. Kamilaris and Prenafeta-Boldú (2018) reported that 40 different studies have been conducted on land use, plant species and soil classification, weed identification, fruit counting, animal growth, weather, product yield index, and moisture content estimation for the purpose of developing

TABLE 8.4

Examples of Agricultural Studies with Artificial Intelligence Techniques

AI techniques	Study subjects	Studies
Fuzzy logic	Irrigation management	Matha et al., 2016; Faridi et al., 2018;
	Detection of water resources	Kurniasih et al., 2018; Kale and Patil, 2018;
	Determination of agricultural potential	Ali et al., 2018
	Decision support system in production	
	Greenhouse systems	
Artificial neural networks	Energy consumption in production	Khoshnevisan et al., 2015; Oppenheim and
	Disease classification	Shani, 2017; Kamilaris and Prenafeta,
	Production management	2018; Liu et al., 2018; Khoshroo et al.,
	Agricultural waste processing	2018
Genetic algorithm	Product drying process	Khawas et al., 2015; Li and Parrott, 2016;
	Land use optimization	Hosseini et al., 2016; Haklı and Uğuz,
	Estimation of soil mechanical resistance	2017; Arya et al., 2018
	Automatic land classification	
	Disease detection	
Expert systems	Weed detection	Montalvo et al., 2013; Romeo et al., 2013;
	Foliage identification	Martinez et al., 2015; Vasquez et al., 2018;
	Due diligence on agricultural machinery	Haklı et al., 2018
	System analysis in animal production	
	Land allocation	
Ant algorithms	Planning contracting works in agriculture	Alaiso et al., 2013; Nguyen et al., 2017;
	Irrigation management	Zhang et al., 2018; Zhang et al., 2019a
	Disease recognition and classification	
	Soil nitrogen content estimation	

Source: Terzi et al. (2019).

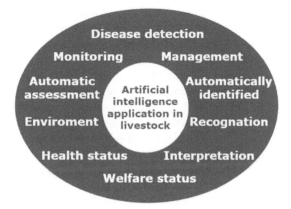

FIGURE 8.20   Artificial intelligence application in livestock.

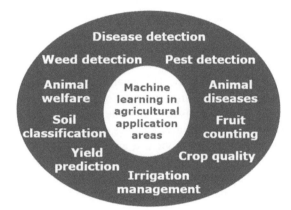

FIGURE 8.21   Machine learning in agricultural application areas.

solutions to various agricultural and food production problems by using deep learning techniques. They also reported that the deep learning method provides higher accuracy and performs better than common image processing methods.

8.12.1   DETECTION OF PLANT DISEASES AND PESTS

Plant diseases and pests are especially active on the young shoots of the plant (Altaş et al., 2019). In the event of a disease, plants exhibit visual signs in the shape of colorful spots with different shapes and sizes according to the type of disease and in the shape of lines seen on stems and different sections or organs of the plants. These symptoms alter color, shape, and size while the disease progresses. With image processing methods, colored objects might be distinguished, and the severity of plant diseases might be determined. Besides image processing methods, expert systems might be improved to allow instant disease diagnosis with machine learning methods (Ozguven, 2020). Recently, the potential use of image processing and machine learning methods for disease detection in the whole plant and or different plant parts (leaves, stem, fruit, and such) has been comprehensively studied by many researchers (Ozguven and Adem, 2019). Monitoring pests and diseases is an extremely important activity to increase productivity in agriculture. Remote sensing combined with machine learning techniques opens new possibilities for tracking and identifying characteristics such as identifying diseases, pests, water, and nutritional stress (Calou et al., 2020).

Ozguven and Adem (2019) conducted a study for the automatic detection of sugar beet leaf spot disease (*Cercospora beticola* Sacc.) with a deep learning technique. For this purpose, the faster R-CNN model and the parameters in the CNN architecture have been changed, and updated faster R-CNN models have been proposed. In this study, a new 1–3 scale was developed by researchers to determine sugar beet leaf spot disease with the help of expert systems. On the new scale, 0 means the whole plant is healthy, 1 means low severity of disease, 2 means severe disease, and 3 means a mixed low and severe disease. The 24-bit 1,024 × 576 resolution images received from the sugar beet leaf images dataset in the research were performed to determine and classify the disease severity as healthy, mild disease, severe disease, or mild and severe mixed disease. The dataset consisted of 155 sugar beet leaves images, including 38 healthy, 20 mildly diseased, 35 severely diseased, and 62 mildly and severely diseased. The faster R-CNN model was preferred in the study due to better determining and classifying highly complicated objects. The faster R-CNN and updated faster R-CNN architectures in the research are seen in Figure 8.22.

The results of applying deep learning methods to sugar beet leaf images are given in Figure 8.23. It was seen that healthy areas were misclassified due to shadows in some images, and diseased areas could not be detected in some images due to reflections. Unlike this, disease detection was better with updated faster R-CNN architecture. This shows that it was crucial to adjust the parameters in the CNN architecture according to the images, to which the faster R-CNN model has been applied. Table 8.5 shows the confusion matrix and Table 8.6 shows the sensitivity, specificity, and accuracy values.

As can be seen in Table 8.5, 111 out of the 117 sugar beet leaf images containing the disease were correctly classified by the proposed model. There was only one incorrect classification in 38 images without the disease. This demonstrates that the specificity values of the updated faster R-CNN approach were higher than the sensitivity values as seen in Table 8.6. The updated faster R-CNN model applied to the present dataset yielded an efficiency of 95.48% in the detection of sugar beet leaf spot disease. As known, to apply deep learning methods successfully, there should be a diversity of samples and the number should be large. Thus, the disease might be classified more effectively. However, with the parameter changes we made in the proposed approach, similar

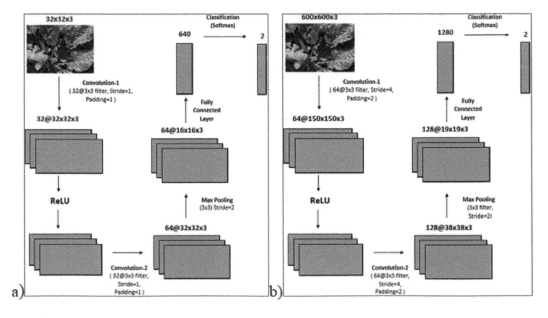

FIGURE 8.22   (a) Faster R-CNN architectures, (b) updated faster R-CNN architectures (Ozguven and Adem, 2019).

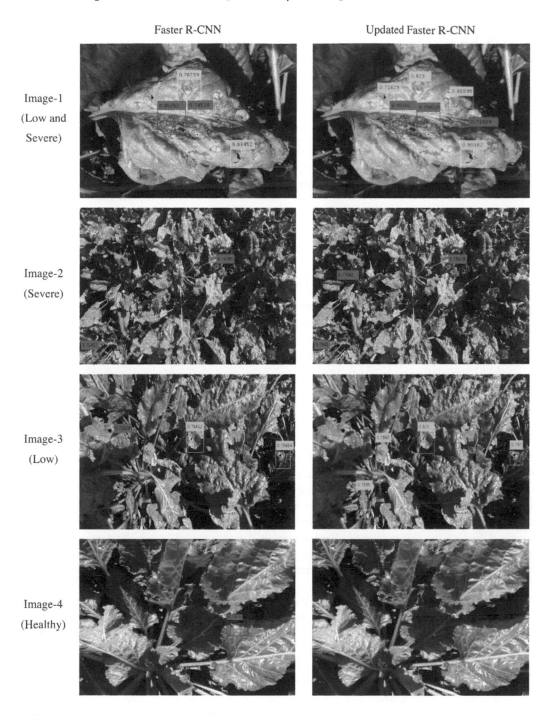

FIGURE 8.23 Application of CNN models to sugar beet leaf images (Ozguven and Adem, 2019)

success rates were achieved with fewer images. It is thought that the model performance will be better when the number of images is increased.

Ozguven (2020) conducted a study for the detection of mildew disease, determination of disease severity, and real-time and automatic monitoring of the course of the disease using cucumber leaf images. The faster R-CNN model, which is a CNN-based object detection and localization method with the latest technology from deep learning methods, was used as a method in the study.

TABLE 8.5

Confusion Matrix of the Updated Faster R-CNN Model Proposed in the Study

		Predict			
		0 (Healthy)	1 (Low)	2 (Severe)	3 (Low and severe)
Actual	0 (Healthy)	37	1	0	0
	1 (Low)	0	19	1	0
	2 (Severe)	0	1	33	1
	3 (Low and severe)	0	1	2	59

Source: Ozguven and Adem (2019).

TABLE 8.6

Success Evaluation of the Updated Faster R-CNN Model Proposed in the Study

	Sensitivity	Specificity	Accuracy
0 (Healthy)	97.37	100	99.36
1 (Low)	95	97.84	97.48
2 (Severe)	94.28	97.6	96.87
3 (Low and severe)	95.16	98.97	97.48
Overall	95.48	95.48	95.48

Source: Ozguven and Adem (2019).

The method proposed by the researcher was trained and tested with a total of 175 images, and a total correct classification rate of 94.86% was obtained as a result of the test studies for disease detection and disease severity in cucumber leaves. It has been reported that the proposed method was found to be successful when compared to the modern methods in the literature in terms of these metrics, and there is no need for sample collection and laboratory analysis since the proposed method has a practical application and is evaluated automatically. In addition, although the proposed method has been studied on cucumber leaves, it has been reported that it is thought to be useful in the rapid detection of other plant diseases and pests, determination of disease severity, and follow-up of the disease process by furthering the study.

The 24-bit 1022 × 764 resolution images received from the cucumber leaf images dataset in the research were performed to determine and classify the disease severity as healthy, mild disease, severe disease, or mild and severe mixed disease. The dataset consisted of 175 cucumber leaf images, including 15 healthy, 45 mildly diseased, 11 severely diseased, and 104 mildly and severely diseased. The results obtained by applying the faster R-CNN model to sample cucumber leaf images for the detection of mildew disease are shown in Figure 8.24. Table 8.7 shows the confusion matrix of the faster R-CNN model, and Table 8.8 shows the sensitivity, specificity, and accuracy values.

As shown in Table 8.7, 152 out of the 160 cucumber leaf images containing the mildew disease were correctly classified by the proposed approach. There was only one incorrect classification out of 15 healthy cucumber leaf images. In the proposed model, as shown in Table 8.8, the sensitivity and specificity values were the same on scales 0 and 1, the sensitivity value was higher on scale 2, and the specificity value was higher on scale 3.

The researcher made the following statements as a result of the study. The faster R-CNN method is considered a method that should be studied more in the detection of plant diseases, pests, and disease severity and progression. In addition to model development studies, increasing the image

Image-1 (Low and Severe)   Image-2 (Severe)

Image-3 (Low)   Image-4 (Healthy)

FIGURE 8.24 Application of the proposed method to cucumber leaf images (Ozguven, 2020).

TABLE 8.7

Confusion Matrix of the Faster R-CNN Model Proposed in the Study

		Predict			
		0 (Healthy)	1 (Low)	2 (Severe)	3 (Low and Severe)
Actual	0 (Healthy)	14	1	0	0
	1 (Low)	1	43	1	0
	2 (Severe)	0	0	10	1
	3 (Low and severe)	0	1	4	99

Source: Ozguven (2020).

quality with platforms, robots, or autonomous vehicles will also contribute to the increase in model performance. In addition, adding spraying mechanisms to these systems may enable the development of autonomous expert systems that can only spray in diseased areas in precision agriculture. Thus, diagnosis of plant diseases and determination of disease severity and course will be faster and more practical, and effective plant protection will be provided with early intervention. In addition, the amount of pesticides will be reduced, the cost of pesticides will be reduced, and adverse environmental effects will be prevented by avoiding unnecessary spraying in site-specific applications.

Calou et al. (2020) used high-spatial-resolution drone images to monitor the extent of Yellow Sigatoka damage in a banana plant, following basic assumptions about the identification,

TABLE 8.8

Evaluation of the Success of the Faster R-CNN Model Proposed in the Study

	Sensitivity	Specificity	Accuracy
0 (Healthy)	93.33	93.33	93.33
1 (Low)	95.56	95.56	95.56
2 (Severe)	90.9	76	90.9
3 (Low and severe)	95.19	99	95.19
Overall	94.86	94.86	94.86

Source: Ozguven (2020).

classification, quantification, and estimation of phenotypic factors. It has been reported that the monthly flights were conducted on a commercial banana plantation using a drone equipped with a 16-megapixel RGB camera, five classification algorithms were used to identify and quantify damage, and field assessments were made according to traditional methodology. The researchers found that the SVM algorithm achieved the best performance (99.28% overall accuracy and 97.13 Kappa index), followed by the ANN and minimum distance algorithms, in quantifying the disease; the SVM algorithm was more effective than other algorithms compared to the traditional methodology used to estimate the extent of Yellow Sigatoka, demonstrating that the tools used for monitoring leaf spots can be handled with RS, machine learning, and high spatial-resolution RGB images. Figure 8.25 shows the damage levels of Yellow Sigatoka pests on banana leaves.

Tetila et al. (2020) developed a computer vision system for the acquisition of soybean pests by UAV images and their identification and classification using deep learning (Figure 8.26). Researchers used a dataset of 5,000 images captured under real growing conditions and evaluated the performance of five deep learning architectures, Inception-v3, ResNet-50, VGG-16, VGG-19, and Xception, for the classification of soybean pest images. The experimental results showed that the deep learning architectures trained with fine-tuning can lead to higher classification rates in comparison to other approaches, reaching accuracies of 93.82%. Deep learning architectures outperformed traditional feature extraction methods, such as SIFT and SURF with the bag-of-visual-words approach, the semi-supervised learning method OPFSEMImst, and supervised learning methods used to classify images, such as SVM, K-NN, and random forest.

The proposed approach uses a set of 5,000 images divided into 13 classes: (1) *Acrididae*, (2) *Anticarsia gemmatalis*, (3) *Coccinellidae*, (5) *Diabrotica speciosa*, (5) *Edessa meditabunda*, (6) *Euschistus heros* adult, (7) *Euschistus heros* nymph, (8) *Gastropoda*, (9) *Lagria villosa*, (10) *Nezara viridula* adult, (11) *Nezara viridula* nymph, (12) *Spodoptera* spp., and (13) without the presence of pests. Figure 8.27 shows examples of superpixel images from the dataset. It has been reported that the images were collected in real field conditions providing various lighting conditions such as sun reflection, cloud cover, shadow, size and positioning of objects, overlap, background variations, and mating and development stages.

8.12.2 DETECTION OF WEEDS

Correct detection and control of weeds in agricultural areas is a necessary procedure to increase plant yield and prevent herbicide pollution. Identifying weeds and precision spraying only weeds are the ideal solution for weed eradication. For this purpose, superior features of imaging systems are used. With the advent of UAVs, the ability to acquire images of the entire agricultural area at very high spatial resolution and at low cost becomes possible, and the resulting input data meets high standards for weed localization and weed management. The similarity of weeds to plants can sometimes be a problem in the automatic detection of weeds. Color information alone is not sufficient to distinguish

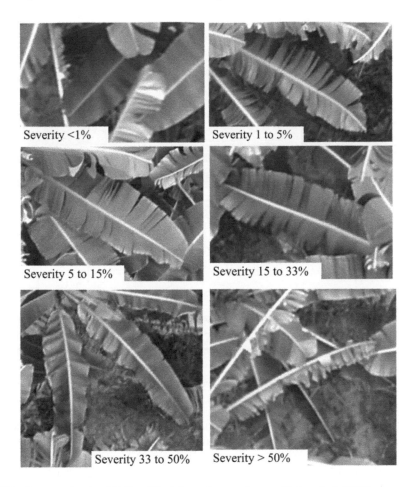

Severity <1%   Severity 1 to 5%

Severity 5 to 15%   Severity 15 to 33%

Severity 33 to 50%   Severity > 50%

FIGURE 8.25   Damage levels of Yellow Sigatoka on banana leaves (Calou et al., 2020).

between plants and weeds. Multispectral cameras provide luminance images with high spectral resolution and spectral reflectance is estimated in several narrow spectral bands for detection. Supervised and unsupervised learning methods can be used to solve this problem automatically.

Gašparović et al. (2020) tested four independent classification algorithms for the creation of weed maps, combining automatic and manual methods, as well as object-based and pixel-based classification approaches, which were used separately on two subsets (Figure 8.28). The classification algorithms were based on a random forest machine learning algorithm for weed and bare soil extraction, followed by an unsupervised classification with the K-means algorithm to further predict weeds and bare soil presence in non-weed and non-soil areas. Of the four classification algorithms tested, the automatic object-based classification method achieved the highest classification accuracy, resulting in an overall accuracy of 89.0% for subset A and 87.1% for subset B. The researchers concluded that automatic classification methods were robustly developed, using at least 0.25% of the scene size as the training dataset in all conditions foreseen for the random forest classification algorithm to work, the use of the algorithm consists of zonal classes and covers areas with similar biological characteristics. They stated that maps can be created and thus made ready for use for weed control.

Alam et al. (2022) developed CNN-based faster R-CNN and YOLOv5 models for real-time detection and classification of tobacco crops/weeds. The researchers presented an open-access image dataset (TobSet) of tobacco plants and weeds obtained from local fields at different growth stages and varying lighting conditions. TobSet consists of 7,000 tobacco plants and 1,000 weed and bare soil images taken manually with digital cameras periodically over two months. The faster R-CNN-based model

FIGURE 8.26   System developed to detect soybean pests with UAV images using deep learning (Tetila et al., 2020).

outperformed the YOLOv5-based model in terms of accuracy and robustness, while the YOLOv5-based model showed faster inference. Success evaluation of the system was carried out in tobacco fields with a four-wheeled mobile robot sprayer controlled using a computer equipped with NVIDIA GTX 1650 GPU (Figure 8.29). At the end of the study, it was determined that faster R-CNN and YOLOv5-based vision systems were able to analyze plants at 10 and 16 frames per second (fps) with 98% and 94% classification accuracy, respectively. In addition, the selective application of pesticides to weeds with the proposed system resulted in a 52% reduction in pesticide use. Figure 8.30 shows the study of tobacco crop and weed detection with real-time faster R-CNN detection.

Fawakherji et al. (2019) presented a deep learning-based method for predicting the condition of the field in terms of crop and weed distribution from images captured by a UAV. The proposed approach runs on an embedded card equipped with a GPU (Figure 8.31). The study was performed on real multispectral images of sugar beet and maize plants from two different public datasets. The segmentation process for plant and weed classification (Figure 8.32) has been reported to be based on an encoder-decoder model architecture for pixel-based segmentation, allowing the model to distinguish between plants that are slightly different or even partially overlapping, the average accuracy obtained at the end of the study was 95% and 99% was reached when the NDVI index was used as an input for vegetation detection.

8.12.3   MONITORING THE GROWTH OF PLANTS

Counting the number of plants grown in a certain area in plant production, that is, obtaining the plant population, may vary according to the effect of various factors during and before the season. Early season plant numbers are often the result of seed quality, planting performance, and response of seedlings to soil and air at emergence. Mid- to late-season plant numbers are often the result of weather, soil, fertilizer, pest pressure, and other management practices. Plant numbers are one of the most common ways for farmers to evaluate crop growth conditions and management

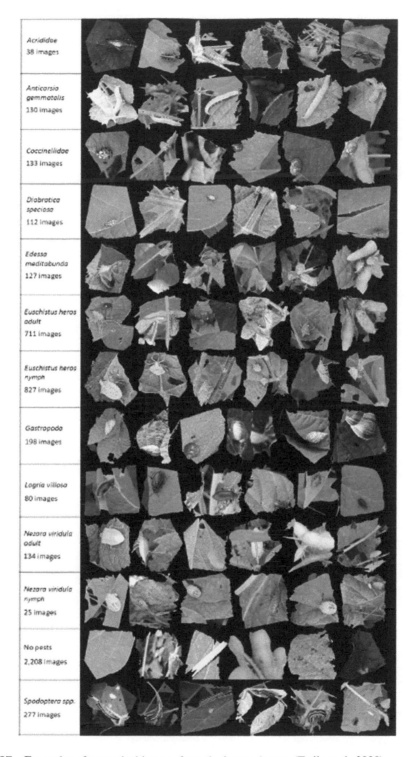

Acrididae
38 images

Anticarsia
gemmatalis
130 images

Coccinellidae
133 images

Diabrotica
speciosa
112 images

Edessa
meditabunda
127 images

Euschistus heros
adult
711 images

Euschistus heros
nymph
827 images

Gastropoda
198 images

Lagria villosa
80 images

Nezara viridula
adult
134 images

Nezara viridula
nymph
25 images

No pests
2,208 images

Spodoptera spp.
277 images

FIGURE 8.27 Examples of superpixel images from the image dataset (Tetila et al., 2020).

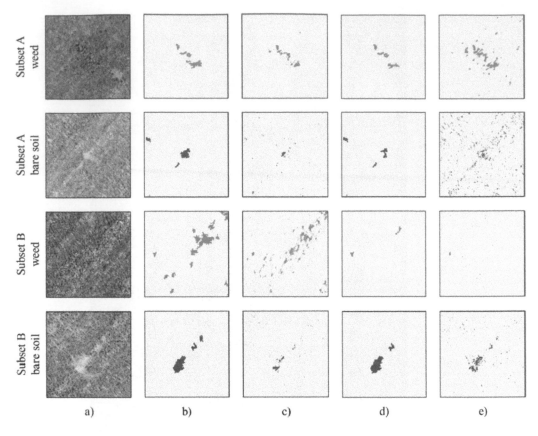

FIGURE 8.28 Comparative visual analysis of classification methods in sample subsets (10 m × 10 m): (a) Digital orthophoto, (b) automated object-based method, (c) automated pixel-based method, (d) manual object-based method, (e) manual pixel-based method (Gašparović et al., 2020).

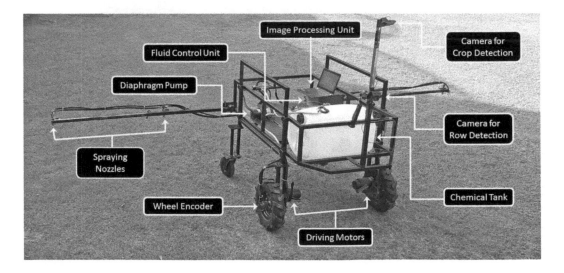

FIGURE 8.29 Developed prototype of the agricultural robotic sprayer (Alam et al., 2022).

FIGURE 8.30 Real-time faster R-CNN detection of tobacco crop and weeds in scenarios with (a) low intra-row plant distance, and (b) high intra-row plant distance (Alam et al., 2022).

practices throughout the season and can be used to make management decisions, such as determining whether replanting is necessary. However, the traditional method for early season plant counting is manual inspection, which is time-consuming, laborious, and spatially limited in scope. In recent years, UAV-based RS has been used to provide low-altitude, high-spatial-resolution images to aid decision-making in crop counting and in many different applications in agriculture. Pang et al. (2020) designed a system that uses deep learning and geometric descriptive information to determine early season maize plant numbers from relatively low spatial resolution (10–25 mm) UAV images covering an area of 10–25 hectares (Figure 8.33). The researchers reported that instead of detecting individual crops in a row, they processed the entire row at once, which significantly reduced their requirement for clarity of the crops. In addition, the researchers stated that their newly developed MaxArea Mask Scoring R-CNN algorithm can separate the rows of plants in each patch image regardless of field conditions, and the system was tested on data collected in two different areas in different years, and the accuracy of the estimated output speed reached 95.8%. In addition, the newly developed MaxArea Mask Scoring R-CNN algorithm is able to segment crop rows in each patch image regardless of terrain conditions, and the robustness of our plan was tested on data collected in two different areas in different years. It has been reported that the accuracy of the estimated emergence rate reaches 95.8% and the system has the potential for real-time applications in the future due to its high processing speed.

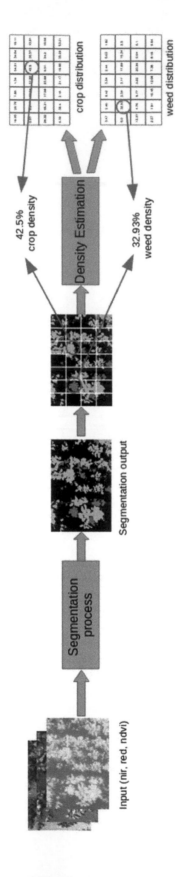

FIGURE 8.31 Flow chart of the proposed approach (Fawakherji et al., 2019).

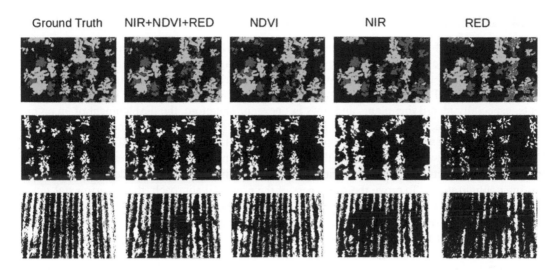

FIGURE 8.32 Qualitative results obtained by CNN with different input types. Rows 1 and 2 represent the output for three classes of segmentation in the sugar beet dataset, and row 2 represents the output for segmentation in the three-class maize dataset (Fawakherji et al., 2019).

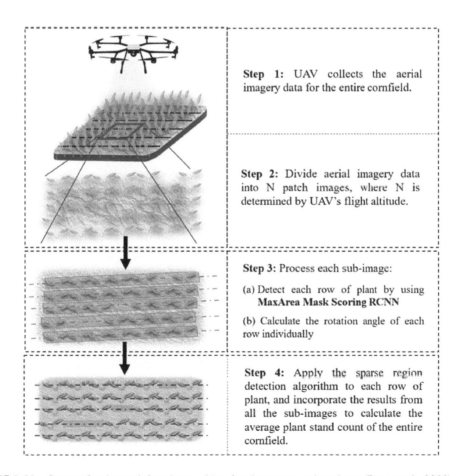

FIGURE 8.33 System for determining the number of early season maize plants (Pang et al., 2020).

8.12.4 Detection of Plant and Tree Locations

Accurate mapping of cropland is an important prerequisite for precision agriculture, as it aids in field management, yield estimation, and environmental management. Optical imaging with sensors mounted on UAVs is today a cost-effective option for capturing images covering cultivated areas. However, only visual inspection of such images can lead to both difficult and inaccurate assessments, especially identifying plants and rows in one step. Therefore, the development of an architecture that can extract individual plants and plantation rows simultaneously from UAV images is an important need to support the management of agricultural systems. For this purpose, Osco et al. (2021) proposed a new deep learning method based on a convolutional neural network (CNN) that simultaneously detects and locates plantation rows while counting plants, considering high-density planting configurations. To prove the robustness of the proposed approach, the experimental setup was evaluated in (a) a corn field (*Zea mays* L.) with different growth stages (Figure 8.34) and (b) a citrus orchard (*Citrus sinensis*) (Figure 8.35). Both datasets characterize different plant density scenarios at different locations, different types of crops, and different sensors and dates. It can be concluded that the CNN method results outperform the results from other deep networks (HRNet, faster R-CNN, and RetinaNet) evaluated with the same task and dataset and that the proposed method can be used to count and position plants and plant rows in UAV images from different plant species. It has also been reported that it can contribute to the sustainable management of agricultural systems by applying it to decision-making models and decision-making models.

8.12.5 Detection of Real-Time Fruit

Establishing an accurate, fast, and reliable fruit detection system is the most important element for fruit yield estimation and automatic harvesting. Fruit segmentation is an important step in distinguishing fruit from the background (leaves and stems). This task is difficult due to variations in fruit color and lighting, as well as a high level of overlap. Sa et al. (2016) developed a fruit detection system for real-time fruit detection with an autonomous agricultural robotic platform using a pre-trained faster R-CNN. The researchers reported that the model is easily adaptable to different fruit species with a minimum number of training images, and they offer approaches in which RGB and NIR images are combined with early and late fusion. As a result of the study, melon F1 = 0.848, strawberry F1 = 0.948, apple F1 = 0.938, avocado F1 = 0.932, mango F1 = 0.942, orange F1 = 0.915, and sweet pepper F1= 0.828 scores were obtained. In addition to this accuracy, it is noted that annotating bounding boxes will be faster instead of annotating at the pixel level. In Figure 8.36 and Figure 8.37, examples of fruit determination made in the study are given.

8.12.6 Detection of Grape Yield

Grape yields in vineyards can vary significantly from year to year as well as spatially due to differences in climate, soil conditions, and pests. Accurate information on grape yield will enable vineyards to better manage their vineyards. The currently applied conventional method for yield estimation is difficult and expensive. In this method, small samples are taken from vineyards during the growing season and estimated to determine the overall yield. For this reason, many CNN models have been developed for accurate and rapid recognition of grapes in vineyards. The irregular shapes and very dense fruiting of the grape bunches can reduce the accuracy of the models.

Wang et al. (2021) proposed the Swin transformer and DETR models to increase the success of grape bunch detection (Figure 8.38). In the study, the optimum number of stages for a Swin transformer was selected through experiments. This model has been compared with traditional CNN models such as faster-R-CNN, SSD, YOLO, and YOLOX. Models were trained under datasets of red grapes collected under natural light. In overexposure, overdarkness, and occlusion situations,

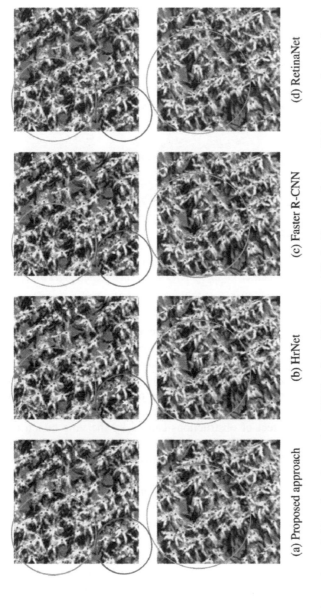

FIGURE 8.34 Comparison of object detection methods HRNet, faster R-CNN, and RetinaNet on a dataset of higher growth stage (mature with cobs) corn plants (orange and blue circles highlight usual and challenging detections, respectively) (Osco et al., 2021).

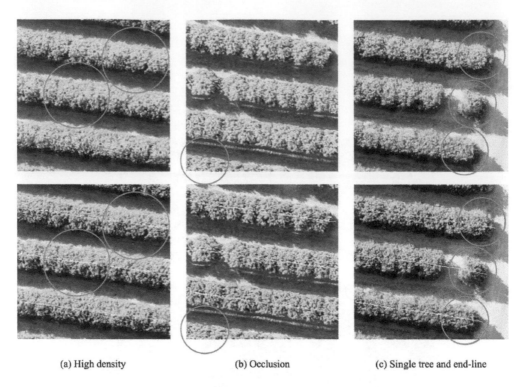

(a) High density   (b) Occlusion   (c) Single tree and end-line

FIGURE 8.35 Plant and plantation row determinations examples of (a) high density, (b) occlusion, and (c) single tree and end-line in the citrus dataset. Orange circles highlight the challenges overcome by the approach in each scene (Osco et al., 2021).

SwingGD can recognize more accurately and robustly compared to other models. The accuracy of the Swin transformer is 91.5%. To confirm the universality of SwingGD, they conducted a test under images of green grapes. Experimental results show that SwingGD has a good effect in practical applications. Figure 8.39 shows examples of bunches of grape detection in the dark.

Millan et al. (2018) developed a new system for the non-destructive, objective, and automatic evaluation of yield in vineyards using image analysis and a Boolean model (Figure 8.40). The researchers reported that the effect of obstructions caused by the vine bunch or other organs in image acquisition has a detrimental effect on the quality of the results, and fruit numbers were evaluated using the Boolean model to reduce the effect of obstructions in the estimation. It has been noted that the Boolean model greatly improves the results when compared to a simpler estimator based on the relationship between cluster area and weight. To evaluate the methodology, three different datasets were studied as cluster images, manually acquired vine images, and vine images captured in motion. As a result of the study, the proposed algorithm estimated the number of berries in cluster images with an RMSE of 20 and a coefficient of determination (R^2) of 0.80. Manually captured vine images were evaluated, providing a mean error of 310 grams and $R^2 = 0.81$. Finally, images captured using a quad equipped with artificial light and automatic camera triggering were also analyzed. The prediction obtained by applying the Boolean model had an average error of 610 grams per segment (three vines) and $R^2 = 0.78$.

An example of cluster partitioning is given in Figure 8.41 for the image captured while in motion. Here, the original image is used to obtain four MPMs (membership probability maps), MPMcolor, MPMcluster affinity, MPMcable, and MPMlinear occurrence region. These MPMs are combined to classify pixels corresponding to clusters for images captured in motion. In Figure 8.42, cluster segmentation is seen in manually captured vine images.

FIGURE 8.36 Example images of detection for two fruits. (a) and (b) show a color (RGB) and near-infrared (NIR) image of sweet pepper detection, shown as red bounding boxes, respectively. (c) and (d) are the determination of rock melon (Sa et al., 2016).

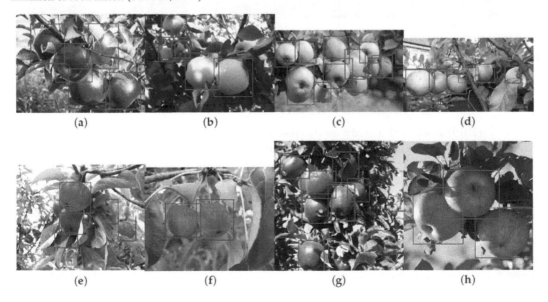

FIGURE 8.37 Eight examples of detection of red (a, e–h) and green (b–d) apples in different varieties (Sa et al., 2016).

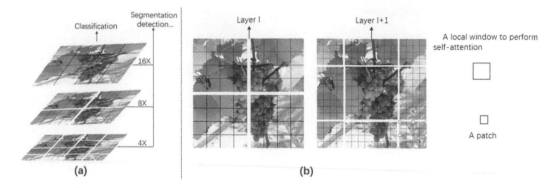

FIGURE 8.38 (a) The proposed Swin transformer model generates hierarchical feature maps. (b) A demonstration of the shifted window approach for computing self attention in the proposed Swin transformer architecture (Wang et al., 2021).

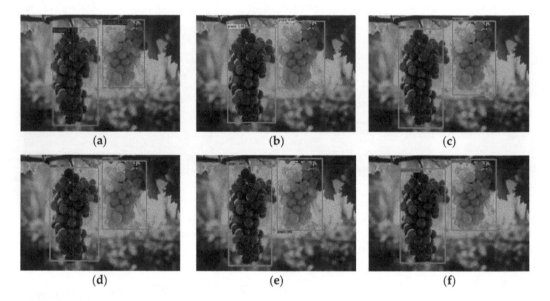

FIGURE 8.39 Examples of grape bunches detection in the dark. (a) SwinGD, (b) DETR, (c) YOLOv3-SPP, (d) faster-R-CNN, (e) SSD, (f) YOLOX. (Wang et al., 2021).

8.12.7 IDENTIFICATION OF PLANT LEAVES

Yigit et al. (2019) conducted a study on the visual automatic identification of plant leaves using artificial intelligence techniques such as artificial neural network, naive Bayes algorithm, random forest algorithm, K-nearest neighbor, and support vector machine. In the study, data from 637 healthy leaves from 32 different plant species were used. Twenty-two visual features of each leaf were obtained by image processing technique and these 22 visual features were evaluated in four groups as size, color, texture, and pattern. To investigate the effects of these groups on classification performance, 15 possible different combinations from four groups were created and then the models were trained with 510 leaf data and tested for accuracy with 127 leaf data. According to the test results, the SVM model with 92.91% accuracy was found to be the most successful descriptor (Figure 8.43). In addition, the researchers tested the SVM model to identify diseased and defective leaves. Five hundred and thirty-six randomly selected leaves corresponding to 80% of 637 healthy and 33 diseased-defective leaves were used for training, and the remaining 134 leaves were used for

FIGURE 8.40 (a) On-the-go image capture system: Automatic camera triggering, LED lighting and structure for easy position adjustment. (b) An example of a vine image taken in motion (Millan et al., 2018).

testing. It was reported that at the end of the test, it was defined with a precision of 92.53%, and the most effective factor in the result was tissue.

8.12.8 Monitoring of Plant-Growing Environments

Adverse weather conditions often cause agricultural damage, reduce crop yield and quality, and ultimately affect profitability. For this reason, it is necessary for the farmers to create the desired optimum environmental conditions, maintain this in a controlled manner, and take the necessary precautions in adverse situations, especially in order to produce quality production in greenhouses. For this purpose, information such as transpiration, photosynthesis, and respiration about plant growth and development is constantly collected by using various sensors and imaging systems in greenhouses, and AI-based systems can be used to evaluate this information. Chang et al. (2021)

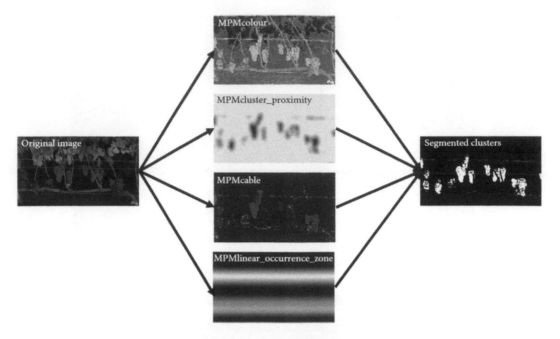

FIGURE 8.41 Cluster segmentation for image captured in motion (Millan et al., 2018).

(a)   (b)

FIGURE 8.42 Cluster segmentation in manually captured vine images: (a) A vine image taken under uncontrolled lighting conditions; (b) segmentation result using RGB and HSV representations (Millan et al., 2018).

presented a new AI methodology to predict harvest time and growth quality of lettuce (*Lactuca sativa* L.) grown in a hydroponic system. In the study, fuzzy logic, neural networks, and models that combine these two methods were used to estimate various physiological parameters related to lettuce plant growth (leaf number, contour area of leaves, and dry mass), net photosynthesis rate, and transpiration rate measurements in real time (Figure 8.44). A small-scale hydroponic growing system using the IoT for environmental data collection and imaging systems was used to measure and predict plant growth parameters in a commercial lettuce-growing greenhouse.

The AI models developed in the study are three fuzzy logic (FL) (FL #1, FL #2, and FL #3), a neural network (NN), and a combination of these two methods (neural-fuzzy, NF) (NF #1 and NF #2) (Figure 8.45). Average hourly temperature (AhT) and average hourly light (AhL) factors are inputs to FL #1 and FL #2. Hourly average CO_2 (AhC) and hourly average humidity (AhH) are input factors for FL #1 and FL #2, respectively. The daily light integral (DLI) is an input for FL #3 and NN. Average daily humidity (AdH) and mean daily temperature (AdT) are inputs for the NN model. FL #1 and FL

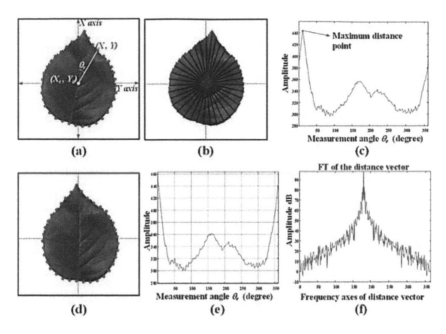

FIGURE 8.43 (a) The center and edges of the non-oriented leaf, (b) distances from origin to edges for non-oriented leaf, (c) 1-D distance vector of non-oriented leaf, (d) the edges of the oriented leaf, (e) 1-D distance vector of oriented leaf, (f) FT of the oriented leaf in logarithmic scale (Yigit et al., 2019).

#2 estimate parameters including net photosynthesis rate (Pn) and transpiration rate (En). NN estimates the daily dry mass (DW) of lettuce. Pn, En, and DLI are inputs for the FL #3 model. FL #3 is used to estimate the contour area (LA) and leaf number (LN) of leaves. LA, LN, and DW were used as model inputs for NF #1 and NF #2 used to predict harvest time (Ht) and plant quality (Qt). The design process of each component of the AI models is shown in detail in Figure 8.45. The straight lines in the figure are data streams. Plots of time scales and parameters for FL, NN, and NF model input/output ports are described in more detail in Table 8.9.

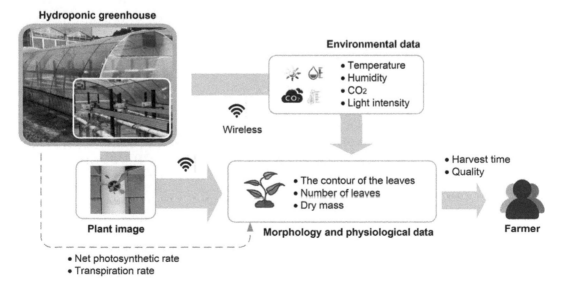

FIGURE 8.44 Conceptual framework of the proposed approach (Chang et al., 2021).

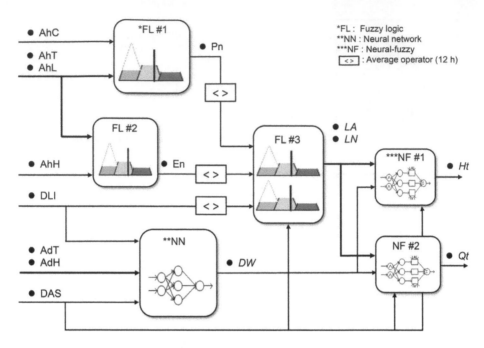

FIGURE 8.45	Components of AI models (Chang et al., 2021).

TABLE 8.9

Clear Range of Membership Functions (MF) Language Variables and Parameters

Linguistic variables	Crisp interval [p, q]	The linguistic label of MFs and its parameters (specified as the vector [a, b, c, d])			
		L	M	H	VH
Temp, °C	[15, 50]	–	[15, 15, 25, 28]	[25, 30, 37.5, 43]	[38, 44.5, 50, 50]
Ligt, $umol \cdot s^{-1} \cdot m^{-2}$	[0, 1,400]	[0, 0, 200, 300]	[200, 300, 700, 900]	[800, 1,000, 1,500, 1,500]	–
CO_2, ppm	[0, 600]	[0, 0, 400, 430]	–	[380, 450, 600, 600]	–
Ht, %	[0, 90]	[0, 0, 30, 55]	–	[35, 60, 90, 90]	–
Pn	[0, 25]	[0, 0, 3, 5]	[4.5, 6, 19, 22]	[19.5, 23, 25, 25]	–
En	[2.5, 4.5]	[0, 0, 2.8, 3.2]	[3, 3.3, 3.6, 3.7]	[3.5, 3.8, 4.5, 4.5]	–
DLI, $umol \cdot s^{-1} - m^{-2}$	[−1, 12,000]	[−1, −1, 3,000, 4,100]	[2,000, 5,000, 8,000, 11,000]	[7,000, 9,800, 12,000, 12,000]	–
DAS, day	[14, 50]	–	[14, 14, 25, 33]	[27, 33, 50, 50]	–
ΔLN, pieces	[0, 0.4]	[0, 0, 0.02, 0.05]	[0.01, 0.08, 0.14, 0.2]	[0.15, 0.22, 0.4, 0.4]	–
ΔLA, pixels	[0, 2,500]	[0, 0, 300, 520]	[490, 700, 1,750, 2,000]	[1,700, 2,000, 2,500, 2,500]	–

Source:	Chang et al. (2021).

Note:	Defining the linguistic label of MFs as low (L), medium (M), high (High), and very high (VH).

At the end of the study, the following was reported by the researchers. Plant growth models, which incorporate knowledge of physics and biology, are effective tools for environmental control. Modeling plant physiology can provide a reference for physiological regulation and maintenance of a particular plant yield and quality. In particular, the proper design of the rules for plant physiological interactions can effectively and accurately predict the physiological parameters of different crops. This approach can also be used to explain the good or bad quality of plant growth and the reasons for the long or short harvest day. Also, the use of the proposed approach was accurate in identifying and interpreting the impact of the interaction. Capable of learning when new scenarios arise, datasets obtained through the IoT and image processing techniques to characterize lettuce can model the growth responses of lettuce plants, thereby increasing extra application possibilities.

8.12.9 Detection of Real-Time Object

Autonomous farming robots need to provide real-time feedback, but this goal is inherently complex due to the wide variety of objects to be detected and the lack of landscape uniformity. Tang et al. (2018) present a multi-view object detection approach based on deep learning in their study. In addition, the object retrieval ability and object detection accuracy of both the multi-view methods (Figure 8.46) and related classical modalities were evaluated and compared based on a test on a small object dataset (Figure 8.47 and Figure 8.48). The validity of the proposed approach was validated in the study and when compared to other methods, it was found that multi-view detection methods were faster while obtaining mAPs that were approximately the same in small object detection.

8.12.10 Identification of Agricultural Machinery

Zhang et al. (2019b) created an image description dataset consisting of seven types of machines (Figure 8.49) and six types of abnormal images (Figure 8.50) from multiple machine images in the agricultural machinery operation monitoring system. In addition, a network called AMTNet was designed and trained for the automatic recognition of agricultural machinery images with the Inception_v3 network. In the study, the generated image dataset contains 125,000 images, with the training set being 100,000 images and the validation set being 25,000 images. As a result of the study, under the same experimental conditions, 97.83% and 100% recognition accuracy were obtained in the AMTNet Top_1 and Top_5 validation sets, respectively. It is reported to outperform classical ResNet_50 and Inception_v3 networks (Figure 8.51). The average area under the curve of the network and the F1 score for image recognition of various machines reached 92% and 96%, respectively. According to the test results, AMTNet is reported to have good robustness to lighting, environmental changes, and small space occlusion, meeting the practical application requirements of intelligent control over agricultural machinery operation.

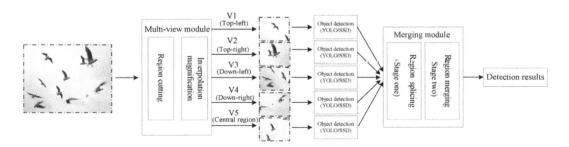

FIGURE 8.46 Model for multi-view object detection (Tang et al., 2018).

FIGURE 8.47 Comparison of small object detection results between the YOLO method (first and third columns) and the multi-view YOLO method (second and fourth columns) (Tang et al., 2018).

8.12.11 Smart Sprayer

In Partel et al. (2021), a low-cost smart sensing system was developed to control the airblast tree crop sprayers (Figure 8.52), and its performance was evaluated by field trials in a commercial citrus orchard (Figure 8.53). The prototype developed by the researchers consists of a lidar, machine vision, GPS, flow meters, sensor fusion, and AI to scan trees for tree height, tree classification, and fruit counting. This intelligent sensing system can detect and classify objects as trees or non-trees (e.g., human, field structures). It can measure tree height and canopy density, detect and count fruits, and control spray nozzles to optimize spraying applications (Figure 8.54). In the study, new software was written in C++ and run on an Nvidia Jetson Xavier NX embedded computer to manipulate and control data using data aggregation and AI techniques. The tree height estimation results of the smart detection system showed a relatively low average error of 6% (Figure 8.55). A CNN was used to perform tree classification with an average accuracy of 84% in classifying the collected images as mature, young, dead, and non-tree objects. Using another CNN for the fruit count module, it has been reported that mature and immature citrus fruits have an F1 score of 89% compared to ground truth labeled images, and this new detection method reduces spray volume by 28% compared to conventional spraying applications.

8.12.12 Use in Agricultural Product Drying

Intelligent control is a multidisciplinary field in control theory, expert systems, automation, computer vision, sensor fusion, operations research, and artificial intelligence interface. In recent years, smart control applications have been seen in the drying of agricultural products. The drying process

FIGURE 8.48 Comparisons of small object detection results between SSD method (the first and third columns) and the multi-view SSD method (the second and fourth columns) (Tang et al., 2018).

is carried out in batches or continuously, depending on the nature of the product. No matter how the drying process is done, it requires careful attention to the sensors and instrumentation for the system observer. In old drying systems, only environmental variables (temperature, humidity, air velocity) were observed and controlled regardless of product quality. In the drying systems used in recent years, controlling the product quality is seen as the main concept in drying (Martynenko and Bück, 2019).

New-generation smart drying technologies require advanced instrumentation for real-time and continuous monitoring of changes in environmental variables such as temperature, velocity, humidity, pressure, and changes in product quality characteristics (moisture content, shrinkage, color, texture, physicochemical properties, etc.) (Figure 8.56). These systems also include tools, technologies, resources, and practices that will contribute to energy savings and environmental sustainability during the drying process. Thus, smart drying systems should include not only the dryer but also smart sensors, converters, and smart control systems in order to change process conditions, increase product quality, and increase energy efficiency. These systems are called smart drying systems; they are special sensors and instruments that reflect the basic quality characteristics of the product and its changes during drying. Biomimetic sensors (electronic nose, electronic tongue), computer vision technology, spectroscopy (microwave/dielectric spectroscopy, NIR reflection spectroscopy, etc.), magnetic resonance imaging (MRI), ultrasound techniques, and electrostatic sensors are included in this group. While computer vision and biomimetic sensors provide information about the quality characteristics perceived by consumers, spectroscopic applications mostly reflect the nutritional value of the product (Özgüven et al., 2020).

With the application of high-efficiency physical space in the drying process of fruits and vegetables with AI technology, problems such as large energy consumption, uneven drying, poor sensory evaluation, and large nutrient loss problems can be solved. By using artificial intelligence technology, you can better observe the drying process and determine the optimum drying scheme. It can also

Subsoiler with a shovel

Subsoiler with a curved surface shovel

Subsoiling preparation machine

Subsoiling combined seed and fertilizer drill

Turnover plow

Rotary cultivator

Seeder

FIGURE 8.49   Images of seven types of agricultural machinery (Zhang et al., 2019b).

Black and White   Blurred   Pointing to the sky   Road Photographed   Object Photographed   Occlusion

FIGURE 8.50   Six types of abnormal images (Zhang et al., 2019b).

ensure the efficient production of dried fruit and vegetable products with high quality. For example, ultrasonic pretreatment and microwave technologies can increase drying speed, but drying uniformity is not good enough. RF technology can dry deeply, but fruits and vegetables are easy to lose color and nutrients. In the smart drying process, a computer vision system and sensor technology are used to observe the drying process, and an artificial intelligence system is used for observation and analysis such as ANN, fuzzy logic, and genetic algorithm for control. It continuously adjusts the drying technologies (microwave, ultrasound, radio frequency, infrared radiation, etc.) at the desired time and manner. Finally, AI technology records, examines, and analyzes products (Chen et al., 2020).

Liu et al. (2022) proposed the mechanism and data dual-drive with equivalent accumulated temperature mutual-window AI-control method for continuous grain drying and established a control system. The researchers reported that grain drying is a complex heat and mass transfer process with significant delay, multidisturbance, nonlinearity, strong coupling, and parameter uncertainty. In addition, researchers reported that AI-control technology is suitable for solving such complex control problems. Experimental validation of this developed system was carried out on a continuous grain drying test platform. As a result, it was determined that the method has implicit prediction, high accuracy, strong stability, and self-adaptive ability, and the maximum humidity control deviation at the outlet of the dryer is −0.58–0.3%. Figure 8.57 is a schematic diagram of the smart drying process control algorithm.

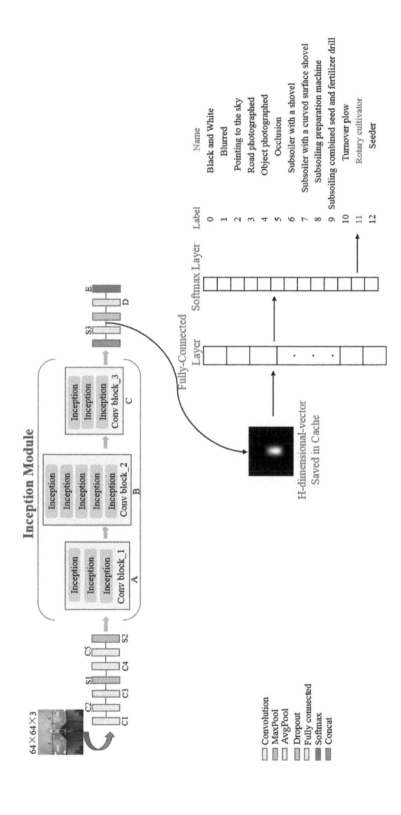

FIGURE 8.51 AMTNet (Zhang et al., 2019b).

FIGURE 8.52   Air flow tree sprayer (Partel et al., 2021).

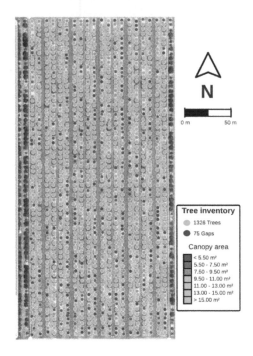

FIGURE 8.53   Selected commercial citrus orchard area to test the success of the developed spraying system (Partel et al., 2021).

8.12.13   IDENTIFICATION OF ANIMALS

Livestock animal identification is of great importance to achieve precision animal production, as it is a prerequisite for modern livestock management and automated behavioral analysis. Regarding cow identification, computer vision-based methods have been widely considered due to their non-contact and practical advantages. Hu et al. (2020) propose a new non-contact cow identification method based on the fusion of deep part features. In the study, first, a set of side-view images of the

FIGURE 8.54 Detecting and counting fruits for which detections are pink squares (Partel et al., 2021).

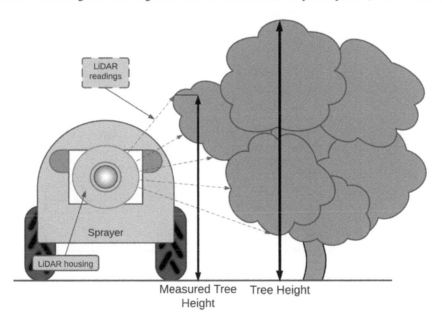

FIGURE 8.55 Lidar reading error in tree height measurement (Partel et al., 2021).

cows were captured, and the YOLO object detection model was applied to locate the cow object in each original image, which was then divided into three parts, head, trunk. and legs, with a part segmentation algorithm using frame differentiation and segmentation span analysis. Next, three independent CNNs were trained to extract deep features from these three parts and a feature fusion strategy was designed to fuse the features. Finally, an SVM classifier trained with the fused features was used to identify each cow. The proposed method achieved 98.36% cow identification accuracy in a dataset containing side-view images of 93 cows, outperforming existing studies. Experimental results showed the effectiveness of the proposed cow identification method and good potential for this method in the individual identification of other farm animals. Figure 8.58 shows the flow chart of the proposed cow identification method.

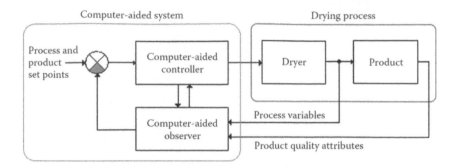

FIGURE 8.56 The structure of intelligent control, focused on product quality (Martynenko and Bück, 2019).

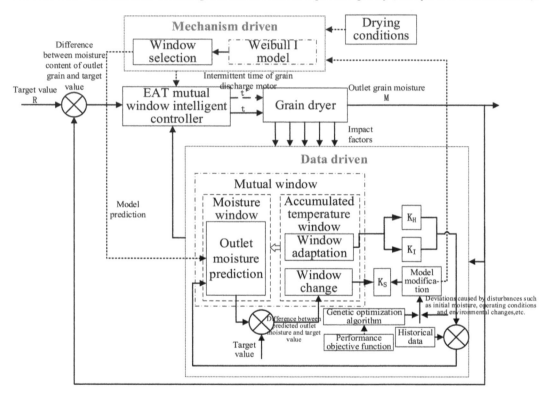

FIGURE 8.57 Schematic diagram of smart drying process control algorithm (Liu et al., 2022).

8.12.14 Use in Feeding Applications in Livestock

The usage of PLF can help farmers to improve management tasks such as monitoring animal performance and health and optimizing feeding strategies. A key component of PLF is precision livestock nutrition, which consists of providing individuals or groups of animals with the amount of nutrients in real time that maximizes nutrient utilization without loss of performance. Precise feeding of animals can reduce protein intake by 25%, nitrogen release into the environment by 40%, and increase profitability by approximately 10%. Feeding success depends on the automatic and continuous collection of data, data processing and interpretation, and control of farm processes. With the advancement of PLF feeding, new nutritional concepts and the development of mathematical models can predict individual animal nutritional requirements in real time. Further progress in these technologies will require the coordination of different experts and stakeholders such as nutritionists, researchers, engineers, technology suppliers, economists, farmers, and consumers (Pomar et al., 2019).

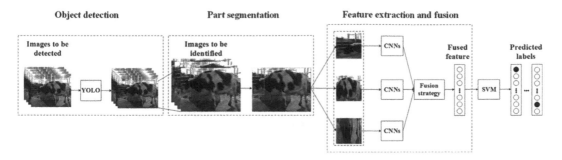

FIGURE 8.58   Flow chart of the proposed cow identification method (Hu et al., 2020).

There are many factors that affect how much feed an individual animal consumes and feed efficiency. These factors can be divided into two important categories as individual animal variations and environmental factors including management decisions. Figure 8.59 shows the system developed for the automatic determination of individual feed consumption. In addition to determining the amount of feed, these systems can also monitor eating habits and eating frequency with software developed based on artificial intelligence. The system developed by Yu et al. (2022) for the automatic determination of individual feed consumption is shown in Figure 8.60. In this study, dairy cow feeding behavior was captured in real time using an edge computing device and processed using DRN-YOLO deep learning algorithms. The research results have demonstrated the ability to effectively provide solutions in the analysis of dairy cow feeding behavior in complex breeding environments.

Campos et al. (2018) presented a method for classifying different grass-eating behaviors by the surface electromyography (sEMG) signal of the masseter muscle. The main hypothesis tested in the study is whether rumination and food eaten can be distinguished from sEMG signal features using machine learning techniques. The three scenarios examined were differentiation (IR) between ruminant and grazing, food identification (FC) for four different foods, and both conditions combined (FCR). The entire data collection, instrumentation, and pattern recognition system is summarized in Figure 8.61. In the study, a new segmentation technique was developed and applied to automatically subdivide the chewing motion signal and evaluated by eight features extracted from seven classifier signals (LDA, QDA, SVM, MLP-NN, RBF-NN, KNN, and MLP-NN) and combined into five sets. Despite the similar characteristics of grasses, it has been reported that the results of food recognition were found to be reasonable and that feeding and rumination were distinguished with relatively high accuracy.

8.12.15   IDENTIFICATION OF ANIMAL SOUNDS

Because animal sounds in breeding contain information about animal welfare and behavior, automatic sound detection has the potential to facilitate a continuous acoustic monitoring system for use in a range of PLF applications. Bishop et al. (2019) presented a multipurpose livestock voice classification algorithm using voice-specific feature extraction techniques and machine learning models. In the study, to test the multipurpose nature of the algorithm, three separate datasets of farm animals consisting of sheep, cattle, and shepherd dogs were created, voice data were extracted from continuous recordings made in situ at three different operational farming operations reflecting the actual distribution conditions, Mel-frequency kepstral coefficients and discrete wavelet transform–based (DWT) features were made, and the classification was determined using an SVM model. At the end of the study, high accuracy was obtained for all datasets (sheep 99.29%, cattle 95.78%, dogs 99.67%). The classification performance alone was insufficient to determine the most appropriate feature extraction method for each dataset; computational timing results of DWT-based features were significantly faster (14.81% to 15.38% reduction in execution time) and a highly sensitive livestock vocalization classification algorithm has been developed which forms the basis of an

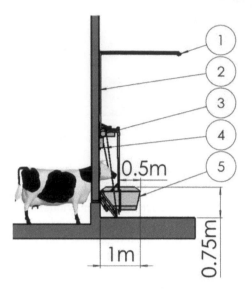

FIGURE 8.59 Developed system for automatic determination of feed consumption: (1) Camera, (2) supporting beam, (3) load cell, (4) electronics box, (5) feeding container (Halachmi et al., 2019).

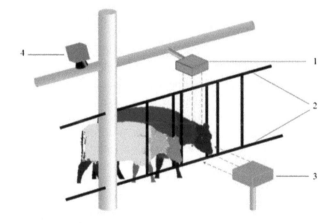

FIGURE 8.60 An improved system for automatic determination of feed consumption (Yu et al., 2022).

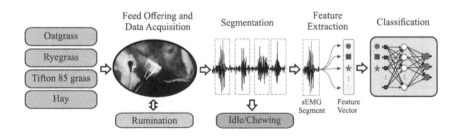

FIGURE 8.61 Animal sEMG data acquisition system (Campos et al., 2018).

automated livestock vocalization detection system. Figure 8.62 shows the spectrogram examples of each class and subclass for the sheep dataset, and Figure 8.63 shows an overview of the proposed multipurpose live animal vocalization detection algorithm.

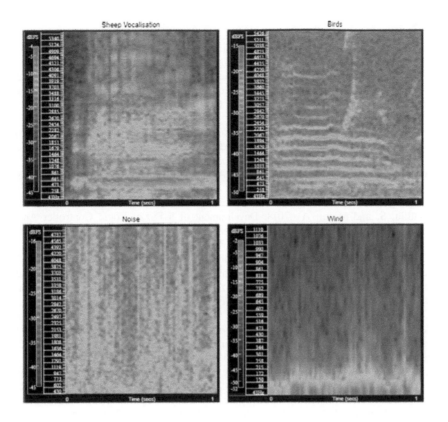

FIGURE 8.62 Spectrogram examples of each class and subclass for the sheep dataset (Bishop et al., 2019).

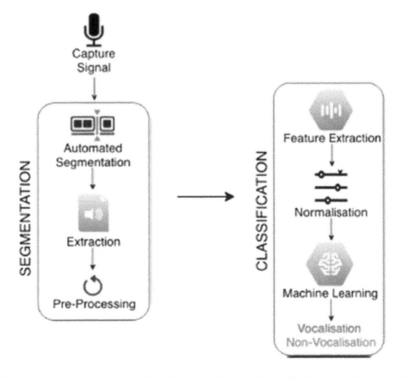

FIGURE 8.63 Overview of the proposed multipurpose livestock vocalization detection algorithm (Bishop et al., 2019).

REFERENCES

Akçetin, E. ve Çelik, U., 2015. Karınca Kolonisi Optimizasyonu Sınıflandırma Algoritması Yöntemi İle Telefon Bankacılığında Doğrudan Pazarlama Kampanyası Üzerine Bir Sınıflandırma Analizi. *İnternet Uygulamaları ve Yönetimi Dergisi*, 6(1), s.5–19. https://doi.org/10.5505/iuyd.2015.20592. (Turkish).

Akgul, Y., Bratieres, S., Ebrahimi, S., Gorichanaz, C., Loudermilk, B., Morris, E., Pârvulescu, C. and Solano, A., 2019. Introduction. *Deep Learning*. (Editors: Goodfellow, I., Bengio, Y. and Courville, A.). www.deeplearningbook.org.

Alaiso, S., Backman, J. and Visala, A., 2013. Ant Colony Optimization for Scheduling of Agricultural Contracting Work. 4th IFAC Conference on Modelling and Control in Agriculture, Horticulture and Post Harvest Industry, August 27–30, 2013. Espoo, Finland.

Alam, M.S., Alam, M., Tufail, M., Khan, M.U., Güneş, A., Salah, B., Nasir, F.E., Saleem, W. and Khan, M.T., 2022. TobSet: A New Tobacco Crop and Weeds Image Dataset and Its Utilization for Vision-Based Spraying by Agricultural Robots. *Applied Sciences*, 12, 1308. https://doi.org/10.3390/app12031308.

Ali, R.B., Bouadila, S. and Mami, A., 2018. Development of a Fuzzy Logic Controller Applied to an Agricultural Greenhouse Experimentally Validated. *Applied Thermal Engineering*, 141(2018), 798–810.

Alpaydın, E., 2004. *Introduction to Machine Learning*. The MIT Press, Cambridge.

Alpaydın, E., 2016. *Machine Learning: The New AI*. The MIT Press, Cambridge. ISBN: 109-8-765-432-1.

Altaş, Z., Ozguven, M.M. ve Yanar, Y., 2019. Bitki Hastalık ve Zararlı Düzeylerinin Belirlenmesinde Görüntü İşleme Tekniklerinin Kullanımı: Şeker Pancarı Yaprak Leke Hastalığı Örneği. International Erciyes Agriculture, Animal & Food Sciences Conference 24–27 April 2019- Erciyes University – Kayseri, Türkiye. (Turkish).

Arya, M.S., Anjali, K. and Unni, D., 2018. *Detection of Unhealthy Plant Leaves Using Image Processing and Genetic Algorithm with Arduino.* . 2018 International Conference on Power, Signals, Control and Computation (EPSCICON), 2018, pp. 1–5, doi: 10.1109/EPSCICON.2018.8379584.

Atalay, M. ve Çelik, E., 2017. Büyük Veri Analizinde Yapay Zekâ ve Makine Öğrenmesi Uygulamaları. Mehmet Akif Ersoy Üniversitesi Sosyal Bilimler Enstitüsü Dergisi. Cilt:9 Sayı:22. s.155–172. (Turkish).

Bao, J. and Xie, Q., 2022. Artificial Intelligence in Animal Farming: A Systematic Literature Review. *Journal of Cleaner Production*, 331, 129956. https://doi.org/10.1016/j.jclepro.2021.129956.

Bishop, J.C., Falzon, G., Trotter, M., Kwan, P. and Meek, P.D., 2019. Livestock Vocalisation Classification in Farm Soundscapes. *Computers and Electronics in Agriculture*, 162, 531–542. https://doi.org/10.1016/j.compag.2019.04.020.

Calou, V.B.C., Teixeira, A.D.S., Moreira, L.C.J., Lima, C.S., De Oliveira, J.B. ve De Oliveira, M.R.R., 2020. The use of UAVs in Monitoring Yellow Sigatoka in Banana. *Biosystems Engineering*, 193, 115–125. https://doi.org/10.1016/j.biosystemseng.2020.02.016.

Camastra, F. and Vinciarelli, A., 2008. *Machine Learning for Audio, Image and Video Analysis: Theory and Applications*. Springer-Verlag London Limited, London. ISBN: 978-1-84800-006-3.

Campesato, O., 2020. *Artificial Intelligence Machine Learning and Deep Learning*. Mercury Learning and Information LLC, Dulles. ISBN: 978-1-68392-467-8.

Campos, D.P., Abatti, P.J., Bertotti, F.L., Hill, J.A.G. and da Silveira, A.L.F., 2018. Surface Electromyography Segmentation and Feature Extraction for Ingestive Behavior Recognition in Ruminants. *Computers and Electronics in Agriculture*, 153, 325–333. https://doi.org/10.1016/j.compag.2018.08.033.

Chang, C.-L., Chung, S.-C., Fu, W.-L. and Huang, C.-C., 2021. Artificial Intelligence Approaches to Predict Growth, Harvest Day, and Quality of Lettuce (Lactuca sativa L.) in a IoT-Enabled Greenhouse System. *Biosystems Engineering*, 212, 77–105. https://doi.org/10.1016/j.biosystemseng.2021.09.015.

Chen, J., Zhang, M., Xu, B., Sun, J., Mujumdar, A.S., 2020. Artificial Intelligence Assisted Technologies for Controlling the Drying of Fruits and Vegetables using Physical Fields: A Review. *Trends in Food Science & Technology*, 105, 251–260. https://doi.org/10.1016/j.tifs.2020.08.015.

Deng, L. and Yu, D., 2014. Deep Learning: Methods and Applications. *Found Trends Signal Process*, 7(3–4), 197–387.

Elgendy, M., 2020. *Deep Learning for Vision Systems*. Manning Publications Co, Shelter Island. ISBN: 978-1-617-29619-2.

Elmas, Ç., 2010. *Yapay Zeka Uygulamaları*. Seçkin Yayıncılık, Ankara. ISBN: 978-975-02-1696-1. (Turkish).

Elmas, Ç., 2018. *Yapay Zeka Uygulamaları*. Seçkin Yayıncılık, 4. Baskı. (Turkish).

Eraslan, G., Hickson, S., Pascanu, R., Ritter, L., Rodrigues, R., Serdyuk, D., Shi, D. and Yang, K., 2019. Sequence Modeling: Recurrent and Recursive Nets. *Deep Learning*. (Editors: Goodfellow, I., Bengio, Y. and Courville, A.). www.deeplearningbook.org.

Faridi, M., Verma, S. and Mukherjee, S., 2018. Integration of GIS, Spatial Data Mining, and Fuzzy Logic for Agricultural Intelligence. *Soft Computing: Theories and Applications: Advances in Intelligent Systems and Computing.* (Editors: Pant, M., Ray, K., Sharma, T., Rawat, S., Bandyopadhyay, A.) vol. 583. Springer, Singapore. 171–183

Fawakherji, M., Potena, C., Bloisi, D.D., Imperoli, M., Pretto, A. and Nardi, D., 2019. UAV Image Based Crop and Weed Distribution Estimation on Embedded GPU Boards. International Conference on Computer Analysis of Images and Patterns, 100–108.

Flach, P., 2012. *Machine Learning: The Art and Science of Algorithms that Make Sense of Data.* Cambridge University Press, Cambridge. ISBN: 978-1-107-09639-4.

Gašparović, M. Zrinjski, M., Barković, D. and Radočaj, D., 2020. An Automatic Method for Weed Mapping in Oat Fields Based on UAV Imagery, *Computers and Electronics in Agriculture*, 173, 105385. https://doi.org/10.1016/j.compag.2020.105385.

Haklı, H. and Uğuz, H., 2017. A Novel Approach for Automated Land Partitioning using Genetic Algorithm. *Expert Systems with Applications*, 82, 10–18.

Haklı, H., Uğuz, H. and Çay, T., 2018. Genetic Algorithm Supported by Expert System to Solve Land Redistribution Problems. *Expert Systems*, 35, e12308. https://doi.org/10.1111/exsy.12308.

Halachmi, I., Levit, H. ve Bloch, V., 2019. Current Trends and Perspective of Precision Livestock Farming (PLF) with Relation to IoT and Data Science Tools. FFTC Agricultural Policy Platform.

Hosseini, M.-P., 2016. A Cloud-Based Brain Computer Interface to Analyze Medical Big Data for Epileptic Seizure Detection. The 3rd Annual New Jersey Big Data Alliance (NJBDA) Symposium.

Hosseini, M., Naeini, S.A.M. and Dehghani, A.A., 2016. Estimation of Soil Mechanical Resistance Parameter by Using Particle Swarm Optimization, Genetic Algorithm and Multiple Regression Methods. *Soil and Tillage Research*, 157, 32–42.

Hosseini, M.-P., Pompili, D., Elisevich, K. and Soltanian-Zadeh, H., 2017. Optimized Deep Learning for EEG Big Data and Seizure Prediction BCI Via Internet of Things. *IEEE Transactions on Big Data*, 3(4), 392–404.

Hosseini, M.-P., 2018. Brain-Computer Interface for Analyzing Epileptic Big Data. Ph.D. thesis, Rutgers University-School of Graduate Studies.

Hosseini, M.-P., Lu, S., Kamaraj, K., Slowikowski, A. and Venkatesh, H.C., 2020. Deep Learning Concepts and Architectures. (Editors: Pedrycz, W. and Chen, S.-M.). Springer Nature Switzerland AG. ISBN: 978-3-030-31755-3.

Hu, H., Dai, B., Shen, W., Wei, X., Sun, J., Li, R. and Zhang, Y., 2020. Cow Identification Based on Fusion of Deep Parts Features. *Biosystems Engineering*, 192, 245–256. https://doi.org/10.1016/j.biosystemseng.2020.02.001.

Imbalzano, E., Orlando, L., Sciacqua, A., Nato, G., Dentali, F., Nassisi, V., Russo, V., Camporese, G., Bagnato, G., Cicero, A.F.G., Dattilo, G., Vatrano, M., Versace, A.G., Squadrito, G. and Di Micco, P., 2022. Machine Learning to Calculate Heparin Dose in COVID-19 Patients with Active Cancer. *Journal of Clinical Medicine*, 11, 219. https://doi.org/10.3390/jcm11010219.

Kale, S.S. and Patil, P.S., 2018. Data Mining Technology with Fuzzy Logic, Neural Networks and Machine Learning for Agriculture. *Data Management, Analytics and Innovation*, 839, 79–87. https://doi.org/10.1007/978-981-13-1274-8_6.

Kamilaris, A. and Prenafeta, S.X., 2018. A Review of the Use of Convolutional Neural Networks in Agriculture. *The Journal of Agricultural Science*, 156, 312–322. doi.org/10.1017/S0021859618000436.

Karakuş, C., 2021. Makine Öğrenmesi Temelleri Ders Notu. https://ckk.com.tr/ders/ML/ML%2000%20Makine%20%C3%96%C4%9Frenmesi%20Ders%20Notu.html. (Turkish).

Kelleher, J.D., 2019. *Deep Learning.* The MIT Press, Cambridge. ISBN: 978-026-2537-55-1.

Khawas, P., Dash, K.K., Das, A.J. and Deka, S.C., 2015. Modeling and Optimization of the Process Parameters in Vacuum Drying of Culinary Banana (Musa ABB) Slices by Application of Artificial Neural Network and Genetic Algorithm. *Drying Technology an International Journal*, 34(4), 2016. 491–503. https://doi.org/10.1080/07373937.2015.1060605

Khoshnevisan, B., Rafiee, S., Iqbal, J., Shamshirband, S., Omid, M., Anuar, N.B. and Abdulwahab, A.W., 2015. A Comparative Study between Artificial Neural Networks and Adaptive Neuro-Fuzzy Inference Systems for Modeling Energy Consumption in Greenhouse Tomato Production: A Case Study in Isfahan Province. *The Journal of Agricultural Science*, 17, 49–62.

Khoshroo, A., Emrouznejad, A., Ghaffarizadeh, A., Kasraei, M. and Omid, M., 2018. Sensitivity Analysis of Energy Inputs in Crop Production using Artificial Neural Networks. *Journal of Cleaner Production*, 197, 992–998.

Krishnamoorthy, C.S. and Rajeev, S., 1996. *Artificial Intelligence and Expert Systems for Engineers*. CRC Press Taylor & Francis Group LLC. ISBN: 0849391253.

Krizhevsky, A., Sutskever, I. and Hinton, G.E., 2012. Imagenet Classification with Deep Convolutional Neural Networks. NIPS'12: Proceedings of the 25th International Conference on Neural Information Processing Systems - Volume 1. December 2012, 1097–1105.

Kulkarni, S.A., Gurupur, V.P. and Fernandes, S.L., 2020. *Introduction to IoT with Machine Learning and Image Processing using Raspberry Pi*. CRC Press Taylor & Francis Group LLC, Boca Raton. ISBN: 978-1-351-00666-8.

Kurniasih, D., Jasmi, K.A., Basiron, B., Huda, M. and Maseleno, A., 2018. The Uses of Fuzzy Logic Method for Finding Agriculture and Livestock Value of Potential Village. *International Journal of Engineering & Technology*, 7(3), 1091–1095.

Li, X. and Parrott, L., 2016. An Improved Genetic Algorithm for Spatial Optimization of Multi-Objective and Multi-Site Land Use Allocation. *Computers, Environment and Urban Systems*, 59, 184–194.

Liakos, K.G., Busato, P., Moshou, D., Pearson, S. and Bochtis, D., 2018. Machine Learning in Agriculture: A Review. *Sensors*, 18, 2674. https://doi.org/10.3390/s18082674.

Liu, Z.W., Liang, F.N. and Liu,Y.Z. 2018. Artificial Neural Network Modeling of Biosorption Process using Agricultural Wastes in a Rotating Packed Bed. *Applied Thermal Engineering*, 140, 95–101.

Liu, Z., Xu, Y., Han, F., Zhang, Y., Wang, G., Wu, Z. and Wu, W., 2022. Control Method for Continuous Grain Drying Based on Equivalent Accumulated Temperature Mechanism and Artificial Intelligence. *Foods*, 11, 834. https://doi.org/10.3390/foods11060834.

Marsland, S., 2015. *Machine Learning: An Algorithmic Perspective*. CRC Press Taylor & Francis Group LLC, Boca Raton. ISBN: 978-1-4665-8333-7.

Matha, T.W., Jati, A.N. and Azmi, F., 2016. Fuzzy Logic as a Method of Decision Making in Automatic Watering Plants. *Journal of Measurement, Electronic, Communication, and Systems*,2(1). 15–21. https://doi.org/10.25124/jmecs.v2i1.1830.

Martinez, V.M., Gil, F.J.G., Gil, J.G. and Gonzalez, R.R., 2015. An Artificial Neural Network Based Expert System Fitted with Genetic Algorithms for Detecting the Status of Several Rotary Components in Agro-Industrial Machines using A Single Vibration Signal. *Expert Systems with Applications*, 42, 6433–6441.

Martynenko, A. and Bück, A., 2019. *Advances in Drying Science and Technology: Intelligent Control in Drying*. CRC Press, Taylor & Francis Group LLC, Boca Raton. ISBN: 13:978-1-4987-3275-8.

Michelucci, U., 2019. *Advanced Applied Deep Learning: Convolutional Neural Networks and Object Detection*. Apress Media, LLC, Dübendorf. ISBN: 978-1-4842-4975-8.

Millan, B., Velasco-Forero, S., Aquino, A. and Tardaguila, J., 2018. On-the-Go Grapevine Yield Estimation Using Image Analysis and Boolean Model. *Hindawi Journal of Sensors*, 2018, 14. https://doi.org/10.1155/2018/9634752.

Mohri, M., Rostamizadeh, A. and Talwalkar, A., 2018. *Foundations of Machine Learning*. The MIT Press, Cambridge. ISBN: 978-026-20-3940-6.

Montalvo, M., Guerrero, J.M., Romeo, J., Emmi, L., Guijarro, M. and Pajares, G., 2013. Automatic Expert System for Weeds/Crops Identification in Images from Maize Fields. *Expert Systems with Applications*, 40, 75–82.

Mueller, J.P. and Massaron, L., 2019. *Deep Learning for Dummies*. John Wiley & Sons, Inc, Hoboken. ISBN: 978-1-119-54303-9.

Müller, A.C. and Guido, S., 2017. *Introduction to Machine Learning with Python: A Guide for Data Scientists*. O'Reilly Media Inc, Sebastopol. ISBN: 978-1-449-36941-5.

Nabiyev, V., 2016. Yapay Zeka. Stratejili Oyunlar, Örüntü Tanıma, Doğal Dil İşleme. Seçkin Yayıncılık 5. Baskı. Sertifika no: 12416, ISBN: 978-975-02-3727-0. (Turkish).

Nguyen, D.C.H., Ascough, J.C., Maier, H.R., Dandy, G.C. and Andales, A.A., 2017. Optimization of Irrigation Scheduling Using Ant Colony Algorithms and an Advenced Cropping System Model. *Environmental Modelling & Software*, 97, 32–45.

Omomule, T.G., Ajayi, O.O. and Orogun, A.O., 2020. Fuzzy Prediction and Pattern Analysis of Poultry Egg Production. *Computers and Electronics in Agriculture*. 171, 105301.

Oppenheim, D. and Shani, G., 2017. Potato Disease Classification Using Convolution Neural Networks. *Advances in Animal Biosciences*, 8, 244–249. https://doi.org/10.1017/S2040470017001376.

Osco, L.P., De Arruda, M.S., Gonçalves, D.N., Dias, A., Batistoti, J., De Souza, M., Gomes, F.D.G., Ramos, A.P.M., De Castro Jorge, L.A., Liesenberg, V., Li, J., Ma, L., Marcato, J. and Gonçalves, W.N., 2021. A CNN Approach to Simultaneously Count Plants and Detect Plantation-Rows from UAV Imagery. *ISPRS Journal of Photogrammetry and Remote Sensing*, 174, 1–17. https://doi.org/10.1016/j.isprsjprs.2021.01.024.

Ozguven, M.M. and Adem, K., 2019. Automatic Detection and Classification of Leaf Spot Disease in Sugar Beet using Deep Learning Algorithms. *Physica A-Statistical Mechanics and Its Applications*, 535, 1–8. https://doi.org/10.1016/j.physa.2019.122537.

Ozguven, M.M., 2020. Deep Learning Algorithms for Automatic Detection and Classification of Mildew Disease in Cucumber. *Fresenius Environmental Bulletin*, 29(08/2020), 7081–7087.

Özgüven, M.M., 2019. Teknoloji Kavramları ve Farkları. International Erciyes Agriculture, Animal & Food Sciences Conference 24–27 April 2019- Erciyes University – Kayseri, Turkiye. (Turkish).

Özgüven, M.M., Beyaz, A., Ormanoğlu, N., Aktaş, T., Emekci, M., Ferizli, A.G., Çilingir, İ. ve Çolak, A., 2020. Hasat Sonrası Ürünlerin Korunmasına Yönelik Mekanizasyon Otomasyon ve Mücadele Teknikleri. Türkiye Ziraat Mühendisliği IX. Teknik Kongresi. Ocak 2020, Ankara. Bildiriler Kitabı-1, s.301–324. (Turkish).

Öztemel, E., 2006. Yapay Sinir Ağları. Papatya Yayıncılık, 2. Baskı, ISBN: 978-975-6797-39-6. (Turkish).

Pang, Y., Shi, Y., Gao, S., Jiang, F., Veeranampalayam-Sivakumar, A-N., Thompson, L., Luck, J. and Liu, C., 2020. Improved Crop Row Detection with Deep Neural Network for Early-Season Maize Stand Count in UAV Imagery. *Computers and Electronics in Agriculture*, 178, 105766. https://doi.org/10.1016/j.compag.2020.105766.

Partel, V., Costa, L. and Ampatzidis, Y., 2021. Smart Tree Crop Sprayer Utilizing Sensor Fusion and Artificial Intelligence. *Computers and Electronics in Agriculture*, 191, 106556. https://doi.org/10.1016/j.compag.2021.106556.

Patterson, J. and Gibson, A., 2017. *Deep Learning: A Practitioner's Approach*. O'Reilly Media, Inc, Sebastopol. ISBN: 978-1-491-91425-0.

Pomar, C., van Milgen, J. and Remus, A., 2019. Precision Livestock Feeding, Principle and Practice. 10.3920/978-90-8686-884-1_18.

Ren, Y., Yang, J., Zhang, Q. and Guo, Z., 2019. Multi-Feature Fusion with Convolutional Neural Network for Ship Classification in Optical Images. *Applied Sciences*, 9, 4209. https://doi.org/10.3390/app9204209.

Romeo, J., Pajares, G., Montalvo, M., Guerrero, J.M., Guijarro, M. and de la Cruz, J.M., 2013. A New Expert System for Greennees Identification in Agriculture Images. *Expert Systems with Applications*, 40, 2275–2286.

Rothman, D., 2018. *Artificial Intelligence by Example*. Packt Publishing, Birmingham. ISBN: 978-1-78899-054-7.

Sa, I., Ge, Z., Dayoub, F., Upcroft, B., Perez, T. and McCool, C., 2016. DeepFruits: A Fruit Detection System Using Deep Neural Networks. *Sensors*, 16(8), 1222. https://doi.org/10.3390/s16081222.

Singh, R., Gehlot, A., Prajapat, M.K. and Singh, B., 2022. *Artificial Intelligence in Agriculture*. CRC Press, Taylor & Francis Group LLC, Boca Raton. ISBN: 978-1-032-15810-5.

Song, H.A. and Lee, S.-Y., 2013. Hierarchical Representation using NMF. International Conference on Neural Information Processing, 466–473.

Tang, C., Ling, Y., Yang, X., Jin, W. and Zheng, C., 2018. Multi-View Object Detection Based on Deep Learning. *Applied Sciences*, 8, 1423. https://doi.org/10.3390/app8091423.

Terzi, İ., Özgüven, M.M., Altaş, Z., Uygun, T., 2019. Tarımda Yapay Zeka Kullanımı. International Erciyes Agriculture, Animal & Food Sciences Conference 24–27 April 2019 - Erciyes University – Kayseri, Turkiye. (Turkish).

Tetila, E.C., Machado, B.B., Astolfi, G., De Souza Belete, N.A., Amorim, W.P., Roel, A.R. and Pistori, H., 2020. Detection and Classification of Soybean Pests using Deep Learning with UAV Images. *Computers and Electronics in Agriculture*, 179, 105836. https://doi.org/10.1016/j.compag.2020.105836.

Vasquez, R.P., Aguilar, A.A., Lopez, M.V., Rivero, L.G., Rodriguez, A.A. and Rojas, A.A., 2018. Expert System Based on a Fuzzy Logic Model for the Analysis of the Sustainable Livestock Production Dynamic System. *Computers and Electronics in Agriculture*. https://doi.org/10.1016/j.compag.2018.05.015.

Wang, J., Zhang, Z., Luo, L., Zhu, W., Chen, J. and Wang, W., 2021. SwinGD: A Robust Grape Bunch Detection Model Based on Swin Transformer in Complex Vineyard Environment. *Horticulturae*, 7, 492. https://doi.org/10.3390/horticulturae7110492.

Wani, M.A., Bhat, F.A., Afzal, S. and Khan, A.I., 2020. *Advances in Deep Learning*. Springer Nature Singapore Pte Ltd, Singapore. ISBN: 978-981-13-6793-9.

Yang, X-S., 2019. *Introduction to Algorithms for Data Mining and Machine Learning*. Academic Press is an imprint of Elsevier, London. ISBN: 978-0-12-817216-2.

Yao, M., Jia, M. and Zhou, A., 2018. *Applied Artificial Intelligence: A Handbook for Business Leaders*. (Editor: Zhang, N.) Topbots Inc., USA ISBN: 978-0-9982890-2-1.

Yigit, E., Sabanci, K., Toktas, A. and Kayabasi, A., 2019. A Study on Visual Features of Leaves in Plant Identification using Artificial Intelligence Techniques. *Computers and Electronics in Agriculture*, 156, 369–377. https://doi.org/10.1016/j.compag.2018.11.036.

Yu, Z., Liu, Y., Yu, S., Wang, R., Song, Z., Yan, Y., Li, F., Wang, Z. and Tian, F., 2022. Automatic Detection Method of Dairy Cow Feeding Behaviour Based on YOLO Improved Model and Edge Computing. *Sensors*, 22, 3271. https://doi.org/10.3390/s22093271.

Zhang, Y., Lee, W.S., Li, M., Zheng, L. and Ritenour, M.A., 2018. Non-Destructive Recognition and Classification of Citrus Fruit Blemishes Based on Ant Colony Optimized Spectral Information. *Postharvest Biology and Technology*, 143, 119–128.

Zhang, Y., Li, M., Zheng, L., Qin, Q. and Lee, W.S., 2019a. Spectral Features Extraction for Estimation of Soil Total Nitrogen Content Based on Modified Ant Colony Optimization Algorithm. *Geoderma*, 333, 23–34.

Zhang, Z., Liu, H., Meng, Z. and Chen, J., 2019b. Deep Learning-Based Automatic Recognition Network of Agricultural Machinery Images. *Computers and Electronics in Agriculture*, 166, 104978. https://doi.org/10.1016/j.compag.2019.104978.

Zheng, A. and Casari, A., 2018. *Feature Engineering for Machine Learning: Principles and Techniques for Data Scientists*. O'Reilly Media Inc., Sebastopol ISBN: 978-1-491-95324-2.

Index